精通 Android 应用开发

王治国　王　捷　著

清华大学出版社
北　京

内 容 简 介

近年来 Android 的兴起和对移动设备开发领域的冲击，已成热门话题。Android 作为最受欢迎的智能手机操作系统，具有广阔的发展前景，而 Android 应用选择了 Java 作为其开发语言，对于 Java 来说，也是一次极好的机会。

本书内容深入浅出、语言通俗易懂，便于读者自学。对于一些较难以理解的概念采用实例进行说明，以帮助读者更好地理解各知识点在实际开发中的应用。本书共分为 15 章，全面介绍了 Android 应用开发的相关知识，其内容覆盖了 Android 概述、开发环境的搭建、界面布局、Widget 组件及事件处理机制、Activity、Intent、Android 的管理员 Service、Android 资源访问、Android 输入/输出处理、音频/视频多媒体应用开发、网络编程、地位服务和地图服务等。最后通过一个实例对书中各章节的知识点进行综合应用。

本书基础翔实，实例丰富，图文并茂、案例真实。从基础到案例覆盖了 android 应用开发的各领域，可作为本科院校、高等职业院校及软件学院计算机类、通信类专业的教材，也适合作为相关培训学校的 Android 培训教材及从事 Android 移动编程和应用开发人员参考用书。

本书封面贴有清华大学出版社防伪标签，无标签者不得销售。
版权所有，侵权必究。侵权举报电话：010-62782989　13701121933

图书在版编目（CIP）数据

精通 Android 应用开发 / 王治国，王捷著. —北京：清华大学出版社，2014(2019.2 重印)
ISBN 978-7-302-35651-6

Ⅰ. ①精…　Ⅱ. ①王…　②王…　Ⅲ. ①移动终端－应用程序－程序设计　Ⅳ. ①TN929.53

中国版本图书馆 CIP 数据核字（2014）第 050770 号

责任编辑：袁金敏
封面设计：刘新新
责任校对：徐俊伟
责任印制：刘海龙

出版发行：清华大学出版社
    网　　址：http://www.tup.com.cn, http://www.wqbook.com
    地　　址：北京清华大学学研大厦 A 座　　　邮　　编：100084
    社 总 机：010-62770175　　　　　　　　　邮　　购：010-62786544
    投稿与读者服务：010-62776969, c-service@tup.tsinghua.edu.cn
    质量反馈：010-62772015, zhiliang@tup.tsinghua.edu.cn
印 装 者：北京九州迅驰传媒文化有限公司
经　　销：全国新华书店
开　　本：185mm×260mm　　　印　张：24　　　字　数：599 千字
版　　次：2014 年 7 月第 1 版　　　　　　　　印　次：2019 年 2 月第 3 次印刷
定　　价：49.00 元

产品编号：054078-01

前　言

当今社会已经进入了信息移动时代，手机功能越来越智能，越来越开放，为了实现这些需求，必须有一个好的开发平台来支持，Android 是 2007 年 11 月由 Google 公司宣布的基于 Linux 平台的开源手机操作系统，任何公司及个人都可以免费获得源代码及开发 SDK。由于 Android 平台的开放性和优异性，得到了业界广泛的支持，是目前最受欢迎的嵌入式操作系统之一，其发展趋势势不可挡，Android 移动软件开发已成为当今最为流行的移动终端开发技术。

移动终端的快速发展使得 Android 系统应用的需求激增，很多在校生和广大开发者都加入了 Android 开发阵营。为了帮助开发者更快地进入 Android 开发行列，笔者精心编写了本书。本书具有以下几个特点：结构合理，从读者的实际需求出发，科学安排知识结构，内容由浅入深，循序渐进，逐步展开，反映了当前 Android 技术的发展和应用水平；浅显易懂，条理清晰、语言简洁，通过大量简单易懂的实例帮助读者快速掌握知识点，每部分既相互连贯又自成体系，使读者既可以按照本书编排的章节顺序进行学习，也可以根据自己的需求对某一章节进行有针对性的学习；实用性强，注重实用性和可操作性，通过实例使读者在掌握相关技能的同时，学习相应的基础知识。所有的实例都调试运行通过，读者可以直接参照使用。

全书分 15 章，各章内容介绍如下。

第 1 章　Android 概述，简单地介绍手机操作系统、Android 的发展及其优越性。

第 2 章　重点介绍如何搭建 Android 开发环境。

第 3 章　介绍了 Android 应用程序的构成及程序的内部执行流程。

第 4～5 章　对 Android 的界面布局和 Widget 组件及事件处理机制进行详细介绍。

第 6 章　主要讨论 Android 的门面 Activity 及其之间的跳转和数据传递。

第 7 章　详细介绍了 Android 系统中的 Intent 功能和用法。

第 8 章　对如何创建、配置 Service 组件及如何启动、停止 Service 进行详细阐述。

第 9 章　对实现消息异步处理的组件 BroadcastReceiver 进行深入探讨。

第 10 章　详细介绍 Android 的 Preferences、文件和数据库 SQLite 三种数据存储方式的使用。

第 11 章　主要讨论 Android 系统中 ContentProvider 组件的功能和用法。

第 12 章　对基于 Android 平台的音视频录制和播放功能进行具体介绍。

第 13～14 章　主要介绍 Android 平台下进行网络编程的方法、如何进行定位及如何实现 Google 提供的地图服务。

第 15 章　介绍如何使用 Android 技术开发一个移动版同学簿，该系统综合运用了本书各章节的知识和技术，包括 Android 如何获取网络数据进行数据的绑定，实现实时网络图片加载、Android UI 布局、UI 界面的动态更新、数据全局共享处理和界面数据交互等。

本书知识点全面，结构合理，重难点突出，实例丰富，语言简洁，适用于 Android 移动软件开发初中级用户。

本书由郑州轻工业学院王治国、王捷编著，参加本书编写的还有钱慎一、胡东华、黄永丽等，此外，白亚东、白永刚、王国胜、刘松云、张丽、张班班、胡文华、尼春雨、蒋军军、聂静等也参与了本书部分内容的编写工作，对此表示衷心的感谢。特别感谢郑州轻工业学院教务处的大力支持。

目　　录

第 1 章　**Android 概述** ··· 1
　1.1　智能手机操作系统简介 ·· 1
　1.2　Android 的基本概念 ·· 2
　　　1.2.1　Android 的前世 ·· 2
　　　1.2.2　Android 的优点 ·· 3
　1.3　Android 系统架构 ·· 3
　1.4　本章小结 ·· 5
第 2 章　**搭建 Android 开发环境** ··· 6
　2.1　开发前的准备工作 ·· 6
　2.2　搭建开发环境 ·· 6
　　　2.2.1　Android SDK 的安装 ·· 7
　　　2.2.2　Eclipse 和 ADT 安装 ·· 8
　　　2.2.3　创建和启动 AVD（Android Virtual Device） ·· 10
　2.3　构建 Android 应用程序 ·· 14
　　　2.3.1　使用 Eclipse 创建 Android 应用程序 ·· 14
　　　2.3.2　运行 Android 应用程序 ·· 16
　　　2.3.3　通过 DDMS 调试 Android 应用程序 ·· 18
　2.4　本章小结 ·· 19
第 3 章　**Android 应用程序剖析** ··· 20
　3.1　Android 应用程序目录结构 ··· 20
　　　3.1.1　gen/目录下的 R.java 文件详解 ··· 23
　　　3.1.2　组件标识符 ·· 24
　　　3.1.3　AndroidMainfest.xml 详细介绍 ·· 25
　3.2　Android 应用程序的执行流程 ··· 29
　3.3　Android 应用程序的基本组件 ··· 30
　　　3.3.1　Activity ··· 31
　　　3.3.2　Service ·· 32
　　　3.3.3　BroadcastReceiver ·· 33
　　　3.3.4　ContentProvider ··· 33
　　　3.3.5　Intent 和 IntentFileter ·· 33
　3.4　本章小结 ·· 34
第 4 章　**界面布局** ··· 35
　4.1　UI 概述 ··· 35

4.2　线性布局 ·· 35
　　4.3　相对布局 ·· 38
　　4.4　绝对布局 ·· 41
　　4.5　表格布局 ·· 43
　　4.6　本章小结 ·· 45
第 5 章　**事件处理及 Widget 组件** ··· 46
　　5.1　基本 Widget 组件 ··· 46
　　　　5.1.1　文本框（TextView）和编辑框（EditText） ·· 46
　　　　5.1.2　Button（按钮）和 ImageButton（图片按钮） ·· 49
　　　　5.1.3　单选框（RadioButton）和复选框（ChekBox） ···································· 51
　　　　5.1.4　AnalogClock 和 DigitalClock ·· 53
　　5.2　高级 Widget 组件 ··· 54
　　　　5.2.1　ListView（列表视图） ··· 54
　　　　5.2.2　Spinner（下拉列表） ··· 59
　　　　5.2.3　ProgressBar（进度条） ··· 63
　　　　5.2.4　SeekBar（拖动条） ·· 66
　　　　5.2.5　DatePicker（日期选择器）和 TimePicker（时间选择器） ······················ 68
　　5.3　对话框 ·· 72
　　　　5.3.1　提示对话框 ·· 72
　　　　5.3.2　多选对话框 ·· 73
　　　　5.3.3　内容输入对话框 ·· 74
　　　　5.3.4　单选对话框 ·· 74
　　　　5.3.5　复选对话框 ·· 75
　　　　5.3.6　列表对话框 ·· 75
　　5.4　消息提示 ·· 76
　　5.5　事件处理机制 ··· 78
　　　　5.5.1　基于监听的事件处理 ·· 78
　　　　5.5.2　基于回调机制的事件处理 ··· 84
　　　　5.5.3　Handler ··· 86
　　5.6　本章小结 ·· 88
第 6 章　**Android 的门面——Activity** ·· 89
　　6.1　Activity 生命周期 ··· 89
　　6.2　Activity 管理栈 ·· 93
　　6.3　创建、配置和使用 Activity ··· 94
　　　　6.3.1　创建 Activity ·· 94
　　　　6.3.2　配置 Activity ·· 96
　　　　6.3.3　启动关闭 Activity ·· 97
　　　　6.3.4　需要传递参数的 Activity 启动 ··· 100
　　　　6.3.5　启动其他 Activity 并返回结果 ··· 104

6.4	启动模式	108
6.5	本章小结	110

第 7 章 Android 的邮递员——Intent ... 111
- 7.1 Intent 概述 ... 111
  - 7.1.1 Intent 属性 ... 111
  - 7.1.2 Intent 解析 ... 113
- 7.2 Intent Filter ... 113
  - 7.2.1 动作检测 ... 114
  - 7.2.2 种类检测 ... 114
  - 7.2.3 数据检测 ... 115
  - 7.2.4 通用情况 ... 116
  - 7.2.5 使用 intent 匹配 ... 117
- 7.3 Intent 的调用 ... 117
  - 7.3.1 显式调用 ... 117
  - 7.3.2 隐式调用 ... 118
  - 7.3.3 在 Intent 中传递数据 ... 121
  - 7.3.4 在 Intent 中传递复杂对象 ... 125
  - 7.3.5 实现 Activity 之间的协同 ... 130
- 7.4 常用 Intent 组件的使用 ... 134
- 7.5 本章小结 ... 137

第 8 章 Android 的隐形管理员——Service ... 138
- 8.1 Service 概述 ... 138
- 8.2 Service 的生命周期 ... 138
  - 8.2.1 startService 启动服务 ... 140
  - 8.2.2 bindSerivce 启动服务 ... 142
- 8.3 Service 的使用方法 ... 144
  - 8.3.1 编写不需和 Activity 交互的本地服务 ... 144
  - 8.3.2 编写本地服务和 Activity 交互 ... 145
  - 8.3.3 编写传递基本型数据的远程服务 ... 149
  - 8.3.4 编写传递复杂数据类型的远程服务 ... 154
- 8.4 IntentService ... 162
- 8.5 本章小结 ... 167

第 9 章 Android 的接收员——BroadcastReceiver ... 168
- 9.1 BroadcastReceiver 概述 ... 168
- 9.2 广播消息 ... 169
  - 9.2.1 自定义 BroadcastReceiver ... 169
  - 9.2.2 普通广播 ... 172
  - 9.2.3 有序广播 ... 177
- 9.3 处理系统广播消息 ... 180

9.4	BroadcastReceiver 的生命周期	186
9.5	本章小结	186

第 10 章 Android 的数据存储 187

10.1	数据存储概述	187
10.2	SharedPreferences	187
	10.2.1 使用 SharedPreferences	187
	10.2.2 PreferenceActivity	193
10.3	文件	199
	10.3.1 应用程序文件读写	200
	10.3.2 操作资源文件	203
	10.3.3 操作 SD 卡上的文件	204
10.4	数据库	211
	10.4.1 SQLite 简介	212
	10.4.2 使用 SQLite 数据库	216
10.5	本章小结	225

第 11 章 Android 的图书馆——ContentProvider 226

11.1	ContentProvider 概述	226
11.2	自定义 ContentProvider	229
11.3	系统 ContentProvider	236
	11.3.1 使用 Contacts Contract Content Provider	237
	11.3.2 读取短信	243
11.4	本章小结	247

第 12 章 Android 多媒体应用开发 248

12.1	音频录制	248
	12.1.1 使用 Intent 录制音频	248
	12.1.2 使用 MediaRecorder 录制音频	250
12.2	音频播放	255
	12.2.1 常见的音频格式	256
	12.2.2 使用 Intent 播放音频	256
	12.2.3 使用 MediaPlayer 播放音频	257
12.3	视频录制	260
	12.3.1 使用 Intent 录制视频	261
	12.3.2 使用 MediaRecorder 录制视频	263
12.4	视频播放	269
	12.4.1 常见的视频格式	269
	12.4.2 使用 Intent 播放视频	269
	12.4.3 使用 VideoView 播放视频	270
	12.4.4 使用 MediaPlayer 播放视频	272
12.5	本章小结	276

第 13 章 Android 的网络编程 277
13.1 Android 网络编程基础 277
13.2 基于 HTTP 协议的网络编程 278
13.2.1 HTTP 介绍 278
13.2.2 使用 HttpURLConnection 访问网络 279
13.2.3 使用 HttpClient 访问网络 286
13.3 基于 Socket 的网络编程 292
13.3.1 套接字 Socket 293
13.3.2 Socket 编程 293
13.4 基于 WebView 的简单浏览器 296
13.5 本章小结 302

第 14 章 定位服务和地图服务 303
14.1 定位服务相关类 303
14.2 定位实例 305
14.3 Google Map 使用 308
14.3.1 申请 Map API KEY 308
14.3.2 开发和测试环境搭建 311
14.4 地图定位 315
14.5 本章小结 317

第 15 章 移动同学簿 318
15.1 系统概述 318
15.1.1 移动同学簿的应用背景 318
15.1.2 移动同学簿的总体需求 318
15.1.3 移动同学簿的功能分析 319
15.1.4 移动同学簿的设计思路 319
15.2 系统功能模块设计 319
15.3 系统数据分析与设计 320
15.4 物理网站的设计与实现 321
15.5 Android 移动端的设计与实现 328
15.5.1 手机端软件结构 328
15.5.2 移动端数据的创建与初始化 329
15.5.3 首页模块的设计与实现 334
15.5.4 信息列表展示页的设计与实现 356
15.5.5 搜索页面的设计及实现 358
15.5.6 个人详细信息页的设计与实现 360
15.5.7 删除功能的设计与实现 373
15.6 本章小结 374

第 1 章 Android 概述

智能手机正快速走入人们的生活,已经有越来越多的人把智能手机当作娱乐、办公的首选设备。智能手机系统显示尤为重要,本章将从智能手机操作系统的分类及其优缺点、Android 操作系统发展和优越性及其系统架构等几个方面进行介绍,使读者对手机操作系统及其 Android 有总体的了解。

1.1 智能手机操作系统简介

2012 年 7 月,中国互联网信息中心发布第 30 次《中国互联网络发展统计报告》,报告显示我国手机网民规模继续稳步增长,截至 2012 年 6 月底,我国手机网民达到 3.88 亿,较 2011 年年底增加了约 3270 万人,占总体网民比例的 72.1%。由这些统计数据不难看出,智能手机和人们的生活息息相关,因此,学习和研究智能手机软件开发,具有广阔的社会需求和工程实践意义。

智能手机是指"像个人电脑一样,具有独立的操作系统,可以由用户自行安装软件、游戏等第三方服务商提供的程序,通过此类程序不断对手机的功能进行扩充,并可以通过移动通讯网络来实现无线网络介入的这样一类手机的总称"。由于智能手机多使用 ARM 而非 X86 的 CPU 体系架构,因此,智能手机操作系统和开发环境与普通计算机有很大区别。目前,主流的智能手机操作系统有 Android、iOS、Symbian、Windows Phone 和 BlackBerry OS 等,它们占据了智能手机市场 99%以上的份额。下面对这些手机操作系统进行逐一简介。

(1) Symbian

Symbian 系统是塞班公司为手机设计的操作系统。2008 年 12 月 2 日,塞班公司被诺基亚收购。2011 年 12 月 21 日,诺基亚官方宣布放弃塞班(Symbian)品牌。由于缺乏新技术支持,塞班的市场份额日益萎缩。截止至 2012 年 2 月,塞班系统的全球市场占有量仅为 3%,中国市场占有率则降至 2.4%。2012 年 5 月 27 日,诺基亚宣布,彻底放弃继续开发塞班系统,取消塞班 Carla 的开发,但是服务将一直持续到 2016 年。2013 年 1 月 24 日晚间,诺基亚宣布,今后将不再发布塞班系统的手机,意味着塞班这个智能手机操作系统,在长达 14 年的历史之后,终将谢幕。

(2) Windows Phone

Windows Phone 是微软发布的一款手机操作系统,2010 年 2 月,微软正式向外界展示 Windows Phone 操作系统。2010 年 10 月,微软公司正式发布 Windows Phone 智能手机操作系统的第一个版本 Windows Phone 7(以下简称 WP 7),并于 2010 年年底发布了基于此平台的硬件设备。主要生产厂商有诺基亚、三星和 HTC 等,从而宣告了 Windows Mobile

系列彻底退出了手机操作系统市场。全新的 WP 7 完全放弃了 WM5，6X 的操作界面，而且程序互不兼容，并且微软完全重塑了整套系统的代码和视觉，但由于担心移动产品和整体品牌的连续性，一开始微软将其命名为"WP 7"。WP 7 曾于 2010 年 2 月 16 日更名为"Windows Phone 7 Series"，其后于 4 月 2 日取消"Series"，改回"Windows Phone 7"。

2011 年 9 月 27 日，微软发布了 Windows Phone 系统的重大更新版本"Windows Phone 7.5"，首度支持中文。Windows Phone 7.5 是微软在 WP 7 的基础上大幅优化改进后的升级版，其中包含了许多系统修正和新增的功能，以及繁体中文和简体中文在内的 17 种新的显示语言。

2012 年 6 月 21 日，微软在美国旧金山召开发布会，正式发布全新操作系统 Windows Phone 8（以下简称 WP8）。WP 8 放弃 WinCE 内核，改用与 Windows 8 相同的 NT 内核。WP 8 系统也是第一个支持双核 CPU 的 WP 版本，宣布 WP 进入双核时代，同时宣告着 WP 7 退出历史舞台。由于内核变更，WP 8 将不支持市面上所有的 WP 7.5 系统手机升级，而 WP7.5 手机只能升级到 WP 7.8 系统。WP 8 于 2012 年 10 月 11 日上市。

（3）iOS

苹果 iOS 是由苹果公司开发的手持设备操作系统。苹果公司最早于 2007 年 1 月 9 日的 Macworld 大会上公布这个系统，最初是设计给 iPhone 使用的，后来陆续套用到 iPod touch、iPad 及 Apple TV 等苹果产品上。iOS 与苹果的 Mac OS X 操作系统相同，它也是以 Darwin 为基础的，因此，同样属于类 Unix 的商业操作系统。原本这个系统名为 iPhone OS，直到 2010 年 6 月 7 日 WWDC 大会上宣布改名为 iOS。截止至 2011 年 11 月，根据 Canalys 的数据显示，iOS 已经占据了全球智能手机系统市场份额的 30%，在美国的市场占有率为 43%。

（4）Android

Android 是一种基于 Linux 的自由及开放源代码的操作系统，主要用于移动设备，如智能手机和平板电脑，由 Google 公司和开放手机联盟领导及开发。Android 操作系统最初由 Andy Rubin 开发，主要支持手机。2005 年 8 月由 Google 收购注资。2007 年 11 月，Google 与 84 家硬件制造商、软件开发商及电信营运商组建开放手机联盟共同研发改良 Android 系统。随后 Google 以 Apache 开源许可证的授权方式，发布了 Android 的源代码。第一部 Android 智能手机发布于 2008 年 10 月。Android 逐渐扩展到平板电脑及其他领域上，如电视、数码相机和游戏机等。2011 年第一季度，Android 在全球的市场份额首次超过塞班系统，跃居全球第一。2012 年 11 月数据显示，Android 占据全球智能手机操作系统市场 76% 的份额，中国市场占有率为 90%。

1.2　Android 的基本概念

Android 操作系统市场占有率越来越高，这取决于它较其他智能手机操作系统有无法比拟的优越性，本节主要介绍 Android 的发展历史及其优越性。

1.2.1　Android 的前世

Android 的诞生还要从 Andy Rubin 说起。Rubin 是硅谷著名的极客（对计算机和网络技术有狂热兴趣并投入大量时间钻研的人），他家的门铃是硅谷最昂贵的玩具——视网膜扫描仪。Rubin 很喜欢机器人，这也就是为什么他为创立的新公司取名 Android 的原因。Rubin 最初的目标是想把 Android 打造成一个可以对任何软件设计人员开放的移动终端平台。很快这个公司就获得了众人的青睐，很多人表示打算买下他的公司。而 Rubin 唯独向 Google 抛出了橄榄枝，他发了一封邮件给拉里·佩奇，表示要跟他合伙。几周之后，Google 就抢先把 Rubin 的公司买下。Google 收购 Android 时没有宣布任何计划，只是向《商业周刊》表示："我们收购 Android 是因为它拥有天才般的工程师，这些工程师具有非常棒的技术。我们非常兴奋让他们加入 Google。"

随着 Rubin 加入 Google，2007 年网络上就盛传全球最大的在线搜索服务商 Google 将进军移动通信市场，并推出资助品牌的移动终端。Google 手机的图片更是满天飞，光外形就有翻盖、滑盖、旋屏和触控等多种版本。更有人将其与苹果公司于 2007 年初推出的 iPhone 相提并论。

2007 年 11 月 5 日，Google 终于揭开谜底。Google 宣布与其他 33 家手机厂商（包括摩托罗拉、华为、宏达电、三星、LG 等）、手机芯片供应商、软硬件供应商、移动运营商联合组成开发手机联盟（Open Handset Alliance，OHA），供发布了名为 Android 的开放手机软件平台。参与开放手机联盟的这些厂商，都会基于 Android 平台来开发新的手机业务。Android 向手机厂商和移动运营商提供一个开放的平台，供他们开发创新性的应用软件。

Android 作为 Google 企业战略的重要组成部分，将进一步推进"随时随地为每个人提供信息"这一企业目标的实现。Google 的目标是让移动通信不依赖于设备甚至平台，基于此，Android 将进一步补充 Google 长期以来的移动发展战略：通过与全球各地的手机厂商和移动运营商结成合作伙伴，开发既实用又有吸引力的移动服务，并推广这些产品。

1.2.2　Android 的优点

与其他智能手机操作系统相比，Android 具有以下几个无可比拟的优点。

（1）开放性。Google 与开放手机联盟合作开发了 Android，Google 通过与运营商、设备制造商、开发商和其他有关各方结成深层次的合作伙伴关系，希望通过建立标准化、开放式的移动电话软件平台，在移动产业内形成一个开放式的生态系统。

（2）应用程序无界限。Android 上的应用程序可以通过标准 API 访问核心移动设备功能。通过互联网，应用程序可以产生它们的功能，可供其他应用程序使用。

（3）应用平等。移动设备商的应用程序可以被替换或扩展，即使是拨号程序或主屏幕这样的核心组件。

（4）快速方便的应用开发。Android 平台为开发人员提供了大量的使用库和工具，开发人员可以快速创建自己的应用。

1.3　Android 系统架构

通过上一节的介绍，可以对 Android 的优点有了初步的了解，那么，这些优越性是由什么来保证的呢？这取决于 Android 的体系架构，其体系架构如图 1.1 所示。

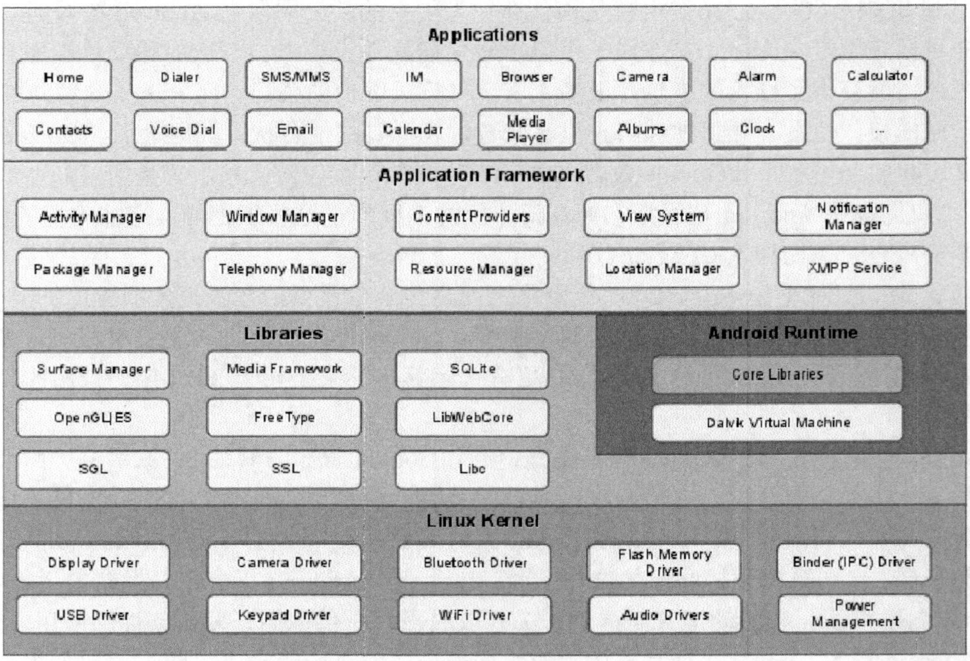

图 1.1　Android 体系架构图

由图 1.1 可以看出，Android 体系架构采用了软件叠层技术，整个架构由应用层、应用程序框架层、Android 运行时、库及 Linux 内核五层构成。

（1）应用层

Android 平台缺省包含了一系列的核心应用程序，包括电子邮件、短信、日历、地图、浏览器和联系人管理程序等，这些应用程序都是用 Java 语言编写运行在虚拟机上的。当然，作为程序员也可以用自己写的程序来替换 Android 提供的应用程序，这需要应用程序框架层来保证。

（2）应用程序框架层

这一层是进行 Android 开发的基础，开发人员可以使用这些框架来开发自己的应用程序，从而简化了程序开发的架构设计，但是必须遵守其框架的开发原则。应用程序框架层包含视图系统、内容提供器、资源管理器、通知管理器、活动管理器、窗口管理器、电话管理器和包管理器九大部分，如图 1.1 所示。

（3）Android 运行时

Android 虽然采用 Java 语言开发、编写应用程序，但却不使用 J2ME 执行 Java 程序，而是用 Android 自有的 Android 运行时（Android Runtime）来执行。Android 运行时包括核心库和 Dalvik 虚拟机两部分，如图 1.1 所示。

核心库包含两部分内容：一部分提供 Java 编程语言核心库的大多数功能；另一部分为 Android 的核心库。与标准的 Java 不同，Android 不是用一个 Dalvik 虚拟机来同时执行多个 Android 应用程序，而是每个 Android 应用程序都用一个自有的 Dalvik 虚拟机来执行。

Dalvik 虚拟机（Dalvik Virtual Machine）是一种基于寄存器的 Java 虚拟机。它是专为移动设备而设计的，它在编写时就已经设想用最少的内存资源来执行，以及支持同时执行多个虚拟机的特性。在设计方面，Dalvik 虚拟机有许多地方参考了 Java 虚拟机，不过 Dalvik 虚拟机执行的中间码并非是 Java 虚拟机执行的 Java 字节码，同时也不直接执行 Java 的类，而是依靠转换工具 dx 将 java 字节码转换为 Dalvik 虚拟机执行时特有的 dex（Dalvik Excutable）格式。

（4）系统库

　　应用程序框架层是贴近于应用程序的软件组件服务，而更底层则是 Android 的库函数（c/c++），这一部分是应用程序框架的支撑，这一层主要包括以下功能。

- 多媒体库：Android 的媒体库函数是以 PacketVideo 公司的 OpenCORE 为基础发展的，该库函数可以播放、录制多种普遍常见的影音格式。
- S 同时执行多个应用程序时，Surface Manager 会负责管理器显示与存取操作间的互动，另外，也负责将 2D 绘图与 3D 绘图进行显示上的合成。
- WebKit：它是一套网页浏览器的软件引擎，该引擎的功能不仅可供 Android 内建的网页浏览器所调用，也可以提供内嵌性网页显示效果。
- SGL：提供 Android 在 2D 绘图方面的绘图引擎。
- OpenGL ES：Android 是依据 OpenGL ES 1.0 API 标准来实现的其 3D 绘图函数库，该函数库可以用软件方式执行也可以用硬件加速方式执行，其中，3D 软件光栅处理方面已进行高度优化。
- FreeType：提供点阵字、向量字的描绘显示。
- 媒体框架：提供了对各种音、视频的支持。Android 支持多种音频、视频、静态图像格式，如 MPEG-4、H.264、MP3、AAC、ARM、JPG、PNG、GIF 等。
- SQLite：SQLite 是一套轻量级的数据库引擎，可供其他应用程序调用。
- Libc：提供了针对移动设备而优化了的 C 库。

（5）Linux 内核层

　　之前提到了 Android 平台的一个主要优点就是开放性，采用 Linux 内核则是 Android 平台开放性的基础。在 Android 平台中操作系统采用了 Linux 2.6 版的内核，它包括了显示驱动、摄像头驱动、Flash 内存驱动、Binder（IPC）驱动、键盘驱动、Wifi 驱动、Audio 驱动和电源管理等。Linux 内核层为我们在软件层和硬件层建立了一个抽象层，使得应用开发人员无需关心硬件细节。不过对手机开发商而言，如果想要 Android 平台运行到自己的硬件平台上就必须对 Linux 内核层进行修改，通常要做的工作就是为自己的硬件编写驱动程序。

1.4　本章小结

　　本章首先介绍了智能手机操作系统的分类及其优缺点，然后对 Android 操作系统及其发展和优越性进行介绍，最后讲述了 Android 如何通过系统架构来体现其优越性，使我们对手机操作系统及其 Android 有总体的了解。

第 2 章 搭建 Android 开发环境

本章主要讲述如何搭建 Android 开发环境，包括获取 SDK、Eclipse 及进行 Android Eclipse 插件设置等。在搭建好开发环境后，如何创建一个 Android 程序，并对这个程序进行简单分析。

2.1 开发前的准备工作

工欲善其事，必先利其器。在开发 Android 应用之前，必须搭建开发环境，本节主要对开发 Android 应用程序所需要的软件进行详细介绍。

（1）JDK6.0 或 JDK7.0

JDK 的版本只要是 5 以上即可，本书采用 JDK7.0，到 java 的官方网站 http://www.oracle.com/technetwork/java/index.html 下载即可获得。

（2）Eclipse

使用 MyEclipse 也可以，但是由于 MyEclipse 是收费的并且插件较多影响运行速度，因此，不建议使用。Eclipse 是一个开发源代码的、基于 Java 的可扩展的集成开发环境。Eclipse 可以集成多种插件，已完成特定语言的开发。下载地址：http://www.eclipse.org/downloads/。

（3）Android SDK

Android SDK 是 Android 应用程序开发工具包，类似于 Java 的 JDK，可以到 Android 的官方网站下载，地址：http://developer.android.com/sdk/index.html。

（4）Eclipse 的插件 ADT（Android Development Tools）

ADT 是一个专门为 Eclipse IDE 设计的旨在提供一个强大的、集成的环境来建立 Android 应用程序的插件。ADT 扩展了 Eclipse 的功能，可以快速建立 Android 项目，创建一个应用程序界面。它添加了基于 Android 框架 API 的组件，使用 Android SDK 工具调试所创建的应用程序，甚至导出签名（或未签名）APKs 以分发应用程序。在 Eclipse 中强烈建议使用 ADT 进行开发，ADT 提供了令人难以置信的提高开发应用程序的效率。下载地址和 Android SDK 相同。

准备好这些工具，就可以安装这些软件来搭建 Android 开发环境了。注意，以上提供的下载地址会由于官方的更新而产生变动，有时下载到的版本会不同，但下载方式相同，如遇问题可以参考官方的帮助文档。

2.2 搭建开发环境

在学习 Android 应用程序开发之前，我们假定读者已经有了一定的 java 开发基础，所以 JDK 的安装过程在此不再赘述。安装完 JDK，配置完环境变量后，就可以进行下一步的安装了。

2.2.1 Android SDK 的安装

在指定地址下载可得到一个 adt-bundle-windows-x86_64-20130522.zip 文件，将该文件解压缩到任意路径，作者解压到 D:\Program Files\。解压缩后在目录下得到一个 android-sdk-windows 文件夹，该文件夹下包含如下文件。

- add-ons：该目录下存放额外的附件软件。刚解压缩时为空。
- platforms：该目录下存放不同版本的 Android 版本。刚解压缩时为空。
- tools：该目录下存放了大量 Android 开发、调试工具。
- SDK Manager.exe：该程序就是 Android SDK 和 AVD（Android 虚拟设备）管理器，通过该工具可以管理 Android SDK 和 AVD。在联网的情况下，运行 SDK Manager.exe，即可看到如图 2.1 所示窗口。

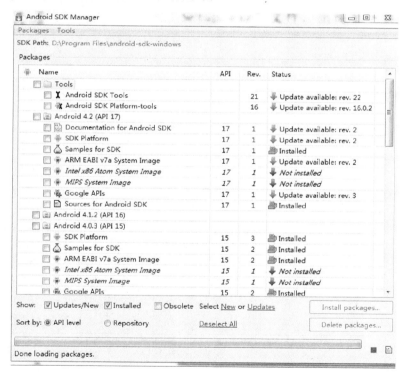

图 2.1　Android SDK Manager

该图左边为可以下载的 SDK 平台列表，选中自己需要的版本，选中后，单击【install packages...】按钮，进入下载页面，下载所需时间与网速有关，请耐心等待。

安装完成后，可以看到在 android-sdk-windows 目录下增加了如下几个文件夹。

- docs：该文件夹下存放了 Android SDK 开发文件和 API 文档等。
- platform-tools：该文件夹下存放了和 Android 平台相关的工具。

- samples：该文件夹下存放了不同 Android 平台的示例程序。

为了可以在命令行窗口使用 Android SDK 的各种工具，建议将 android-sdk-windows 目录下的 tools 子目录、platform-tools 子目录添加到系统的 Path 环境变量中，方法和设置 JDK 环境变量相同。

2.2.2 Eclipse 和 ADT 安装

Eclipse 的安装比较简单，直接找到 eclipse-java-juno-SR2-win32-x86_64 文件，将其解压缩到指定的目录即可。打开解压缩后的文件夹，双击 eclipse.exe 即可运行 Eclipse。

启动 Eclipse，即可为其安装 ADT 插件。选择主菜单上的【Help】→【Install New Software】命令，出现如图 2.2 所示对话框，单击【Add】按钮，在对话框的 name 一栏中输入"ADT"，然后单击【Archive...】按钮，浏览和选择已经下载的 ADT 插件压缩文件的路径，如图 2.3 所示。

图 2.2　安装新插件窗口

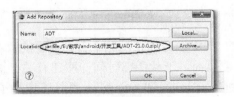

图 2.3　定位 ADT 所在目录

单击【OK】按钮，会看到这个插件的信息，选中 Developer Tools，然后单击【Next】按钮，出现如图 2.4 所示界面，耐心等待几分钟，出现如图 2.5 所示界面。

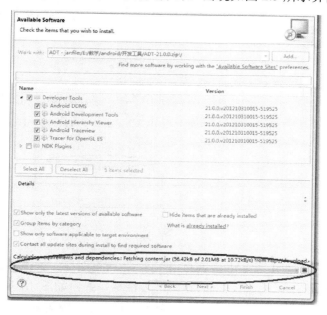

图 2.4　安装进度

图 2.5　所安装工具列表

一直单击【Next】按钮，直到出现如图 2.6 所示界面，选择【I accept the term of the license agreements】复选框，然后单击【Finish】按钮，等待安装完成，中间可能会出现警告信息，点击【OK】按钮继续即可，安装完成，会提示重启 Eclipse，至此，ADT 插件安装完毕。

图 2.6　许可协议

重启 Eclipse 后，会在工具条上看到多出了 两个图标，这说明 ADT 安装成功了，然后还要指定 Android SDK 的路径。首次单击 图标，弹出如图 2.7 所示界面。

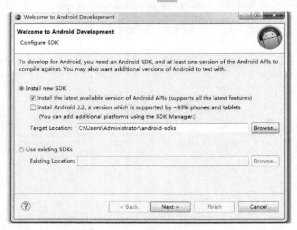

图 2.7　定位 SDK

Install new SDK 下面选中第一项，然后在 Target Location 一栏定位到之前解压缩得到的 android-sdk-windows 文件，Use existing SDKs 也定位到同样的位置，如图 2.8 所示。一直单击【Next】按钮，直到单击【Finish】按钮为止。至此，Android 开发环境搭建完成，下面就可以创建一个 Android 项目来小试牛刀了！

2.2.3　创建和启动 AVD（Android Virtual Device）

搭建好开发环境就可以创建 Android 应用程序了，当完成一个 Android 应用程序后需要测试一下程序的运行结果，而 Android 应用程序必须在 3G 手机上测试，如果没有支持

Android 平台的 3G 手机该怎么办呢？SDK 为用户提供了方便，在 SDK 中集成了 Android 虚拟设备 AVD，利用 AVD 可以创建各种模拟手机，利用模拟手机可以获得跟实际手机一样的体验。通常有两种方式管理 AVD：一种是图形界面方式；另一种是命令行方式。

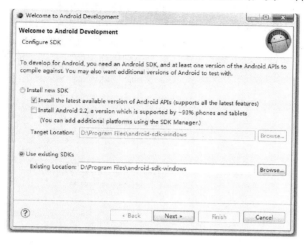

图 2.8　使用已有的 SDK

1．利用图形界面管理 AVD

利用图形界面管理 AVD，首先需要利用 AVD 管理工具创建一个 AVD。打开 Eclipse，单击工具栏上的 图标，弹出如图 2.9 所示对话框。其次，单击右侧的【New】按钮，弹出如图 2.10 所示对话框，在此对话框中需要对创建的 AVD 属性进行相应的设置，其中重要属性含义极其取值如表 2.1 所示。最后，设置好属性值后，单击【OK】按钮，返回到图 2.9 所示的 AVD 管理器界面，在此列出了所有创建的 AVD。选中一个 AVD，单击管理器界面右侧的【Start】按钮，即可启动 AVD，启动所耗时间根据机器配置有所差别。本书创建的 AVD 启动之后的界面如图 2.11 所示。

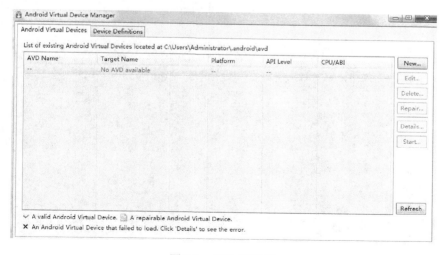

图 2.9　AVD 管理器

表 2.1　AVD 重要属性及其含义

属性	含义及取值
AVD Name	该属性代表创建的 AVD 名字，可以随意指定
Device	指定模拟器的屏幕尺寸和型号，可以在下拉列表中选择
Target	指创建的 AVD 搭载的 Android 版本，可以在下拉列表中选择
SD Card	Size 选项，指该模拟器所安装的 SDCard 的大小。创建的 SDCard 是以镜像文件的形式存放，file 属性指定该镜像文件存放于电脑硬盘上的位置，一般保留默认值即可

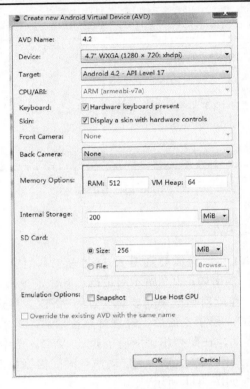

图 2.10　创建新的 AVD

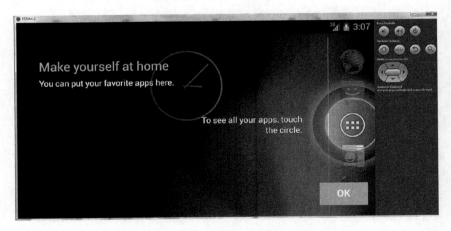

图 2.11　启动完成的模拟器

2. 通过命令行管理 AVD

在命令行下管理 AVD 需要借助 android 命令（位于 Android SDK 安装目录的 tools 子目录下），如果直接执行 android 命令将会启动 Android SDK 和 AVD 管理器。此外，该命令还支持如下子命令。

- list：列出机器上所有已经安装的 Android 版本和 AVD 设备。
- list avd：列出机器上所有已经安装的 AVD 设备。
- list target：列出机器上所有已经安装的 Android 版本。
- Create avd：创建一个 AVD 设备。
- Move avd：移动或重命名一个 AVD 设备。
- Delete avd：删除一个 AVD 设备。
- Update avd：升级一个 AVD 设备使之复合新的 SDK 环境。
- Create project：创建一个新的 Android 项目。
- Update project：更新一个已有的 Android 项目。
- Create test-project：创建一个新的 Android 测试项目。
- Update test-project：更新一个已有的 Android 测试项目。

如果希望查看系统上已经安装的 AVD 设备，则在命令行窗口运行 android list 或 android list avd 命令即可，如图 2.12 所示。

图 2.12　运行 android list avd 命令后的结果

如果要创建一个新的 AVD，可执行如下命令。

```
Android create avd -n <avd 名称> -t <android 版本> -p <avd 设备保存位置> -s <选择 AVD 皮肤>
```

在上面的 create avd 子命令中，只有-n 和-t 选项是必须选的，其余都是可选的。如果不设置-p 选项，创建的 AVD 设备默认保存在 C:\Users\Administrator\.android 目录下（以 Windows 7 为例）。下面使用命令创建一个名为 test，运行 Android4.0.3 的模拟器。

```
Android create avd -n test -t 8
```

上面的命令中 8 是 Android 4.03 的代号，执行命令，系统会提醒用户是否需要定制 AVD 硬件，可以选择 yes 或 no，如果输入 yes，即可开始定制 AVD 硬件的各种选项，定制完成后系统开始创建 AVD 设备，如果选择 no，则直接开始创建 AVD 设备。创建后通过命令查看当前创建的所有 AVD，如图 2.13 所示。

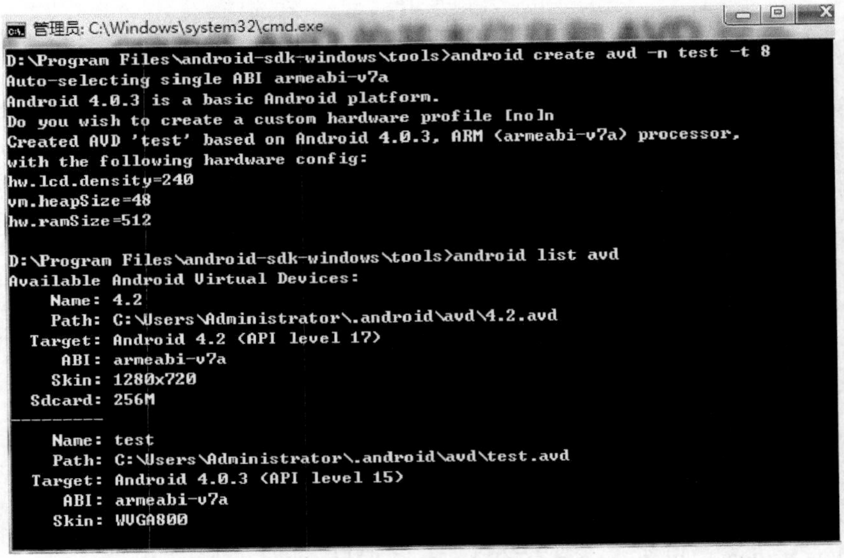

图 2.13 创建新的 AVD 设备并列出当前所有 AVD 设备

从图中可以看到，当前系统中有两个 AVD 设备（之前通过图形界面创建了一个名字为 4.2 的 AVD 设备），我们可以在 C:\Users\Administrator\.android 看到一个 avd 子目录，该子目录下包含了两个文件和两个文件夹。

- 4.2.avd 和 4.2.ini：其中，4.2.avd 目录下有一个 userdata.img 文件，它是 AVD 中用户数据的镜像。还有一个 sdcard.img，它是该 AVD 所使用的虚拟 SD 卡的镜像。
- test.avd 和 test.ini：文件的含义同上。

当然也可以通过命令行方式启动 AVD 设备，在 Android SDK 安装目录的 tools 子目录下有一个 emulator.exe 文件，它就是 Android 模拟器。这个模拟器做的十分出色，几乎可以模拟真实手机的绝大部分功能，当然它只是模拟，不要指望用模拟器与你现实中的朋友发短信或者打电话。使用 emulator.exe 启动模拟器有两种方法。

- emulator -avd <avd 名称>。
- emulator -data 镜像文件名称。

2.3 构建 Android 应用程序

完成了 Android 开发环境的设置及模拟器的创建和启动，相信大家很想体验一下 Android 开发的魔力所在。本节将带领大家逐步构建第一个 Android 应用程序。

2.3.1 使用 Eclipse 创建 Android 应用程序

在 Eclipse 中安装 ADT 插件后，开发 Android 应用就会变得非常方便，因为 Eclipse 会自动完成许多工作，具体步骤如下。

打开 Eclipse，选择主菜单中的【File】→【Project...】命令，弹出如图 2.14 所示窗口，单击 Android 选项，选中下面的 Android Application Project，单击【Next】按钮，弹出如图 2.15 所示窗口，在此填写应用程序的基本属性信息，属性说明如表 2.2 所示。设置好属性值后，一直单击【Next】按钮，直到出现如图 2.16 所示界面。

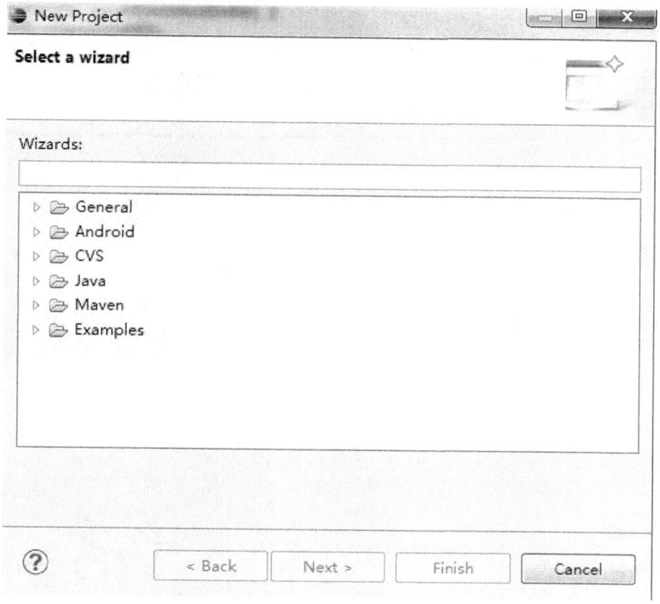

图 2.14　选择项目类别

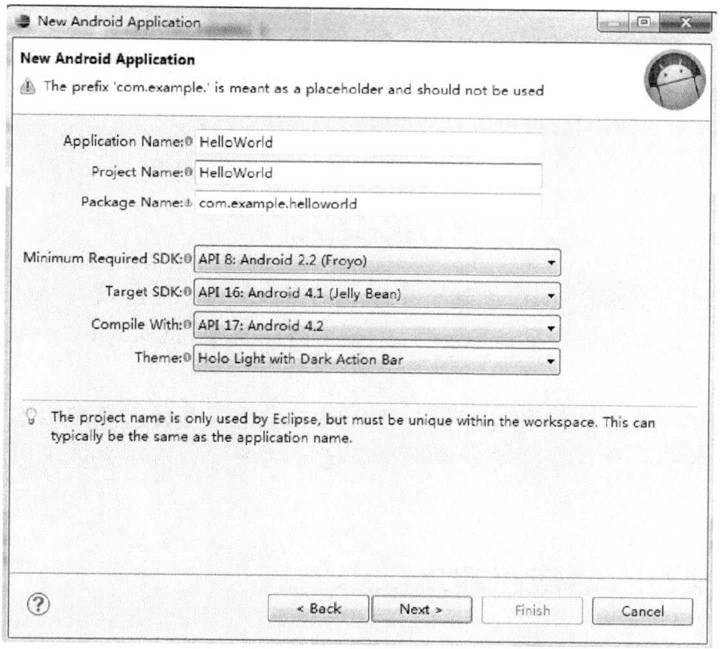

图 2.15　项目属性

表 2.2 项目属性表

名称	属性说明
Application Name	应用程序的名字,它将显示在应用程序的标题栏上
Project Name	包含这个项目的文件夹名称
Package Name	包名称,和 Java 应用程序的包概念相同
Minimum Required SDK	此值对应应用程序要求的最低 API 版本,如果设备上的 API 版本低于此值,则应用程序无法在设备上运行
Target SDK	要运行应用程序的目标模拟器的 SDK 版本
Compile With	编译应用程序所使用的版本

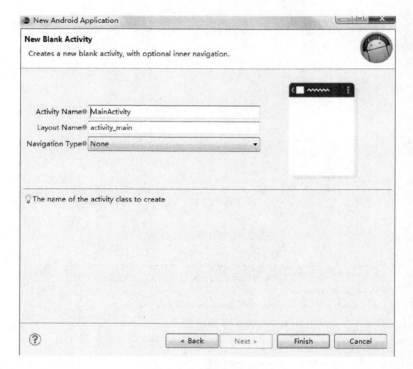

图 2.16　创建 Activity

　　ADT 插件会自动为应用程序创建一个默认的 Activity,在此填写 Activity 的名字及对应布局文件的名字,本例保持默认值不变。单击【Finish】按钮,第一个 Android 应用程序创建完成。

　　截止目前为止,虽然并没有写任何一行代码,但是该项目已经可以运行了,这是由于使用 ADT 生成的每一个项目本身就是一个可运行的项目。接下来就可以在模拟器上执行这个项目了。

2.3.2　运行 Android 应用程序

　　运行 Android 项目是否需要首先启动模拟器呢?运行之前启动模拟器是可以的,这样运行 Android 项目之后会自动发布到已经启动的模拟器上。如果运行之前有多于一个的模拟器已经启动，那么，在运行时会有一个界面提示选择要发布的目标模拟器。如果运行之前没有启动任何模拟器，那么运行代码后，会自动启动一个默认的模拟器。运行 Android 项目最常用的方式如下。

在 Eclipse 的【Package explorer】视图中，右键单击新建的项目【HelloWorld】，选择【Run As】后单击【Android Application】按钮即可运行项目。

项目发布的目标模拟器可以自行设定，右键单击项目，选择【Run As】，然后单击【Run Configuration】按钮，弹出如图 2.17 所示窗口。在窗口右边选择 Target 标签，指定运行的目标模拟器。HelloWorld 项目运行的结果如图 2.18 所示。

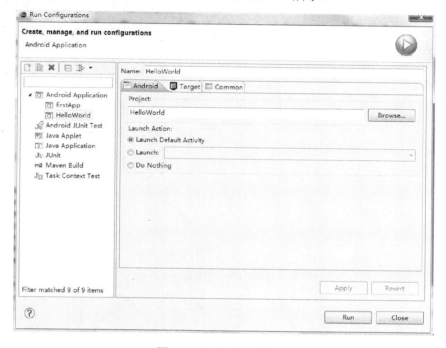

图 2.17 运行时配置窗口

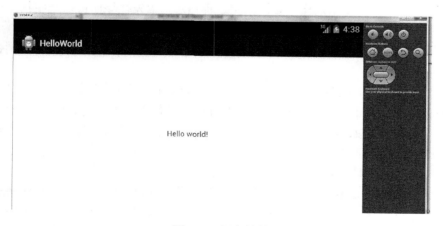

图 2.18 运行结果

2.3.3 通过 DDMS 调试 Android 应用程序

当 Android 应用程序在模拟器上运行时，用户甚至看不到程序运行的过程，在控制台也看不到程序的输出，那么该如何调试 Android 应用呢？不用担心，Android 已经为用户考虑好了这个问题。Android 提供了一个 DDMS 调试环境，DDMS 是 Dalvik Debug Monitor Service 的简称，是一个功能非常强大的调试环境。运行如下命令：monitor.bat 即可看到如图 2.19 所示窗口，该窗口为 Android 调试器窗口。

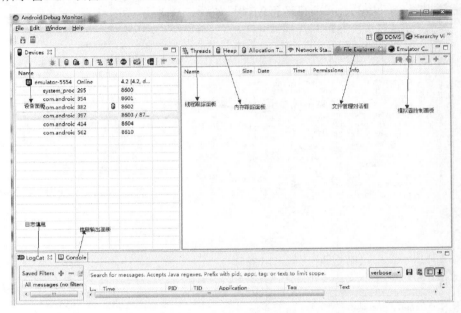

图 2.19　DDMS 调试窗口

DDMS 窗口中有如下几个重要的面板。
- 设备面板：该面板会列出当前所有运行的模拟器，并列出个模拟器内的所有进程信息。如果需要查看指定模拟器或指定进程信息，应先在该面板内选中指定模拟器或进程。
- 信息输出面板：该面板位于 DDMS 窗口下方，相当于传统 Java 应用控制台，因此非常重要。
- 线程跟踪面板：该面板可用于查看指定进程内所有正在执行的线程的状态。
- Heap 内存跟踪面板：该面板可用于查看指定进程内堆内存的分配和回收信息。
- 模拟器控制面板：该面板用于让模拟器拨打电话、发送短信等，还可以虚拟设置模拟器的位置信息等。
- 文件管理对话框：可以在模拟器和本地文件之间进行导入和导出。

实际上，如果在 Eclipse 中安装了 ADT 插件，那么 Eclipse 就会将 DDMS 集成进来，在 Eclipse 中可以直接切换到 DDMS 视图（Perspective）。单击 Eclipse 中右上角的 图 （Open Perspective 按钮，弹出如图 2.20 所示窗口。在弹出窗口中选择 DDMS 就可以打开 DDMS

调试窗口。

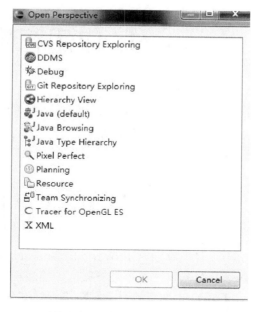

图 2.20 Open Perspective 窗口

2.4 本章小结

要开发 Android 应用程序，首先要做充足的开发前准备工作，本章首先详细阐述了如何获取 SDK、Eclipse 及进行 Android Eclipse 插件设置等，其次讲述如何搭建 Android 开发环境，最后讲述在搭建好平台上，如何创建一个简单的 Android 程序，并对这个程序进行了简单的分析。使读者对 Android 应用程序开发有一个初步的了解。

第 3 章　Android 应用程序剖析

用户要想编写出复杂的应用程序，首先要对 Android 应用程序的构成及程序的内部执行流程有一个清晰的了解。本章通过对一个简单的应用程序的深入剖析，使读者对 Android 应用程序的构成及执行流程有个清晰的了解。

3.1　Android 应用程序目录结构

之前我们已经开发了一个项目名称为 HelloWorld 的 Android 应用程序，也许你很疑惑，好像什么都没做，只是输入了几个名字，点了几下鼠标，应用程序就可以运行。这里面到底发生了什么？本节将对 HelloWorld 程序的目录结构进行详细分析。图 3.1 所示是 HelloWorld 在 Eclipse 中的目录层次结构，下面对其中主要的目录极其文件进行介绍。

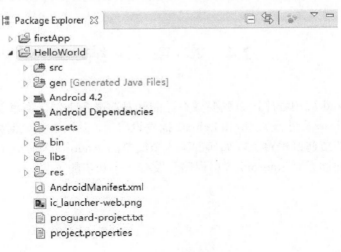

图 3.1　项目目录结构

项目的根下有几个重要的文件（夹），下面详细讲解每个文件夹的功能和作用。

src/——专门存放编写的 java 源代码的包。

android 2.1/——存放 Android 自身的 jar 包。

gen/——该目录不用开发人员维护，但又非常重要的目录，该目录用来存放由 Android 开发工具所生成的目录，该目录下的所有文件都不是用户创建的，而是由 ADT 自动生成的，该目录下的 R.java 文件非常重要，后面会详细介绍。

assets/——该目录用来存放应用中用到的类似于视频文件、MP3 的一些媒体文件。

res/——是 resource 的缩写，为资源目录，该目录可以存放一些图标、界面文件、应用中用到的文字信息。

AndroidManifest.xml——该文件是系 Android 项目的系统清单文件，它用于控制 Android 应用的名称、图标、访问权限等整体属性。

default.properties——该文件一般也不需要手工去更改，文件存放了项目对应的一些环境配置，如应用要求运行的最低 Android 版本。

资源被编译到最终的 APK 文件里。Android 创建了一个被称为 R 的类，这样在 Java 代码中可以通过它关联到对应的资源文件。

接下来对 res/的子目录做更加详细的说明。

（1）res/drawable

res/目录下有三个 dawable 文件夹，区别只是将图标按分辨率高低来放入不同的目录，其中，【drawable-hdpi】用来存放高分辨率的图标，【drawable-mdpi】用来存放中等分辨率的图标，【drawable-ldpi】用来存放低分辨率的图标。程序运行时可以根据手机分辨率的高低选取相应目录下的图标。不过，如果不想准备过多图片，那么也可以只准备一张图标将其放入三个目录的任何一个中去。

（2）res/values 文件夹

① strings.xml

用来定义字符串和数值，在 Activity 中使用 getResources().getString(resourceId)或 getResources().getText(resourceId) 取得资源。打开 helloworld 项目的 string.xml，可以看到如下内容。

```
<?xml version="1.0" encoding="utf-8"?>
<resources>

    <string name="app_name">HelloWorld</string>
    <string name="hello_world">Hello world!</string>
    <string name="menu_settings">Settings</string>

</resources>
```

每个 string 标签声明了一个字符串，name 属性指定其引用名。为什么需要把应用中出现的文字单独放在 string.xml 文件中呢？原因有如下两点。

一是为了国际化，Android 建议将在屏幕上显示的文字定义在 strings.xml 中，如果今后需要进行国际化，如开发的应用本来是面向国内用户的，当然要在屏幕上使用中文，而如今要让应用走向世界，打入日本市场，当然需要在手机屏幕上显示日语，如果没有把文字信息定义在 strings.xml 中，就需要修改程序内容了。但当我们把所有屏幕上出现的文字信息都集中存放在 strings.xml 文件之后，只需要再提供一个 strings.xml 文件，把里面的汉字信息都修改为日语，再运行程序时，Android 操作系统会根据用户手机的语言环境和国家来自动选择相应的 strings.xml 文件，这时手机界面就会显示出日语，这样做国际化非常方便。二是为了减少应用的体积，降低数据冗余。假设在应用中要使用"我们一直在努力"这段文字 10 000 次，如果不将"我们一直在努力"定义在 strings.xml 文件中，而是在每次使用时直接写上这几个字，这样下来程序中将有 70 000 个字，这 70 000 个字占 136KB 的空间。而由于手机的资源有限，其 CPU 的处理能力及内存是非常有限的，136KB 对手机程序来说是个不小的空间，在做手机应用时一定要记住"能省内存就省内存"，而如果将这几个字定义在 strings.xml 中，在每次使用到的地方通过 Resources 类来引用该文字，只占用了 14B，对降低应用体积效果非常有效。当然我们在开发时并不会用到这么多的文字信息，但是"不以善小而不为，不以恶小而为之"，作为手机应用开发人员，一定要养成良好的编程习惯。

② styles.xml

用来定义样式。打开本项目的 style.xml 文件，内容如下。

```xml
<resources>
    <!--
        Base application theme, dependent on API level. This theme is replaced
        by AppBaseTheme from res/values-vXX/styles.xml on newer devices.
    -->
    <style name="AppBaseTheme" parent="android:Theme.Light">
        <!--
            Theme customizations available in newer API levels can go in
            res/values-vXX/styles.xml, while customizations related to
            backward-compatibility can go here.
        -->
    </style>
    <!-- Application theme. -->
    <style name="AppTheme" parent="AppBaseTheme">
        <!-- All customizations that are NOT specific to a particular API-level
             can go here. -->
    </style>
</resources>
```

注意：Android 中的资源文件不要以数字作为文件名，这样会导致错误。

（3）res/layout 目录下的布局文件

本例中的布局文件是 ADT 默认自动创建的 activity_main.xml 文件。可以用两种方式 Graphical Layout 者 xml 清单显示其中的内容，在 Eclipse 中，这两种查看方式可以随意切换。双击打开此 xml 文件，内容如下。

```xml
<RelativeLayout xmlns:android="http://schemas.android.com/apk/res/android"
    xmlns:tools="http://schemas.android.com/tools"
    android:layout_width="match_parent"
    android:layout_height="match_parent"
    tools:context=".MainActivity" >

    <TextView
        android:layout_width="wrap_content"
        android:layout_height="wrap_content"
```

```
            android:layout_centerHorizontal="true"
            android:layout_centerVertical="true"
            android:text="@string/hello_world" />
</RelativeLayout>
```

与在网页中布局中使用 HTML 文件相同，Android 在 XML 文件中使用 XML 元素来设定屏幕布局。每个文件包含整个屏幕或部分屏幕，被编译进一个视图资源，可以被传递给 Activity.setContentView 或被其他布局文件引用。文件保存在工程的 res/layout/ 目录下，它被 Android 资源编辑器编译。

3.1.1 gen/目录下的 R.java 文件详解

R.java 文件中默认有 attr 、drawable、layout、string 等四个静态内部类，每个静态内部类分别对应一种资源，如 layout 静态内部类对应 layout 中的界面文件，其中，每个静态内部类中的静态常量分别定义一条资源标识符，如 "public static final int activity_main=0x7f030000;" 对应的是 layout 目录下的 activity_main.xml 文件。具体的对应关系，如图 3.2 所示。

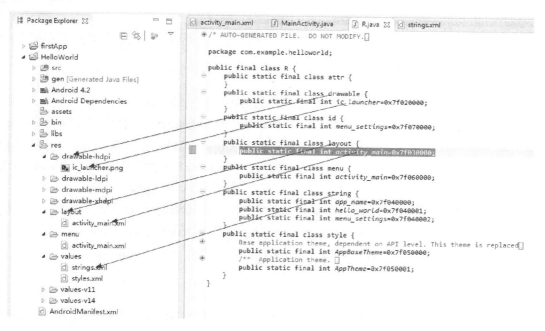

图 3.2　R.java 中的资源的对应关系图

现在已经理解了 R.java 文件中内容的来源，也即当开发者在 res/ 目录中任何一个子目录中添加相应类型的文件后，ADT 会在 R.java 文件中相应的匿名内部类当中自动生成一条静态 int 类型的常量，对添加的文件进行索引。如果在 layout 目录下再添加一个新的界面，那么在 public static final class layout 中也会添加相应的静态 int 常量。相反，在 res 目录下删除任何一个文件，其在 R.java 中对应的记录会被 ADT 自动删除。再比如说，在

strings.xml 添加一条记录，在 R.java 的 string 内部类中也会自动增加一条记录。R.java 文件会给开发程序带来很大的方便，如在程序中使用 "public static final int ic_launcher=0x7f020000;" 就可以找到其对应的 ic_launcher 图片。R.java 文件除了有自动标识资源的"索引"功能外，还有另一个主要的功能，当 res 目录中的某个资源在应用中没有被使用到，在该应用被编译时系统就不会把对应的资源编译到该应用的 APK 包中，这样可以节省 Android 手机的资源。

3.1.2 组件标识符

通过对 R.java 文件的介绍，读者已经了解了 R 文件的索引作用，它可以检索到应用中需要使用的资源。下面介绍如何通过 R.java 文件来引用到所需要的资源。

（1）在 Java 程序当中，可以按照 Java 的语法来引用。

① R.resource_type.resource_name

注意，resource_name 不需要文件的后缀名。

比如，上面的 ic_launcher.png 文件的资源标识符可以通过如下方式获取。

R.drawable.ic_launcher

② android.R.resource_type.resource_name

Android 系统本身自带了很多资源，用户也可以进行引用，只是需要在前面加上 "android." 以声明该资源来自 Android 系统。

（2）在 XML 文件中引用资源的语法如下：

① @[package:]type/name，使用自己包下的资源可以省略 package。

在 xml 文件中，如 activity_main.xml 以及 AndroidMainfest.xml 文件中通过 "@drawable/ic_launcher" 的方式获取。其中 "@" 代表 R.java 类，"drawable" 代表 R.java 中的静态内部类 "drawable"，"/ic_launcher" 代表静态内部类 "drawable" 中的静态属性 "ic_launcher"。该属性可以指向 res 目录下的 "drawable-*dpi" 中的 ic_launcher.png 图标。

其他类型的文件类似，凡是在 R 文件中定义的资源都可以通过 "@ Static_inner_classes_name / resourse_name" 的方式获取。如 "@id/button"，"@string/app_name"。

② 如果访问的是 Android 系统中带的文件，则要添上包名 "android:"。

如 android:textColor="@android:color/red"。

（3）"@+id/string_name" 表达式

顺便说一下，在布局文件当中我们需要为一些组件添加 Id 属性作为标示，可以使用如下的表达式 "@+id/string_name"，其中，"+" 表示在 R.java 的名为 id 的内部类中添加一条记录。如"@+id/button" 的含义是在 R.java 文件中的 id 这个静态内部类添加一条常量名为 button，该常量就是该资源的标识符。如果 id 这个静态内部类不存在，则会先生成它。通过该方式生成的资源标识符，仍然可以以 "@id/string_name" 的方式引用。示例代码片段如下。

```
< RelativeLayout
```

```
android:layout_width = "fill_parent"
android:layout_height = "wrap_content"
>
< Button
android:layout_width = "wrap_content"
android:layout_height = "wrap_content"
android:text = "@string/cancle_button"
android:layout_alignParentRight = "true"
android:id = "@+id/cancle" />
< Button
android:layout_width = "wrap_content"
android:layout_height = "wrap_content"
android:layout_toLeftOf = "@id/cancle"
android:layout_alignTop = "@id/cancle"
android:text = "@string/ok_button" />
</ RelativeLayout >
```

其中，android:id="@+id/cancle" 将其所在的 Button 标识为 cancle，在第二个 Button 中通过 "@id/cancle" 对第一个 Button 进行引用。

3.1.3　AndroidMainfest.xml 详细介绍

每个应用程序都有一个功能清单文件 AndroidManifest.xml（一定是这个名字）在它的根目录里，该清单文件给 Android 系统提供了关于这个应用程序的基本信息，系统在运行任何程序代码之前必须知道这些信息。今后，我们开发 Activity、Broadcast、Service 之后，都要在 AndroidManifest.xml 中进行定义。另外，如果使用到系统自带的服务如拨号服务、应用安装服务、GPRS 服务等都必须在 AndroidManifest.xml 中声明权限。

AndroidManifest.xml 主要包含以下功能。

- 命名应用程序的 Java 应用包，这个包名用来唯一标识应用程序。
  - ➤ 描述应用程序的组件——活动、服务、广播接收者、内容提供者；对实现每个组件和公布其功能（比如，能处理哪些意图消息）的类进行命名，这些声明使得 Android 系统了解这些组件及它们在什么条件下可以被启动。
  - ➤ 决定应用程序组件运行在哪个进程中。
- 声明应用程序所必须具备的权限，用以访问受保护的部分 API，以及和其他应用程序交互。
- 声明应用程序其他的必备权限，用以组件之间的交互。
- 列举测试设备 Instrumentation 类，用来提供应用程序运行时所需的环境配置及其他信息，这些声明只在程序开发和测试阶段存在，发布前将被删除。
- 声明应用程序所要求的 Android API 的最低版本级别。
- 列举 application 所需要链接的库。

下面以 HelloWorld 项目的功能清单文件为例进行讲解。

```xml
<?xml version="1.0" encoding="utf-8"?>
<manifest xmlns:android="http://schemas.android.com/apk/res/android"
    package="com.example.helloworld"
    android:versionCode="1"
    android:versionName="1.0" >

    <uses-sdk
        android:minSdkVersion="8"
        android:targetSdkVersion="16" />

    <application
        android:allowBackup="true"
        android:icon="@drawable/ic_launcher"
        android:label="@string/app_name"
        android:theme="@style/AppTheme" >
        <activity
            android:name="com.example.helloworld.MainActivity"
            android:label="@string/app_name" >
            <intent-filter>
                <action android:name="android.intent.action.MAIN" />

                <category android:name="android.intent.category.LAUNCHER" />
            </intent-filter>
        </activity>
    </application>

</manifest>
```

以下详细讲解各个标签。

（1）<manifest>元素

```xml
<manifest xmlns:android="http://schemas.android.com/apk/res/android"
    package="com.example.helloworld"
    android:versionCode="1"
    android:versionName="1.0" >
```

该元素是 AndroidManifest.xml 文件的根元素，该元素为必选。其中，根据 xml 文件的语法，"xmlns:android"指定该文件的命名空间。功能清单文件会使用"http://schemas.android.com/apk/res/android"所指向的一个文件。"package"属性是指定 Android 应用所在的包。"android:versionCode"指定应用的版本号。如果应用需要不断升级，在升级时应该修改该值。"android:versionName"是版本名称，名称的取定可根据爱好而定。

（2）<application>元素

```xml
<application
    android:allowBackup="true"
    android:icon="@drawable/ic_launcher"
```

```
        android:label="@string/app_name"
        android:theme="@style/AppTheme" >
        <activity
            android:name="com.example.helloworld.MainActivity"
            android:label="@string/app_name" >
            <intent-filter>
                <action android:name="android.intent.action.MAIN" />
                <category android:name="android.intent.category.LAUNCHER" />
            </intent-filter>
        </activity>
    </application>
```

<application> 是非常重要的一个元素，今后，开发的许多组件都会在该元素下定义，该元素为必选元素。<application> 的 "icon" 属性用来设定应用的图标。<application> 的 "label" 属性用来设定应用的名称。指定其属性值所用的表达式 "@string/app_name" 含义与上面的表达式 "@drawable/ic_launcher" 相同，同样指向 R.java 件中的 string 静态内部类中的 app_name 属性所指向的资源。这里它指向的是 "strings.xml" 文件中的一条记录 " app_name"，其值为 "HelloWorld"，因此，这种表达方式等价于 android:label = "HelloWorld"。

（3）<activity>元素

<activity> 元素的作用是注册一个 Activity 信息，当我们在创建 "HelloWorld" 这个项目时，指定了【Create Activity】属性为 "MainActivity"，之后 ADT 在生成项目时帮我们自动创建了一个 Activity 名称就是 "MainActivity.java"，Activity 在 Android 中属于组件，它需要在功能清单文件中进行配置。<activity>元素的 "name" 属性指定的是该 Activity 的类名。<activity>元素的 "label" 属性表示 Activity 所代表的屏幕标题，其属性值的表达式在上面已经介绍过了，不再赘述。该属性值在 AVD 运行程序到该 Activity 所代表的界面时，会在标题上显示该值。

（4）<intent-filter>元素

翻译成中文是 "意图过滤器"。首先简单介绍什么是意图（Intent）。应用程序的核心组件（活动、服务和广播接收器）通过意图被激活，意图代表的是你要做的一件事情，代表你的目的，Android 寻找一个合适的组件来响应这个意图，如果需要会启动这个组件一个新的实例，并传递给这个意图对象，后面会有详细介绍。

组件通过意图过滤器（intent filters）通告它们所具备的功能——能响应的意图类型。由于 Android 系统在启动一个组件前必须知道该组件能够处理哪些意图，那么意图过滤器需要在 manifest 中以<intent-filter>元素指定。一个组件可以拥有多个过滤器，每一个描述该组件所具有的不同能力。一个指定目标组件的显式意图将会激活那个指定的组件，意图过滤器不起作用。但是一个没有指定目标的隐式意图只在它能够通过组件过滤器时才能激活该组件。

第一个过滤器 ——

```
<intent-filter>
```

```xml
            <action android:name="android.intent.action.MAIN" />

            <category android:name="android.intent.category.LAUNCHER" />
        </intent-filter>
```

是最常见的，它表明这个 activity 将在应用程序加载器中显示，就是用户在设备上看到的可供加载的应用程序列表。换句话说，这个 activity 是应用程序的入口，是用户选择运行这个应用程序后所见到的第一个 activity。

（5）权限 Permissions

HelloWorld 项目的功能清单文件中并没有出现<Permissions>元素，但是 Permission 也是一个非常重要的节点，在后面的学习中会经常用到。Permission 是代码对设备上数据的访问限制，这个限制被引入来保护可能会被误用而曲解或破坏用户体验的关键数据和代码。如拨号服务、短信服务等。每个许可被一个唯一的标签所标识。这个标签常常指出了受限的动作。

如申请发送短信服务的权限需要在功能清单文件中添加如下语句。

```xml
<uses-permission android:name="android.permission.SEND_SMS"/>
```

一个功能（feature）最多只能被一个权限许可保护。如果一个应用程序需要访问一个需要特定权限的功能，它必须在 manifest 元素内使用 <uses-permission> 元素来声明这一点。这样，当应用程序安装到设备上之后，安装器可以通过检查签署应用程序认证的机构来决定是否授予请求的权限，在某些情况下，会询问用户。如果权限已被授予，那应用程序就能够访问受保护的功能特性。如果没有，访问将失败，但不会给用户任何通知。因此我们在使用一些系统服务，如拨号、短信、访问互联网、访问 SDCard 时一定要记得添加相应的权限，否则会出现一些难以预料的错误。

应用程序还可以通过权限许可来保护它自己的组件（活动、服务、广播接收器、内容提供者）。它可以利用 Android 已经定义（列在 android.Manifest.permission 中）或其他应用程序已声明的权限许可，或者定义自己的许可。一个新的许可通过<permission>元素声明。比如，一个 Activity 可以用下面的方式保护。

```xml
<manifest . . . >
<permission android:name="com.example.project.DEBIT_ACCT" . . . />
. . .
<application . . .>
<activity android:name="com.example.project.FreneticActivity" . . . >
android:permission="com.example.project.DEBIT_ACCT"
. . . >
. . .
</activity>
</application>
. . .
<uses-permission android:name="com.example.project.DEBIT_ACCT" />
. . .
```

```
</manifest>
```

注意：在这个例子里，这个 DEBIT_ACCT 许可并非仅仅在<permission>元素中声明，如果该应用程序的其他组件要使用到该组件，那么它同样声明在<uses-permission>元素里。

（6）库 Libraries

每个应用程序都链接到缺省的 Android 库，这个库包含了基础应用程序开发包（实现了基础类如活动、服务、意图、视图、按钮、应用程序、内容提供者等）。然而，一些包处于它们自己的库中。如果你的应用程序使用了其他开发包中的代码，它必须显式地请求链接到它们。这个 manifest 必须包含一个单独的 <uses-library> 元素来命名每一个库。如在进行单元测试的时候需要引入其所需要的库。

代码片段如下：

```
<application android:icon="@drawable/icon"
            android:label="@string/app_name">
    <uses-library android:name="android.test.runner" />
</application>
```

3.2 Android 应用程序的执行流程

经过前面对 Android 项目目录结构的介绍，以及相关文件的讲解，我们对许多细节已经有所了解，只是 Android 程序是如何执行呢？下面做一个总结。

发布程序到手机上之后，当双击该应用的图标时，系统会将这个点击事件包装成一个 Intent 该 Intent 包含两个参数。

```
{ action : "android.intent.action.MAIN",
  category : "android.intent.category.LAUNCHER" },
```

这个意图被传递给 HelloWorld 这个应用之后在应用的功能清单文件中寻找与该意图匹配的意图过滤器，如果匹配成功，找到相匹配的意图过滤器所在的 Activity 元素，再根据<activity> 元素的 "name" 属性来寻找其对应的 Activity 类。接着 Android 操作系统创建该 Activity 类的实例对象，对象创建完成之后，会执行到该类的 onCreate 方法，此 OnCreate 方法是重写其父类 Activity 的 OnCreate 方法而实现。onCreate 方法用来初始化 Activity 实例对象。如下是 HelloWorld.java 类中 onCreate 方法的代码。

```
@Override
protected void onCreate(Bundle savedInstanceState) {
    super.onCreate(savedInstanceState);
    setContentView(R.layout.activity_main);
}
```

其中，super.onCreate(savedInstanceState)的作用是调用其父类 Activity 的 OnCreate 方法来实现对界面的画图绘制工作。在实现自己定义的 Activity 子类的 OnCreate 方法时一定要记得调用该方法,以确保能够绘制界面。

setContentView(R.layout. main) 的作用是加载一个界面,该方法中传入的参数是"R.layout. activity_main", 其含义为 R.java 类中静态内部类 layout 的静态常量 activity_main 的值,而该值是一个指向 res 目录下的 layout 子目录下 activity_main 文件的标识符,因此代表着显示 activity_main 所定义的画面。

关于 Activity 类的执行流程及其生命周期会在后面的部分详细讲解。

Android 程序执行的整个序列图如图 3.3 所示。

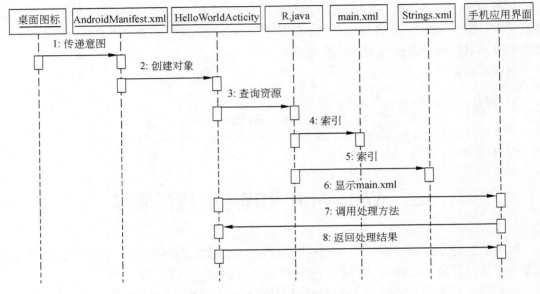

图 3.3　Android 应用执行序列图

3.3　Android 应用程序的基本组件

在第 1 章中我们提到了 Android 平台的几大优点,其中,包括开放性和应用程序平等。的确,Android 最吸引人的特性之一就是应用程序可以利用其他应用程序来完成想要的功能!例如,你的应用程序需要用到图片浏览功能,而这时正好有另一个应用程序已经开发出一个合适的图片浏览程序,太好了!你再也不用自己开发图片浏览程序,而是直接利用已有的程序即可。那么,如何才能利用别的应用程序中的成果呢?只需要在必要时启动那个图片浏览功能。

这听起来简直不可思议,Android 平台是如何完成这样神奇的工作呢?与其他计算机平台上的应用程序不同,Android 应用程序没有唯一的启动入口(如 C 语言中 main()函数入口),一个 Android 应用程序是由多个不同的组件组合而成,组件之间通过 Intent 来实现通信。Android 系统的基本组件包括 Activity、Service、BroadcastReceiver、ContentProvider 等,此外,还包括专门负责在基本组件之间传递消息的 Intent 组件。所有这些组件都必须在 AndroidManifest.xml 文件中声明,这些组件如何协调工作呢?下面通过一个简单的用户与应用程序交互的例子来说明 Android 程序中上述组件是如何配合的,这里包含了两个 Activity,如图 3.4 所示。

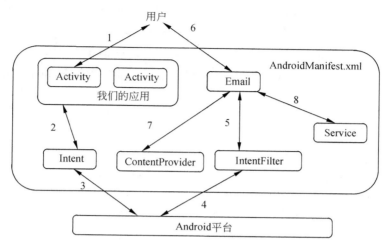

图 3.4　示例应用程序的结构

- 首先,用户通过 Activity 与应用程序交互,如图中步骤 1。
- 应用程序中的 Activity 通过 Intent 来向 Android 平台请求启动一个能处理打开 Email 的应用程序,如图中步骤 2 和步骤 3。
- Android 系统通过 AndroidManifest.xml 中声明的 IntentFilter 找到能处理打开 Email 的应用程序,如图中步骤 4 和步骤 5。
- 用户与 Email 应用程序进行交互,如图中步骤 6。
- Email 应用程序通过 ContentProvider 来使用另一个录音应用程序产生的音频文件,如图中步骤 7。
- 用户播放刚才的音频文件,并返回到了之前的应用程序,此时音频文件仍然会继续播放,因为 Service 将在后台工作,如图中的步骤 8。

当然,并不是每个 Android 应用程序都必须包含这些组件,但是一旦确定了应用程序中需要的组件,就应该在 AndroidManifest.xml 中声明它们。接下来对这些基本组件做简单地介绍,使读者对这些组件建立一个大致的认识,后面章节还会对这些组件详细介绍。

3.3.1　Activity

Activity 是应用程序的表示层。应用程序中的每个屏幕显示都通过集成和扩展基类 Activity 来实现。Activity 利用 View 来实现应用程序的 GUI,而手机用户则直接通过 GUI 和应用程序交互,如应用程序通过 GUI 向用户显示信息,用户通过 GUI 向应用程序发出指令和响应。

例如,一个短信应用程序,需要一个 Activity 来显示联系人列表,同时需要另一个

Activity 显示用户输入的短信内容，甚至还可能需要第三个 Activity 显示已收到短信的内容。虽然这些 Activity 整体形成了一个完整的短信程序用户界面，但实际上每个 Activity 是独立的。当然，它们也有共同点——每个 Activity 都是继承自 Activity 的子类。

应用程序往往由多个 Activity 组成。一个应用程序需要多少个 Activity？每个 Activity 表示什么样的用户界面？这些问题都取决于具体的应用程序设计。通常的原则是，程序启动后显示的第一副画面是应用程序的第一个 Activity，以后根据应用程序的需要从一个 Activity 跳转到一个新的 Activity。下面这段代码展示了之前的项目 HelloWorld 创建 Activity 的方法。

```java
public class MainActivity extends Activity {

    @Override
    protected void onCreate(Bundle savedInstanceState) {
        super.onCreate(savedInstanceState);
        setContentView(R.layout.activity_main);
    }
}
```

对于每个 Activity，系统会分配一个默认的窗口。一般情况下，窗口将沾满整个屏幕。改变默认属性，窗口大小也是可调整的。窗口的显示位置也可以悬浮在其他窗口之上。Activity 同时也能使用别的窗口，为了提醒用户，可以在一个 Activity 中使用弹出对话框。

Activity 窗口内的可见内容通过 View 提供，View 对象继承自 View 类，每个 View 对象控制这窗口内的一个巨型空间。View 是一种层次结构，父 View 包含的布局属性会被子 View 继承。位于 View 层次关系最底部的子 View 对象所代表的矩形空间就是跟用户进行交互的地方。例如，我们可以用一个 View 对象来显示图片，并在用户点击图片时产生相应的动作。Android 自带了很多不同的 View 供开发者使用，如按钮、文本框、滚动条、菜单项等，这些内容会在后续章节进行详细介绍。

既然 Activity 的内容通过 View 来显示，那么如何才能将 View 对象放入 Activity 中呢？可以调用 Activity.setContentView()，如上面代码中的最后一行。

3.3.2 Service

Service 与 Activity 的地位是并列的，它也代表一个单独的 Android 组件。但与 Activity 相反，Service 没有可见的界面，它的特点是能长时间在后台运行，也可以这样理解，Service 是具有一段较长生命周期且没有用户界面的程序。

为什么我们需要长时间在后台运行的 Service？想想音乐播放器！可能在播放音乐的同时去编辑短信或浏览网页，就像笔者现在就一边写书一边听着音乐，这种情况下音乐播放器不可能一直处于前台。为了让音乐一直播放下去，需要将播放音乐的任务放在后台。这样，即使音乐播放器已经不再显示了，用户仍然可以听到音乐。所以，我们需要这样的机制——长时间在后台运行的 Service。与 Activity 组件需要集成 Activity 基类相似，Service 组件需要继承 Service 基类。一个 Service 组件被运行起来后，它将拥有自己独立的生命周期。

3.3.3 BroadcastReceiver

BroadcastReceiver 是用户接受广播通知的组件。广播是一种同时通知多个对象的事件通知机制。Android 中的广播通知要么来自系统，要么来自普通应用程序。很多事件都可能导致系统广播，如手机所在的时区发生变化，电池电量低，用户改变系统语言设置等。当然也有广播来自应用程序，比如，一个应用程序通知其他应用程序某些数据已经下载完毕。

为了响应不同的事件通知，应用程序可以注册不同的 BroadcastReceiver，而所有的 BroadcastReceiver 都继承自基类 BroadcastReceiver。需要说明的是，BroadcastReceiver 自身并不实现图形用户界面，但是当它收到某个通知消息后，BroadcastReceiver 可以启动 Activity 作为响应，或者通过 NotificationManager 提醒用户。

3.3.4 ContentProvider

在 Android 中，每个应用程序都使用自己的用户 id 并在自己的进程中运行。这样做的好处是，可以保护系统及应用程序，避免被其他不正常的应用程序影响，每个进程都拥有独立的进程地址空间和虚拟内存。当应用程序彼此间需要共享资源时，这样的架构必须需要一个妥善的解决方案。例如，Contacts 应用程序内存中保存使用者的联系资料，当你在 Email 中要填写收信人时，希望读取 Contacts 内的联系人资料。由于 Contacts 和 Email 这两个应用程序运行在不同的进程中，因此，它们无法直接通过内存共享联系人资料。为了解决应用程序间数据通信、共享的问题，Android 提供了 ContentProvider 机制。

ContentProvider 能将应用程序特定的数据提供给另一个应用程序使用。数据的存储方式可以是 Android 文件系统，也可以是 SQLite 数据库，或者别的合理方式。

ContentProvider 继承自父类 ContentProvider，并且实现了一组标准的接口，通过这组接口，其他应用程序能对数据进行读写和存储。然而，需要使用数据的应用程序并不是直接调用这组方法，而是通过调用 ContentResolver 对象的方法来完成的。ContentResolver 对象可以与任意 ContentProvider 通信。

3.3.5 Intent 和 IntentFileter

严格地说，Intent 并不是 Android 应用的组件，但是它对于 Android 应用的作用非常大——它是 Android 应用程序内不同组件之间通信的载体。当 Android 运行需要连接不同的组件时，通常需要借助 Intent 实现。Intent 可以启动应用中另一个 Activity，也可以启动一个 Service 组件，还可以发送一条广播消息来触发系统中的 BroadcastReceiver。也就是说，Activity、Service、BroadcastReceiver 三种组件之间的通信都以 Intent 为载体，只是不同组件使用 Intent 的机制略有区别。具体组件间如何通过 Intent 进行通信，后续章节会详细介绍，在此不再赘述。

3.4 本章小结

本章主要对 Android 应用程序进行了深入剖析，首先介绍了 Android 应用程序的构成及程序的内部执行流程，并对一个 Android 应用程序所需要的基本组件功能及其作用进行简单介绍，使读者对 Android 应用程序的内部执行有一个清晰的认识。

第 4 章 界 面 布 局

在 Android 应用中，用户界面是非常重要的，它是人与手机之间传递、交换信息的媒介和对话接口，是 Android 系统的重要组成部分。它实现信息的内部形式与用户可以接受形式之间的转换。iPhone 之所以被人们所推崇，除了其功能强大外，最重要的是完美的 UI（用户界面）设计，在 Android 系统中，我们也可以开发出与 iPhone 同样绚丽多彩的 UI。本章主要介绍 Android 应用程序的几种不同的布局格式。

4.1 UI 概述

一个 Android 应用的用户界面是由 View 和 ViewGroup 对象构建的。它们有很多的种类，并且都是 View 类的子类，View 类是 Android 系统平台上用户界面表示的基本单元。View 类的一些子类被统称为"widgets（工具）"，它们提供了诸如文本输入框和按钮之类的 UI 对象的完整实现。ViewGroup 是 View 的一个扩展，它可以容纳多个子 View。通过扩展 ViewGroup 类，可以创建由相互联系的子 View 组成的复合控件。ViewGroup 类同样可以被扩展用作 layout（布局）管理器，如 LinearLayout（线性布局）、TableLayout（表格布局）及 RelativeLayout（相对布局）等布局架构。并且用户可以通过用户界面与程序进行交互。通过使用这些布局模型的组合、嵌套并设置子控件的布局参数，完全可以构建出各种复杂的用户界面，下面对这几种布局模型分别进行详细介绍。读者也可以通过查看文档来了解所有的布局信息，对于每个 Android 开发者而言，Android 提供的官方文档是必看的。

下面简单介绍读者应该如何查看 Android 文档——这是一种学习方法。实际上，掌握学习方法比记住几个知识点更重要。首先定位到 Android SDK 的安装目录，找到 docs 子目录，打开 docs 子目录下的 index.html 页面，单击该页面上方的 Dev Guide 标签页，用户将看到如图 4.1 所示页面。在此页面，可以通过点击链接查看自己感兴趣的内容，如果具有良好的英文阅读能力加上扎实的 Java 基础，完全可以通过此指南开发出各种 Android 应用程序。

4.2 线 性 布 局

线性布局由 LinearLayout 类来代表，线性布局有点像 AWT 编程里的 FlowLayout，它们会将容器里的组件一个挨一个地排列起来。LinearLayout 不仅可以控制各组件横向排列，也可以控制纵向排列。LinearLayout 以它的垂直或水平的属性值，来排列所有的子元素。所有的子元素都被堆放在其他元素之后，因此一个垂直列表的每一行只会有 一个元素，而不管它们有多宽,而一个水平列表将会只有一个行高(高度为最高子元素的高度加上边框高度)。

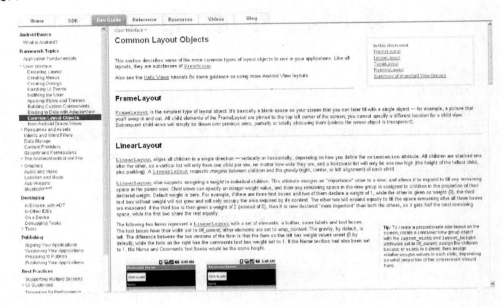

图 4.1 Android 开发指南

线性布局与 AWT 中 FlowLayout 的最大区别在于:Android 的线性布局不会换行,当组件一个挨着一个排到头了,剩下的组件将不会被显示。在 Awt 中 FlowLayout 则会另起一行排列多出来的组件。 LinearLayout 的常用 XML 属性及相关方法见表 4.1。

表 4.1 LinearLayout 常用属性说明

XML 属性	相关方法	说明
Android:gravity	setGravity(int)	设置布局管理器内组件对齐方式,该属性支持 top、bottom、left、center_vertical 等,可以同时指定多种对齐方式,多个属性值用竖线隔开,竖线前后不能有空格
android:orientation	setOrientation(int)	设置布局管理器内组件的排列方式,vertical:垂直,默认 horizontal:水平

LinearLayout 还支持为其包含的 widget 或是 container 指定填充权值。允许其包含的 widget 或是 container 可以填充屏幕上的剩余空间。剩余的空间会按这些 widgets 或者是 containers 指定的权值比例分配屏幕。默认的 weight 值为 0,表示按照 widgets 或者是 containers 实际大小来显示,若高于 0 的值,则将 Container 剩余可用空间分割,分割大小具体取决于每一个 widget 或者是 container 的 layout_weight 及该权值在所有 widgets 或者是 containers 中的比例。例如,如果有三个文本框,前两个文本框的取值一个为 2,另一个为 1,显示第三个文本框后剩余的空间的 2/3 给权值为 2 的,1/3 大小给权值为 1 的。而第三个文本框不会放大,按实际大小来显示。也就是权值越大,重要度越大,显示时所占的剩余空间越大。

例如,修改之前的项目 HelloWorld 的布局文件 activity_main.xml 如下。

```xml
<LinearLayout xmlns:android="http://schemas.android.com/apk/res/android"
    android:layout_width="fill_parent"
    android:layout_height="fill_parent"
    android:orientation="vertical" >
    <LinearLayout
        android:layout_width="fill_parent"
        android:layout_height="fill_parent"
        android:orientation="horizontal"
        android:layout_weight="3" >
        <TextView
            android:layout_width="wrap_content"
            android:layout_weight="1"
            android:layout_height="fill_parent"
            android:background="#aa0000"
            android:text="red"/>
        <TextView
            android:layout_width="wrap_content"
            android:layout_height="fill_parent"
            android:layout_weight="2"
            android:background="#00aa00"
            android:text="green" />
        <TextView
            android:layout_width="wrap_content"
            android:layout_height="fill_parent"
            android:layout_weight="1"
            android:background="#0000aa"
            android:text="blue"
            android:textColor="#aaaaaa" />
    </LinearLayout>
    <LinearLayout
        android:layout_width="fill_parent"
        android:layout_height="fill_parent"
        android:orientation="vertical"
        android:layout_weight="1" >
        <TextView
            android:layout_width="fill_parent"
            android:layout_height="wrap_content"
            android:text="row1"
            android:background="#aaaaaa"
            android:textSize="15pt"
            android:layout_weight="1" />
        <TextView
            android:layout_width="fill_parent"
            android:layout_height="wrap_content"
            android:background="#00aa00"
```

```
                    android:layout_weight="1"
                    android:textSize="15pt"
                    android:text="row2" />
                <TextView
                    android:layout_width="fill_parent"
                    android:layout_height="wrap_content"
                    android:layout_weight="1"
                    android:background="#0000aa"
                    android:textSize="15pt"
                    android:text="row3"
                    android:textColor="#aaaaaa"/>
        </LinearLayout>
</LinearLayout>
```

运行此项目，显示效果如图 4.2 所示。

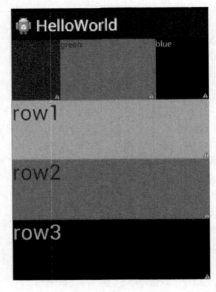

图 4.2　线性布局界面

本例中，最外层布局为垂直的线性布局模型，宽度为占满整个屏幕（fill_parent），高度也为占满整个屏幕。然后在里面又定义了两个线性布局（布局允许嵌套），其中第一个是水平的线性布局，里面放置了三个颜色不同的文本标签，同时为三个标签设置了不同的权重，根据权重大小占据相应大小的空间；第二个线性布局采用垂直排列方式，里面同样放置了三个颜色不同的标签，三个标签的权重相等，所占据的空间也相同。

4.3　相对布局

相对布局由 RelativeLayout 代表，相对布局容器内子组件的位置总是相对兄弟组件、父容器来决定的,因此,这种布局方式被称为相对布局。如果 A 组件的位置是由 B 组件的位置决定的,Android 要求先定义 B 组件,再定义 A 组件。

RelativeLayout 的 XML 属性如表 4.2 所示。

表 4.2 RelativeLayout 属性及相关方法说明

XML 属性	相关方法	说明
android:gravity	setGravity(int)	设置该布局容器内部各子组件的对齐方式
android:ignoreGravity	setIgnoreGravity(int)	设置哪个组件不受 gravity 组件的影响

为了控制该布局容器中个子组件的布局分布,RelativeLayout 提供了一个内部类:RelativeLayout.LayoutParams,该类提供了大量的 XML 属性来控制 RelativeLayout 布局容器中子组件的布局分布。此类中只能设置为 true、false 的 XML 属性如表 4.3 所示。RelativeLayout.LayoutParams 中属性值为其他 UI 组件 ID 的属性如表 4.4 所示。

表 4.3 RelativeLayout.LayoutParams 中取值只能为 Boolean 的属性

属性	说明
android:layout_centerHorizontal	控制该子组件是否位于布局容器的水平居中位置
android:layout_centerVertical	控制该子组件是否位于布局容器的垂直居中位置
android:layout_Inparent	控制该子组件是否位于布局容器的中央位置
android:layout_alignParentBottom	控制该子组件是否位于布局容器低端对齐
android:layout_alignParentLeft	控制该子组件是否位于布局容器左边对齐
android:layout_alignParentRight	控制该子组件是否位于布局容器右边对齐
android:layout_alignParentTop	控制该子组件是否位于布局容器顶端对齐

表 4.4 RelativeLayout.LayoutParams 中属性值为其他 UI 组件 ID 的属性

XML 属性	说明
android:layout_toRightOf	控制该子组件位于给出 ID 组件的右侧
android:layout_toLeftOf	控制该子组件位于给出 ID 组件的左侧
android:layout_above	控制该子组件位于给出 ID 组件的上方
android:layout_below	控制该子组件位于给出 ID 组件的下方
android:layout_alignTop	控制该子组件位于给出 ID 组件的上边界对齐
android:layout_alignBottom	控制该子组件位于给出 ID 组件的下边界对齐
android:layout_alignLeft	控制该子组件位于给出 ID 组件的左边界对齐
android:layout_alignRight	控制该子组件位于给出 ID 组件的右边界对齐

RelativeLayout 允许子元素指定它们相对于其他元素或父元素的位置(通过 ID 指定)。因此,可以左右对齐或上下对齐,或置于屏幕中央的形式来排列两个元素。元素按顺序排列,因此,如果第一个元素在屏幕的中央,那么相对于这个元素的其他元素将以屏幕中央的相对位置来排列。如果使用 XML 来指定这个 layout,在定义它之前,被关联的元素必须定义。

我们再次修改 HelloWorld 项目中的布局文件,如下所示。

```
<RelativeLayout xmlns:android="http://schemas.android.com/apk/res/
android"
    android:layout_width="fill_parent"
    android:layout_height="fill_parent" >
```

```xml
<TextView
    android:id="@+id/textView1"
    android:layout_width="fill_parent"
    android:layout_height="wrap_content"
    android:text="请输入" />
<EditText
    android:id="@+id/editText1"
    android:layout_width="fill_parent"
    android:layout_height="wrap_content"
    android:layout_below="@+id/textView1">
</EditText>
<Button
    android:id="@+id/button1"
    android:layout_width="wrap_content"
    android:layout_height="wrap_content"
    android:layout_alignParentRight="true"
    android:layout_below="@+id/editText1"
    android:layout_marginRight="10dp"
    android:text="取消" />
<Button
    android:id="@+id/button2"
    android:layout_width="wrap_content"
    android:layout_height="wrap_content"
    android:layout_toLeftOf="@id/button1"
    android:layout_alignTop="@id/button1"
    android:text="确定" />
</RelativeLayout>
```

运行效果如图 4.3 所示。

图 4.3 相对布局界面

在此界面布局中，首先定义一个 RelativeLayout 充满整个屏幕，然后定义一个 TextView，接着定义一个 EditText 组件，使其位于 TextView 下方，最后是两个 Button 组件，其中取消按钮位于 EditText 下方，靠屏幕右边放置，右边留出 10 dp 大小的空白，确定按钮位于取消按钮的左边，两个按钮上边缘对齐。

4.4 绝对布局

绝对布局由 AbsoluteLayout 代表。绝对布局就像 Java AWT 中的空布局，就是 Android 不提供任何布局控制，而是由开发人员自己通过 X、Y 两个坐标控制组件的位置。当使用 AbsoluteLayout 作为布局容器时，布局容器不再管理组件的位置、大小——这些都需要开发人员自己控制。AbsoluteLayout 可以让子元素指定准确的 x/y 坐标值，并显示在屏幕上。(0, 0)为左上角，当向下或向右移动时，坐标值将变大。AbsoluteLayout 没有页边框，允许元素之间互相重叠（尽管不推荐）。一般不建议使用相对布局，因为运行 Android 应用的手机往往千差万别，因此，屏幕大小、分辨率都可能存在差异，使用绝对布局会很难兼顾不同屏幕大小、分辨率的问题。

使用绝对布局时，每个自组件都可指定如下两个 XML 属性。
- layout_x：指定该子组件的 X 坐标。
- layout_y：指定该子组件的 Y 坐标。

Android 中定义的距离单位如下。
- px（Pixels，像素）：对应屏幕上的实际像素点。
- in（Inches，英寸）：屏幕物理长度单位。
- mm（Millimeters，毫米）：屏幕物理长度单位。
- pt（Points，磅）：屏幕物理长度单位，1/72 英寸。
- dp（与密度无关的像素）：逻辑长度单位，在每英寸 160 点的屏幕上，1dp=1px=1/160 英寸。随着密度变化，对应的像素数量也在变化，但并没有直接的变化比例。
- dip：与 dp 相同，多用于 Google 示例中。
- sp（与密度和字体缩放度无关的像素）：与 dp 类似，但是可以根据用户的字体大小首选项进行缩放。

在实际应用中，尽量使用 dp 作为空间大小单位，sp 作为和文字相关大小单位。下面通过实例演示绝对布局的使用，同样，修改 HelloWorld 项目中的布局文件 activity_main.xml 如下。

```xml
<AbsoluteLayout xmlns:android="http://schemas.android.com/apk/res/android"
    android:layout_width="fill_parent"
    android:layout_height="fill_parent"
    android:orientation="vertical" >
    <!-- 定义一个文本框，使用绝对定位 -->
    <TextView
        android:layout_width="wrap_content"
        android:layout_height="wrap_content"
        android:layout_x="20dip"
        android:layout_y="20dip"
        android:text="用户名：" />
    <!-- 定义一个文本编辑框，使用绝对定位 -->
```

```xml
<EditText
    android:layout_width="wrap_content"
    android:layout_height="wrap_content"
    android:layout_x="80dip"
    android:layout_y="15dip"
    android:width="200px" />
<!-- 定义一个文本框,使用绝对定位 -->
<TextView
    android:layout_width="wrap_content"
    android:layout_height="wrap_content"
    android:layout_x="20dip"
    android:layout_y="80dip"
    android:text="密　码：" />
<!-- 定义一个文本编辑框,使用绝对定位 -->
<EditText
    android:layout_width="wrap_content"
    android:layout_height="wrap_content"
    android:layout_x="80dip"
    android:layout_y="75dip"
    android:password="true"
    android:width="200px" />
<!-- 定义一个按钮,使用绝对定位 -->
<Button
    android:layout_width="wrap_content"
    android:layout_height="wrap_content"
    android:layout_x="130dip"
    android:layout_y="135dip"
    android:text="登　录" />
</AbsoluteLayout>
```

运行效果如图 4.4 所示。

图 4.4　绝对布局

4.5 表 格 布 局

表格布局由 TableLayout 代表，表格布局采用行列的形式来管理 UI 组件，TableLayout 并不需要明确声明需要多少行、多少列，而是通过添加 TableRow、其他组件来控制表格的行数和列数。

每次向 TableLayout 中添加一个 TableRow，该 TableRow 就是一个表格行，TableRow 也是容器，因此，也可以向其添加其他组件，每添加一个组件，该表格就增加一列。

如果直接向 TableLayout 中添加组件，此组件将直接占用一行。

在表格布局中，列的宽度由该列中最宽的那个单元格决定，整个表格布局的宽度则取决于父容器的宽度（默认总是沾满父容器本身）。

下面实例演示表格布局的使用，修改 HelloWorld 项目的布局文件，如下所示。

```xml
<TableLayout xmlns:android="http://schemas.android.com/apk/res/android"
    android:layout_width="fill_parent"
    android:stretchColumns="0,1,2,3"
    android:layout_height="fill_parent" >
    <TableRow
        android:layout_width="wrap_content"
        android:layout_height="wrap_content" >
        <TextView
            android:gravity="center"
            android:padding="3dip"
            android:text="姓名" />
        <TextView
            android:gravity="center"
            android:padding="3dip"
            android:text="性别" />
        <TextView
            android:gravity="center"
            android:padding="3dip"
            android:text="电话" />
    </TableRow>
    <TableRow
        android:layout_width="wrap_content"
        android:layout_height="wrap_content" >
        <TextView
            android:gravity="center"
            android:padding="3dip"
            android:text="杨过" />
        <TextView
            android:gravity="center"
```

```
                android:padding="3dip"
                android:text="男" />
        <TextView
                android:gravity="center"
                android:padding="3dip"
                android:text="5211314" />
    </TableRow>
    <TableRow
        android:layout_width="wrap_content"
        android:layout_height="wrap_content" >
        <TextView
                android:gravity="center"
                android:padding="3dip"
                android:text="小龙女" />
        <TextView
                android:gravity="center"
                android:padding="3dip"
                android:text="女" />
        <TextView
                android:gravity="center"
                android:padding="3dip"
                android:text="0210210" />
    </TableRow>
</TableLayout>
```

- 表格布局的风格跟 HTML 中的表格比较接近，只是采用的标签不同。
- <TableLayout>是顶级元素，说明采用的是表格布局。
- <TableRow>定义一个行。
- <TextView>定义一个单元格的内容。
- android:stretchColumns="0,1,2,3"该属性指定每行都由"0、1、2、3"列占满空白空间。
- gravity 指定文字对其方式，本例都设为居中对齐。
- gadding 指定视图与视图间的内容空隙，单位为像素。

运行该项目，结果如图 4.5 所示。

图 4.5　表格布局

4.6 本章小结

要想构建出多彩的 Android 应用程序界面,必须对界面中的组件进行合理地放置,本章主要介绍了 Android 常用的几种组件布局管理,并通过简单实例给出各布局的使用方法,使读者对于 Android 的界面设计有一定的了解。

第 5 章　事件处理及 Widget 组件

上一章简单介绍了 UI 和布局管理器的概念,使读者对于 Android 的界面设计有了一定的了解。Android 提供了大量功能丰富的 UI 组件,本章首先为大家介绍常见的 UI 组件,开发者只要使用合适的布局管理器把这些 UI 组件组合起来就可以开发出优秀的图形用户界面,为用户提供完美的体验。同时,为了让这些 UI 组件能响应用户的鼠标、键盘等动作,本章还会为大家讲解 Android 的事件响应机制,这样保证图形界面应用可以响应用户的交互操作。通过本章学习,读者能开发出美观的图形用户界面,这些图形用户界面是 Android 应用开发的基础,也是非常重要的组成部分。

5.1　基本 Widget 组件

Android 当中的 UI 控件种类繁多,初学者的学习进度往往会被这么多的控件所阻碍。为了使读者能快速掌握控件的使用方法,我们首先从最简单的控件开始学习,这些控件是开发 Android 应用程序频繁使用的,而且用法比较容易掌握和理解。

5.1.1　文本框(TextView)和编辑框(EditText)

TextView 直接继承了 View,同时它还是 EditText、Button 两个 UI 组件类的父类,作用就是在界面上显示文本——从这点上看,它类似于 Swing 中的 JLable,不过它比 JLable 的功能更加强大。

从功能上看,TextView 其实就是一个文本编辑器,只是 Android 关闭了它的文字编辑功能。如果想要定义某个可编辑的文本框,可以使用它的子类 EditText。

TextView 提供了大量的 XML 属性,这些 XML 属性大部分既可适用于 TextView,也可适用于 EditText,但有少量 XML 只能适用于其中之一。表 5.1 列出了 TextView 的常见属性及对应方法说明。

表 5.1 中 android:autoLink 属性值是如下几个属性值的一个或几个,多个属性值之间用竖线隔开。

- none:不设置任何超链接。
- web(对应于 Linkfy.WEB_URLS):将文本中的 URL 地址转换为超链接。
- email(对应于 Linkfy.EMAL_ADRESSES):将文本中的 E_mail 地址转换为超链接。
- phone(对应于 Linkfy.PHONE_NUMBERS):将文本中的电话号码转换为超链接。
- map(对应于 Linkfy.MAP_ADDRESSES):将文本中的街道地址转换为超链接。
- all:相当于指定 web|email|phone|map。

Android：ellipsize 属性可支持如下几个属性值。
- none：不进行任何处理。
- start：在文本开头部分进行省略。
- middle：在文本中间部分进行省略。
- end：在文本结尾部分进行省略。
- marquee：在文本结尾处以淡出的方式省略。

表 5.1　TextView 的常见属性及对应方法说明

属性名称	对应方法	说明
android:autoLink	setAutoLinkMask(int)	设置是否将指定格式的文本转换为可单击的超链接显示
android:gravity	setGravity(int)	设置文本框内文本的对齐方式
android:height	setHeight(int)	设置文本的高度，单位为 px
android:minHeight	setMinHeight(int)	设置文本的最小高度，单位为 px
android:maxHeight	setMaxHeight(int)	设置文本的最大高度，单位为 px
android:width	setWidth(int)	设置文本的宽度，单位为 px
android:minWith	setMinWidth(int)	设置文本的最小宽度，单位为 px
android:maxWidth	setMaxWidth(int)	设置文本的最大宽度，单位为 px
android:hint	setHint(int)	设置文本框内容为空时，默认显示的提示文本
android:text	setText(CharSequence)	设置文本框内文本的内容
android:textColor	setTextColor(ColorStateList)	设置文本框内文本的颜色
android:textSize	setTextSize(float)	设置文本框内字体的大小
android:typeface	setTypeFace(Typeface)	设置文本框内文本的字体
android:ellipsize	setEllipse(TextUtils,TruncateAt)	设置当显示的文本超过了文本框的长度时如何处理文本内容
Android:lines	setlines(int)	设置文本框默认占几行
Android:MinLines	setMinLines(int)	设置文本框最少占几行
Android:MaxLines	setMaxLines(int)	设置文本框最多占几行
Android:password	setTransformationMethod(TransformationMethod)	设置文本框是一个密码框
Android:phoneNumber	setKeyListener(KeyListener)	设置文本框只能接受电话号码

可以通过以下两种方法创建 TextView。
- 在程序中创建 TextView 对象，代码段如下。

```
TextView tv=new TextView(this);
tv.setText("hello");
setContentView(tv);
```

- 在 XML 布局文件中创建，如下所示。

```
<TextView
android:id="@+id/myTextView"
android:layout_width="fill_parent"
android:layout_height="wrap_content"
android:text="你好"
```

/>

下面通过实例演示这两种组件的具体使用。新建一个 uiPro 项目，创建布局文件 activity_main.xml，其内容如下。

```xml
<?xml version="1.0" encoding="utf-8"?>
<TableLayout xmlns:android="http://schemas.android.com/apk/res/android"
    android:orientation="vertical"
    android:layout_width="fill_parent"
    android:layout_height="fill_parent"
    >
<TableRow>
<TextView
    android:layout_width="fill_parent"
    android:layout_height="wrap_content"
    android:text="用户名："
    android:textSize="10pt"
    />
<EditText
    android:layout_width="fill_parent"
    android:layout_height="wrap_content"
    android:hint="请填写登录帐号"
    android:selectAllOnFocus="true"
    />
</TableRow>
<TableRow>
<TextView
    android:layout_width="fill_parent"
    android:layout_height="wrap_content"
    android:text="密码："
    android:textSize="10pt"
    />
<EditText
    android:layout_width="fill_parent"
    android:layout_height="wrap_content"
    android:password="true"
    />
</TableRow>
<TableRow>
<TextView
    android:layout_width="fill_parent"
    android:layout_height="wrap_content"
    android:text="电话号码："
    android:textSize="10pt"
    />
<EditText
```

```
            android:layout_width="fill_parent"
            android:layout_height="wrap_content"
            android:hint="请填写您的电话号码"
            android:selectAllOnFocus="true"
            android:phoneNumber="true"
            />
    </TableRow>
</TableLayout>
```

此界面中包含三个 TextView 和三个 EditText，第一个编辑框通过属性 android:hint 设定了编辑框中的默认提示信息，当用户输入前，编辑框中显示默认提示信息。第二个编辑框通过 android:password="true" 设置为密码框，用户在此输入的字符会用点号代替。第三个编辑框通过 android:phoneNumber="true" 设置为电话号码输入框，当用户把输入焦点定位到此处时，系统自动显示数字输入键盘，运行效果如图 5.1 所示。

图 5.1　TextView 和 EditText 用法实例图

5.1.2　Button（按钮）和 ImageButton（图片按钮）

Button 继承自 TextView，ImageButton 继承自 Button，它们的主要功能是在界面上生成一个按钮，用户可以单击，从而触发 OnClick 事件。关于事件处理机本章后续部分会详细讲解。Button 和 ImageButton 的主要区别在于：前者生成的按钮上显示文字，而后者上面则显示图片。

按钮的用法比较简单，可以通过设置相关的属性值为按钮增加背景颜色或图片，但这都是固定的，不会随着用户的动作而改变。如果使用图片按钮，可以通过指定属性值指定图片，但不能指定文字。在实际应用中，用户可能会对按钮有更高的要求，如按钮的形状、颜色、图片和文字等，这就需要对按钮进行更高级的定制。下面通过实例演示按钮的复杂设计。

为了定义图片随用户动作改变的按钮,可以使用 XML 资源文件来定义 Drawable 对象,再将其设为 Button 的 android:background 属性值,或者设为 ImageButton 的 android:src 属性值。

首先,在 res/drawable 下创建 selector.xml,内容如下。

```xml
<?xml version="1.0" encoding="UTF-8"?>
<selector xmlns:android="http://schemas.android.com/apk/res/android">
    <!-- 指定按钮按下时的图片 -->
    <item android:state_pressed="true"
        android:drawable="@drawable/red"
    />
    <!-- 指定按钮松开时的图片 -->
    <item android:state_pressed="false"
        android:drawable="@drawable/purple"
    />
</selector>
```

修改项目 uiPro 的布局文件 activity_main.xml 的内容如下:

```xml
<?xml version="1.0" encoding="utf-8"?>
<TableLayout xmlns:android="http://schemas.android.com/apk/res/android"
    android:orientation="horizontal"
    android:layout_width="fill_parent"
    android:layout_height="fill_parent"
    >
<TableRow>
<!-- 普通文字按钮 -->
<Button
    android:layout_width="wrap_content"
    android:layout_height="wrap_content"
    android:background="@drawable/red"
    android:text="普通按钮"
    android:textSize="10pt"
/>
<!-- 普通图片按钮 -->
<ImageButton
    android:layout_width="wrap_content"
    android:layout_height="wrap_content"
    android:src="@drawable/blue"
    android:background="#0000ff"
/>
</TableRow>
<TableRow>
<!-- 按下时显示不同图片的按钮 -->
```

```
<ImageButton
    android:layout_width="wrap_content"
    android:layout_height="wrap_content"
    android:src="@drawable/selector"
    android:background="#000000"
/>
<!-- 带文字的图片按钮-->

<Button
    android:id="@+id/test"
    android:layout_width="wrap_content"
    android:layout_height="wrap_content"
    android:background="@drawable/selector"
    android:text="带文字的图片按钮" />

</TableRow>
</TableLayout>
```

运行该项目，效果如图 5.2 所示。

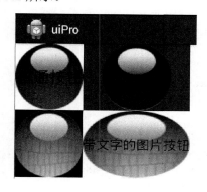

图 5.2 Button 示例图

上图所示界面中，上面两个按钮的背景色和图片都是固定的，用户单击按钮不会产生任何变化；用户按下下面两个按钮时，能看到按钮的图片切换成红色。

5.1.3 单选框（RadioButton）和复选框（ChekBox）

单选框（RadioButton）和复选框（ChekBox）是所有用户界面中最普通的 UI 组件，Android 中这两种组件都继承自 Button 按钮，因此，它们可以直接使用 Button 支持的属性和方法。

RadioButton 和 ChekBox 与普通按钮的区别是，多了一个可选中的功能，它们具有属性 android:checked，该属性用于指定组件初始是否被选中。RadioButton 与 ChekBox 的区别是，一组 RadioButton 只能选中其中一个，因此，RadioButton 通常与 RadioGroup 一起使用，用于定义一组单选按钮。下面通过实例介绍 RadioButton 和 ChekBox 的用法。

修改 uiPro 项目的布局文件 activity_main.xml，内容如下。

```xml
<?xml version="1.0" encoding="utf-8"?>
<LinearLayout xmlns:android="http://schemas.android.com/apk/res/android"
    android:layout_width="match_parent"
    android:layout_height="match_parent"
    android:orientation="vertical" >
    <TextView
        android:id="@+id/textView1"
        android:layout_width="wrap_content"
        android:layout_height="wrap_content"
        android:text="您的性别" />
    <RadioGroup
        android:layout_width="fill_parent"
        android:layout_height="wrap_content"
        android:orientation="horizontal"
        android:checkedButton="@+id/woman"
        android:id="@+id/sex">
        <RadioButton
            android:id="@+id/man"
            android:text="男"/>
        <RadioButton
            android:id="@id/woman"
            android:text="女"/>
    </RadioGroup>
    <TextView
        android:id="@+id/textView2"
        android:layout_width="wrap_content"
        android:layout_height="wrap_content"
        android:text="您的喜好" />
    <CheckBox
        android:id="@+id/checkBox1"
        android:layout_width="wrap_content"
        android:layout_height="wrap_content"
        android:text="上网" />
    <CheckBox
        android:id="@+id/checkBox2"
        android:layout_width="wrap_content"
        android:layout_height="wrap_content"
        android:text="购物" />
    <CheckBox
        android:id="@+id/checkBox3"
        android:layout_width="wrap_content"
        android:layout_height="wrap_content"
        android:text="看电影" />
```

```
        <CheckBox
            android:id="@+id/checkBox4"
            android:layout_width="wrap_content"
            android:layout_height="wrap_content"
            android:text="学习" />
</LinearLayout>
```

其中，在 RadioGroup 的属性中 android:checkedButton="@+id/woman"指定 id 为 woman 的 RadioButton 为默认选定项，注意，"@"后有"+"。被指定为默认选中的 RadionButton 的 id 不用"+"，直接应用前面定义过的标识即可。

运行项目，效果如图 5.3 所示。

图 5.3　单选框和复选框示例图

在性别选项中，只能选中一项，男或女，而在喜好选项中，可以选中多项。

5.1.4　AnalogClock 和 DigitalClock

AnalogClock 和 DigitalClock 是两个非常简单的时钟 UI 组件，前者继承了 View 组件，它重写了 View 的 OnDraw 方法，会在 View 上显示模拟时钟；后者继承了 TextView，即它本身就是文本框，只不过它里面显示的内容是当前时间。这两个组件都会显示当前时间，不同的是，DigitalClock 显示数字时钟，可以显示当前的秒数；而 AnalogClock 显示模拟时钟，不能显示当前描述。下面实例演示这两种组件的用法。

修改 uiPro 项目的布局文件 activity_main.xml，内容如下：

```
<?xml version="1.0" encoding="utf-8"?>
<LinearLayout xmlns:android="http://schemas.android.com/apk/res/android"
    android:layout_width="match_parent"
    android:layout_height="match_parent"
    android:orientation="vertical" >
    <AnalogClock
        android:id="@+id/analogClock1"
```

```
            android:layout_width="wrap_content"
            android:layout_height="wrap_content" />
    <DigitalClock
            android:id="@+id/digitalClock1"
            android:layout_width="wrap_content"
            android:layout_height="wrap_content"
            android:textSize="20pt"
            />
</LinearLayout>
```

运行该项目，效果如图 5.4 所示。

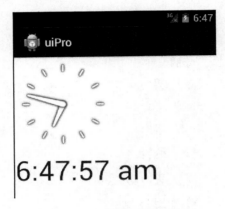

图 5.4 时钟组件示例图

想要设置模拟时钟显示的字体大小和颜色等属性，可以通过相关属性值进行控制，与设置 TextView 上的字体方法相同。

Android 的基本 UI 组件不止以上列举的这些，还有很多其他基本的 UI 组件，因为篇幅限制，在此不再一一列举，在做实际项目开发时，需要用到哪些组件，可以查阅帮助文档，里面有所有组件用法的详细介绍。

5.2 高级 Widget 组件

本节继续介绍 UI 的高级组件，有些读者可能会觉得这么多 UI 组件，真正做开发时不可能每种组件都会用到，要一一的记住太繁琐了！其实这是一种误解，就像玩搭积木玩具，都希望能有一个大大的积木桶，里面积木的种类越多越好，这样就可以搭建出各种各样的造型。UI 组件也是同样的道理，系统提供的组件越多，越有利于用户构造满足用户需求的 UI 界面。

5.2.1 ListView（列表视图）

ListView 类为 AdapterView 的间接子类，可以以列表的形式显示数据，至于显示什么

第 5 章　事件处理及 Widget 组件

数据及如何显示数据，则需要 Adaper（适配器）类及其子类的配合。若以"电视+DVD"的放映流程来比喻，则可以将 ListView 理解为电视屏幕，提供一个数据显示的场所，而 Adaper 可以理解为影碟播放机，对播放进行控制，需要显示的数据理解为碟片，数据在 ListView 中显示的格式，可以理解为播放时指定的屏幕制式，如宽屏和全屏等。播放影片可以使用 DVD 机也可使用 VCD 机进行播放，因此，可以采用不同的 Adaper，也即其子类，如 SimpleAdapter、ArrayAdapter、CursorAdapter 等。当然播放的光盘可以是 DVD 盘或 VCD 盘，即数据的来源可以有多种渠道，因此，数据可以来自自定义的数组、List、数据库、内容提供者（后两种方式本书后面会有详细讲解）。至于指定数据在 ListView 中的显示方式，可以先定义一组元素的预期布局文件，之后，所有的数据项将按此格式显示。这实质上就是 MVC 模式的思想，强调数据与 UI 组件的分离。

ListView 类的常用方法有以下几个。

setAdapter(ListAdapter adapter)：为 ListView 绑定一个 Adapter。

setChoiceMode(int choiceMode)：为 ListView 指定一个显示模式，可选值有三个：CHOICE_MODE_NONE（默认值，没有单选或多选效果）、CHOICE_MODE_SINGLE（单选效果）、CHOICE_MODE_MULTIPLE（多选框效果）。

setOnItemClickListener(AdapterView.onItemClickListener listener)：为其注册一个元素被点击事件的监听器，当其中某一项被单击时，调用其参数 listener 中的 onItemClick()方法。

下面以 ArrayAdapter 为例，演示如何使用 ListView 显示数据。

创建名为 listviewPro 的工程，编写 string.xml，内容如下。

```xml
<?xml version="1.0" encoding="utf-8"?>
<resources>
    <string name="app_name">listview 示例</string>
    <string name="hello_world">Hello world!ListViewActivity</string>
    <string name="menu_settings">Settings</string>
    <string name="name">姓名</string>
</resources>
```

编写布局文件 activity_main.xml，内容如下：

```xml
<LinearLayout xmlns:android="http://schemas.android.com/apk/res/android"
    android:layout_width="fill_parent"
    android:layout_height="fill_parent" >
    <TextView
       android:layout_width="wrap_content"
       android:layout_height="wrap_content"
       android:text="@string/name" />
    <ListView
       android:layout_width="fill_parent"
       android:layout_height="wrap_content"
       android:id="@+id/listview">

    </ListView>
```

```
</LinearLayout>
```

编写 MainActivity.java,内容如下:

```java
package com.example.listviewpro;
import android.os.Bundle;
import android.app.Activity;
import android.view.Menu;
import android.view.View;
import android.widget.ArrayAdapter;
import android.widget.AdapterView;
import android.widget.ListView;
import android.widget.Toast;
public class MainActivity extends Activity {
    private ListView listView;
    private String[] name={"Stone","Tiantian","Cindy","Kimi","Angela"};
    protected void onCreate(Bundle savedInstanceState) {
        super.onCreate(savedInstanceState);
        setContentView(R.layout.activity_main);
        listView=(ListView)findViewById(R.id.listview);
        //创建一个 ArrayAdapter
        ArrayAdapter adapter=new ArrayAdapter(this,
                android.R.layout.simple_list_item_1,name);
        listView.setAdapter(adapter);
        //为 listView 注册一个元素点击事件监听器
        listView.setOnItemClickListener(new AdapterView
                .OnItemClickListener() {
            public void onItemClick(AdapterView<?> arg0,
                    View arg1, int arg2,long arg3) {
                Toast.makeText(MainActivity.this, name[arg2],
                        Toast.LENGTH_LONG).show();
            }
        });
    }
    public boolean onCreateOptionsMenu(Menu menu) {
        getMenuInflater().inflate(R.menu.activity_main, menu);
        return true;
    }
}
```

下面对其中一些代码做下简单解释。

- ArrayAdapter adapter = new ArrayAdapter(Context context, inttextViewResourceId, Object[] objects);

ArrayAdapter 构造方法的参数解释。

context:当前的 Context 对象;

textViewResourceId:一个包含了 TextView 元素的布局文件,用来指定 ListView 中的

每一项的显示格式；

android.R.layout.simple_list_item_1 是 Android 平台自带的一个布局文件，里面只包含一个 TextView 标签。其内容如下。

```xml
<?xml version="1.0" encoding="utf-8"?>
<TextView xmlns:android="http://schemas.android.com/apk/res/android"
android:id="@android:id/text1"
android:layout_width="fill_parent"
android:layout_height="wrap_content"
android:textAppearance="?android:attr/textAppearanceLarge"
android:gravity="center_vertical"
android:gravity="center_vertical"
android:paddingLeft="6dip"
android:minHeight="?android:attr/listPreferredItemHeight"
/>
```

object：要显示的数据，为一个数组。

- onItemClick(AdapterView<?> parent, View view, int position, long id)

参数介绍如下。

parent：被点击的 ListView 对象；

view：被点击的那一项；

position：被点击的那一项在 ListView 中的位置；

id ：被选中的那一行的 id。

运行该项目，效果如图 5.5 所示，点击其中一项，弹出一个 Toast，效果如图 5.6 所示。

如果程序的窗口仅仅需要显示一个列表，没有别的元素，则可以直接让 Activity 集成 ListActivity 来实现，ListActivity 的子类无须调用 setContentView()方法来显示某个界面，而是可以直接传入一个内容 Adapter，ListActivity 的子类就呈现出一个列表。

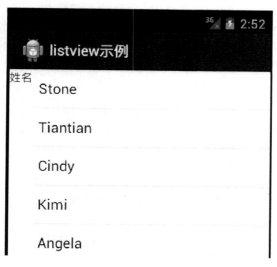

图 5.5　ListView 效果示例图

图 5.6 选中 ListView 其中一项的效果图

下面对 MainActivity 进行修改，内容如下。

```
package com.example.listviewpro;
import android.app.ListActivity;
import android.os.Bundle;
import android.view.View;
import android.widget.ArrayAdapter;
import android.widget.ListView;
import android.widget.Toast;
public class MainActivity extends ListActivity{
    private String[] name={"Stone","Tiantian","Cindy","Kimi","Angela"};
    protected void onCreate(Bundle savedInstanceState){
        super.onCreate(savedInstanceState);
        //创建一个 ArrayAdapter
        ArrayAdapter adapter=new ArrayAdapter(this,
                    android.R.layout.simple_list_item_1,name);
        //直接将 adapter 与 MainActivity 的 ListView 绑定
        setListAdapter(adapter);

    }
//直接重写 ListActivity 类的 onListItemClick()方法对 ListView 中的选项点击事件监听
    protected void onListItemClick(ListView l,View v,int position,long id){
        Toast.makeText(MainActivity.this, name[position], Toast
        .LENGTH_LONG).show();
    }
}
```

执行该项目，运行效果与之前一致。

本书后续内容会学习从数据库及内容提供者获取数据，同时对利用 SimpleAdapter、CursorAdapter 绑定数据进行详细介绍。

5.2.2 Spinner（下拉列表）

手机屏幕较小，当需要用户选择时，可以提供一个下拉列表将所有可选项列出来，供用户选择，以此提高用户的体验。Spinner 与 ListView 一样，也是 AdapterView 的一个间接子类，是一个显示数据的窗口。

Spinner 的数据源和 istView 相同，可以通过绑定适配器获取数据，也可以通过数组获取数据，下面通实过例子讲解通过这两种方式获取数据的方法。

- 使用数组作为数据源。

创建名为 spinnerPro 的工程，编写布局文件 activity_spinner.xml，内容如下。

```
<LinearLayout xmlns:android="http://schemas.android.com/apk/res/android"
    android:layout_width="fill_parent"
    android:layout_height="fill_parent"
    android:orientation="vertical">
    <TextView
        android:id="@+id/spinnerText"
        android:layout_width="fill_parent"
        android:layout_height="wrap_content"
        android:text="TextView" />
    <Spinner
        android:id="@+id/spinner01"
        android:layout_width="fill_parent"
        android:layout_height="wrap_content" />
</LinearLayout>
```

编写 spinnerActivity，内容如下。

```
package com.example.spinnerpro;
import android.os.Bundle;
import android.app.Activity;
import android.view.Menu;
import android.view.View;
import android.widget.AdapterView;
import android.widget.AdapterView.OnItemSelectedListener;
import android.widget.ArrayAdapter;
import android.widget.Spinner;
import android.widget.TextView;
public class SpinnerActivity extends Activity {
    private static final String[] m={"A 型","B 型","AB 型","O 型","其他"};
    private TextView textView;
```

```java
        private Spinner spinner;
        private ArrayAdapter<String> adapter;
        protected void onCreate(Bundle savedInstanceState) {
            super.onCreate(savedInstanceState);
            setContentView(R.layout.activity_spinner);
            textView=(TextView)findViewById(R.id.spinnerText);
            spinner=(Spinner)findViewById(R.id.spinner01);
            //将选项数据与ArrayAdapter适配器连接起来
            adapter=new ArrayAdapter<String>(this,android.R.layout.simple
            _spinner_item,m);
            //设置下拉列表的风格
            adapter.setDropDownViewResource(android.R.layout.simple_spinner_
            dropdown_item);
            //将adapter添加到Spinner中
            spinner.setAdapter(adapter);
            //为spinner添加监听事件
            spinner.setOnItemSelectedListener(new SpinnerSelectdeListener());
            //设置默认值
            spinner.setVisibility(View.VISIBLE);
        }
        //使用数组形式操作数据
        class SpinnerSelectdeListener implements OnItemSelectedListener{

            public void onItemSelected(AdapterView<?> arg0, View arg1, int arg2,
                    long arg3) {
                textView.setText("您的血型是："+m[arg2]);
            }

            public void onNothingSelected(AdapterView<?> arg0) {

            }
        }

        public boolean onCreateOptionsMenu(Menu menu) {
            //Inflate the menu; this adds items to the action bar if it is present.
            getMenuInflater().inflate(R.menu.activity_spinner, menu);
            return true;
        }

}
```

运行该项目，效果如图5.7所示。单击某一选项，则该选项的值添加在TextView的文本后面。

第 5 章 事件处理及 Widget 组件

图 5.7 使用数组的 Spinner 示例图

- 使用 XML 文件作为数组源。

在 spinnerPro 工程的 values 文件夹下新建一个 arrays.xml 文件，内容如下。

```xml
<?xml version="1.0" encoding="utf-8"?>
<resources>
    <string-array name="plantes">
        <item>诺基亚</item>
        <item>联想</item>
        <item>苹果</item>
        <item>三星</item>
        <item >华为</item>
        <item >其它</item>
    </string-array>
</resources>
```

修改 SpinnerActivity，代码如下。

```
package com.example.spinnerpro;
import android.os.Bundle;
import android.app.Activity;
import android.view.Menu;
import android.view.View;
import android.widget.AdapterView;
import android.widget.AdapterView.OnItemSelectedListener;
import android.widget.ArrayAdapter;
import android.widget.Spinner;
```

```java
import android.widget.TextView;
public class SpinnerActivity extends Activity {
    private TextView textView;
    private Spinner spinner;
    private ArrayAdapter adapter;
    protected void onCreate(Bundle savedInstanceState) {
        super.onCreate(savedInstanceState);
        setContentView(R.layout.activity_spinner);
        textView=(TextView)findViewById(R.id.spinnerText);
        spinner=(Spinner)findViewById(R.id.spinner01);
        //将选项数据与ArrayAdapter适配器连接起来
        adapter=ArrayAdapter.createFromResource(this, R.array.plantes,
        android.R.layout.simple_spinner_item);
        //设置下拉列表的风格
        adapter.setDropDownViewResource(android.R.layout.simple_spinner_dropdown_item);
        //将adapter添加到Spinner中
        spinner.setAdapter(adapter);
        //为spinner添加监听事件
        spinner.setOnItemSelectedListener(new SpinnerXMLSelectedListner());
        //设置默认值
        spinner.setVisibility(View.VISIBLE);
    }
    //使用XML形式操作数据
    class SpinnerXMLSelectedListner implements OnItemSelectedListener{
        public void onItemSelected(AdapterView<?> arg0, View arg1, int arg2,
                long arg3) {
            textView.setText("您使用什么牌子的手机: "+adapter.getItem(arg2));
        }
        public void onNothingSelected(AdapterView<?> arg0) {

        }
    }

    public boolean onCreateOptionsMenu(Menu menu) {
        //Inflate the menu; this adds items to the action bar if it is present.
        getMenuInflater().inflate(R.menu.activity_spinner, menu);
        return true;
    }

}
```

再次运行该项目,效果如图 5.8 所示。

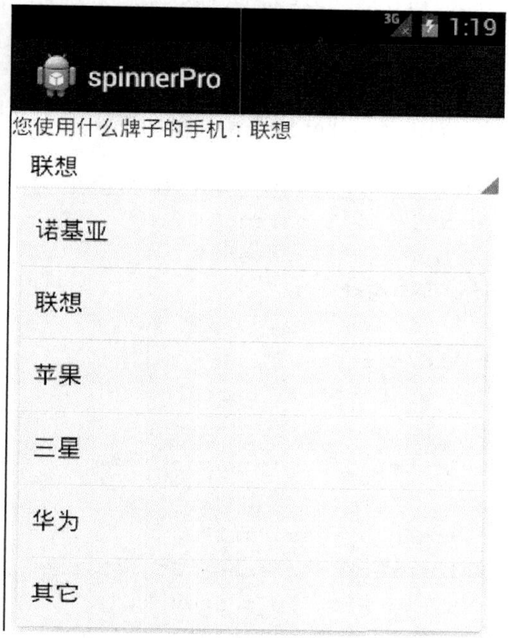

图 5.8 使用 XML 文件的 Spinner 示例图

5.2.3 ProgressBar(进度条)

进度条也是 UI 界面中一种经常使用的组件,通常用于向用户显示某个耗时操作完成的百分比。因此,进度条可以动态显示进度,避免长时间执行某个耗时操作时,让用户感觉程序失去了响应,从而更好地提高用户界面的友好性。

Android 支持几种风格的进度条,最常见的两种进度条是"环形进度条"和"水平进度条",可以通过 style 属性指定风格,水平进度条对应的属性只为:

style="?android:attr/progressBarStyleHorizontal"。

ProgressBar 类中常用的方法如下。

ProgressBar.setMax(int max):设置总长度;

ProgressBar.setProgress(int progress):设置已经开启长度为 0,假设设置为 50,进度条将进行到一半停止。

下面通过实例演示进度条的用法。

创建工程 progressBarDemo,编写 string.xml,内容如下。

```
<?xml version="1.0" encoding="utf-8"?>
<resources>
    <string name="app_name">进度条示例</string>
    <string name="hello_world">Hello world!progressBarActivity</string>
```

```xml
        <string name="progressbar1">环形进度条</string>
        <string name="progressbar">水平进度条</string>
        <string name="menu_settings">Settings</string>
</resources>
```

编写其布局文件 activity_progress_bar.xml,内容如下。

```xml
<LinearLayout xmlns:android="http://schemas.android.com/apk/res/android"
    android:layout_width="fill_parent"
    android:layout_height="fill_parent"
    android:orientation="vertical" >
    <TextView
        android:id="@+id/textView1"
        android:layout_width="fill_parent"
        android:layout_height="wrap_content"
        android:text="@string/progressbar1" />
    <ProgressBar
        android:id="@+id/progress_Bar1"
        style="?android:attr/progressBarStyleLarge"
        android:layout_width="wrap_content"
        android:layout_height="wrap_content" />
    <TextView
        android:id="@+id/textView2"
        android:layout_width="fill_parent"
        android:layout_height="wrap_content"
        android:text="@string/progressbar" />
    <ProgressBar
        android:id="@+id/progress_Bar"
        style="?android:attr/progressBarStyleHorizontal"
        android:layout_width="wrap_content"
        android:layout_height="wrap_content" />
</LinearLayout>
```

编写 ProgressBarActivity 类,代码如下。

```java
package com.example.progressbardemo;
import android.os.Bundle;
import android.os.Handler;
import android.app.Activity;
import android.view.Menu;
import android.widget.ProgressBar;
public class ProgressBarActivity extends Activity {
    private ProgressBar mProgress;
    private int mProgressStatus=0;
    //创建一个Handler对象
    private Handler mHandler=new Handler();
    @Override
```

```java
protected void onCreate(Bundle savedInstanceState) {
    super.onCreate(savedInstanceState);
    setContentView(R.layout.activity_progress_bar);
    mProgress=(ProgressBar)findViewById(R.id.progress_Bar);
    //设置进度条的最大值,其将为该进度条显示的基数
    mProgress.setMax(10000);
    //新开启一个线程
    new Thread(new Runnable(){
        public void run(){
            //循环10000次,不停地更新ProgressStatus的值
            while(mProgressStatus++<10000){
                //将一个Runnable对象添加到消息队列中
                //并且当执行到该对象时执行run()方法
                mHandler.post(new Runnable(){
                    public void run(){
                        //重新设置进度条当前值
                        mProgress.setProgress(mProgressStatus);
                    }
                });
            }
        }
    }).start();
}
@Override
public boolean onCreateOptionsMenu(Menu menu) {
    //Inflate the menu; this adds items to the action bar if it is present.
    getMenuInflater().inflate(R.menu.activity_progress_bar, menu);
    return true;
}
```

代码中开启一个新线程,并在新线程中循环10000次,每次循环更新进度条的当前值,并且刷新界面,执行程序,效果如图5.9所示。

图5.9 进度条示例图

5.2.4　SeekBar（拖动条）

拖动条和进度条非常相似，只是进度条采用颜色填充来表明进度完成的程度，而拖动条则通过滑块的位置来标识数值——而且拖动条允许用户拖动滑块来改变其值，因此，拖动条通常用于对系统的某种数值进行调节，如音量调节、播放进度等。

SeekBar 允许用户改变拖动条的滑块外观，改变滑块外观通过如下属性来指定。

Android:thumb：指定一个 Drawable 对象，该对象将作为自定义滑块。

为了能让程序响应拖动条滑块位置的改变，可以为其绑定一个 OnSeekBarChangeListener 监听器。

下面通过实例演示 SeekBar 的功能和用法，该程序可以实现通过拖动滑块改变图片的透明度。

创建名为"seekBarPro"的工程，其应用程序名字设为"拖动条示例"，改程序界面有两个组件：一个 ImageView 用于显示图片；另一个 SeekBar 用于动态改变图片的透明度，其对应的界面布局文件 activity_seek_bar.xml 的内容如下。

```xml
<LinearLayout xmlns:android="http://schemas.android.com/apk/res/android"
    android:layout_width="fill_parent"
    android:layout_height="fill_parent"
    android:orientation="vertical" >
    <ImageView
        android:id="@+id/image"
        android:layout_width="fill_parent"
        android:layout_height="240px"
        android:src="@drawable/beauty" />
    <SeekBar
        android:id="@+id/seekBar"
        android:layout_width="fill_parent"
        android:layout_height="wrap_content"
        android:max="255"
        android:progress="255"
        android:thumb="@drawable/icon"/>
</LinearLayout>
```

上面的内容中黑体标记的部分定义了该拖动条的最大值、当前值都是 255，并通过 android:thumb 属性来改变拖动条上滑块的外观。

该工程的主程序 SeekBarActivity 的代码如下。

```java
package com.example.seekbardemo;
import android.os.Bundle;
import android.app.Activity;
import android.view.Menu;
import android.widget.ImageView;
```

```java
import android.widget.SeekBar;
import android.widget.SeekBar.OnSeekBarChangeListener;
public class SeekBarActivity extends Activity {
    protected void onCreate(Bundle savedInstanceState) {
        super.onCreate(savedInstanceState);
        setContentView(R.layout.activity_seek_bar);
        final ImageView image=(ImageView)findViewById(R.id.image);
        SeekBar seekBar=(SeekBar)findViewById(R.id.seekBar);
        seekBar.setOnSeekBarChangeListener(new OnSeekBarChange-
        Listener(){
            //当拖动条的滑块位置发生改变时触发该方法
            public void onProgressChanged(SeekBar seekBar, int progress,
                    boolean fromUser) {
                //动态改变图片的透明度
                image.setAlpha(progress);
            }

            public void onStartTrackingTouch(SeekBar seekBar) {
                //TODO Auto-generated method stub
            }
            public void onStopTrackingTouch(SeekBar seekBar) {
                //TODO Auto-generated method stub
            }
        });
    }
    public boolean onCreateOptionsMenu(Menu menu) {
        //Inflate the menu; this adds items to the action bar if it is present.
        getMenuInflater().inflate(R.menu.activity_seek_bar, menu);
        return true;
    }

}
```

运行该项目，效果如图 5.10 所示。

图 5.10　SeekBar 示例图

5.2.5 DatePicker（日期选择器）和 TimePicker（时间选择器）

DatePicker 和 TimePicker 是两个比较易用的控件，它们都是从 FrameLayout 派生出来的，前者供用户选择日期，后者供用户选择时间。

DatePicker 和 TimePicker 在 FrameLayout 的基础上提供了一些方法来获取当前用户所选择的日期、时间；如果程序需要获取用户选择的日期、时间，则可以通过为它们添加相应的监听器来实现。

（1）DatePicker

日期选择器常见方法如下。

- public int getDayOfMonth()：获取选择的天数；
- public int getMonth()：获取选择的月份。（注意：返回数值为 0..11，需要自己+1 来显示）；
- public int getYear()：获取选择的年份；
- public void init (int year, int monthOfYear, int dayOfMonth, DatePicker. OnDateChangedListener onDateChangedListener)：初始化状态（初始化年月日），其中各参数含义如下。
  - ➢ Year：初始年（注意使用 new Date()初始化年时，需要+1900，如下，date.getYear() + 1900）；
  - ➢ monthOfYear：初始月；
  - ➢ dayOfMonth：初始日。
- onDateChangedListener：日期改变时通知用户的事件监听，可以为空（null）。
- public void setEnabled (boolean enabled)：设置视图的启用状态，该启用状态随子类的不同而有不同的解释。其中，enabled 设置为 true 表示启动视图，反之禁用；
- public void updateDate (int year, int monthOfYear, int dayOfMonth)：更新日期。

（2）TimePicker

时间选择器的常用方法如下。

- public int getBaseline ()：返回窗口空间的文本基准线到其顶边界的偏移量。如果这个部件不支持基准线对齐，这个方法返回–1。返回值：基准线的偏移量，如果不支持基准线对齐，则返回–1。
- public Integer getCurrentHour ()：获取当前时间的小时部分（0-23）。
- public Integer getCurrentMinute ()：获取当前时间的分钟部分。
- public boolean is24HourView ()：获取当前系统设置是否是 24 小时制。
- public void setCurrentHour (Integer currentHour)：设置当前小时。
- public void setCurrentMinute (Integer currentMinute)：设置当前分钟（0-59）。
- public void setEnabled (boolean enabled)：设置可用的视图状态，可用的视图状态的解释在子类中改变。
- public void setIs24HourView (Boolean is24HourView)：设置是 24 小时还是上午/下午制。

- public void setOnTimeChangedListener (TimePicker.OnTimeChangedListener onTimeChangedListener)：设置时间调整事件的回调函数，其中参数：onTimeChangedListener 为回调函数，不能为空。

（3）使用实例

下面通过一个具体实例演示这两种组件的用法。

新建一个名为"ChooseDate"的工程，编写布局文件 main.xml，内容如下。

```xml
<?xml version="1.0" encoding="utf-8"?>
<LinearLayout xmlns:android="http://schemas.android.com/apk/res/android"
    android:orientation="vertical"
    android:layout_width="fill_parent"
    android:layout_height="fill_parent"
    >
<TextView
    android:layout_width="fill_parent"
    android:layout_height="wrap_content"
    android:text="您购买此产品为日期为："
    />
<!-- 定义一个 DatePicker 组件 -->
<DatePicker android:id="@+id/datePicker"
    android:layout_width="wrap_content"
    android:layout_height="wrap_content"
    android:layout_gravity="center_horizontal"
    />
<!-- 定义一个 TimePicker 组件 -->
<TimePicker android:id="@+id/timePicker"
    android:layout_width="wrap_content"
    android:layout_height="wrap_content"
    android:layout_gravity="center_horizontal"
    />
<!-- 显示用户输入日期、时间的控件 -->
<EditText android:id="@+id/show"
    android:layout_width="fill_parent"
    android:layout_height="wrap_content"
    android:editable="false"
    android:cursorVisible="false"
    />
</LinearLayout>
```

编写对应的 ChooseDate 类，代码如下。

```
package org.crazyit.choosedate;
import java.util.Calendar;
import android.app.Activity;
import android.os.Bundle;
```

```java
import android.widget.DatePicker;
import android.widget.DatePicker.OnDateChangedListener;
import android.widget.EditText;
import android.widget.TimePicker;
import android.widget.TimePicker.OnTimeChangedListener;
public class ChooseDate extends Activity
{
    //定义5个记录当前时间的变量
    private int year;
    private int month;
    private int day;
    private int hour;
    private int minute;
    @Override
    public void onCreate(Bundle savedInstanceState)
    {
        super.onCreate(savedInstanceState);
        setContentView(R.layout.main);
        DatePicker datePicker = (DatePicker)findViewById(R.id.datePicker);
        TimePicker timePicker = (TimePicker)findViewById(R.id.timePicker);
        //获取当前的年、月、日、小时、分钟
        Calendar c = Calendar.getInstance();
        year = c.get(Calendar.YEAR);
        month = c.get(Calendar.MONTH);
        day = c.get(Calendar.DAY_OF_MONTH);
        hour = c.get(Calendar.HOUR);
        minute = c.get(Calendar.MINUTE);
        //初始化DatePicker组件，初始化时指定监听器
        datePicker.init(year , month ,day
            , new OnDateChangedListener()
        {

            @Override
            public void onDateChanged(DatePicker arg0, int year
                , int month, int day)
            {
                ChooseDate.this.year = year;
                ChooseDate.this.month = month;
                ChooseDate.this.day = day;
                //显示当前日期、时间
                showDate(year, month+1 , day , hour, minute);
            }
```

第 5 章　事件处理及 Widget 组件

```
        });
        //为 TimePicker 指定监听器
        timePicker.setOnTimeChangedListener(new OnTimeChangedListener()
        {
            @Override
            public void onTimeChanged(TimePicker arg0, int hour, int minute)
            {
                ChooseDate.this.hour = hour;
                ChooseDate.this.minute = minute;
                //显示当前日期、时间
                showDate(year, month+1 , day , hour, minute);
            }
        });
    }
    //定义在 EditText 中显示当前日期、时间的方法
    private void showDate(int year, int month , int day
            , int hour , int minute)
    {
        EditText show = (EditText)findViewById(R.id.show);
        show.setText("您的购买日期为：" + year + "年" + month + "月"
                + day + "日 " + hour + "时" + minute + "分");
    }
}
```

运行此项目，单用户通过这两个组件来选择日期和时间时，相应的监听器被触发，从而在文本框中显示对应的日期和时间，效果如图 5.11 所示。

图 5.11　日期时间选择器示例图

5.3 对 话 框

Android 提供了丰富的对话框支持,常用的对话框有下面 4 种。
- AlertDialog:功能丰富、应用较广泛;
- ProgressDialog:进度对话框,这个对话框只是对简单进度条的封装;
- DatePickerDialog:日期选择对话框,这个对话框是对 DatePicker 的包装;
- TimePickerDialog:时间选择对话框,是对 TimePicker 的包装。

这四种对话框中功能最强、用法最灵活的就是 AlertDialog,因此,应用最为广泛,本节主要介绍 AlertDialog 的功能和用法。

AlertDialog 是一个提示窗口,要求用户做出选择,该对话框中一般会有几个选择按钮、标题信息和提示信息。AlertDialog 提供了一些方法来生成四种预定义对话框。
- 带消息、带 N 个按钮的提示对话框。
- 带列表、带 N 个按钮的列表对话框。
- 带多个单选列表项,带 N 个按钮的对话框。
- 带多个多选列表项,带 N 个按钮的对话框。

AlertDialog 的创建方式有两种:

一是直接 new 一个 AlertDialog 对象,然后调用 AlertDialog 对象的 show 和 dismiss 方法来控制对话框的显示和隐藏;二是在 Activity 的 onCreateDialog(int id)方法中创建 AlertDialog 对象并返回,然后调用 Activty 的 showDialog(int id)和 dismissDialog(int id)来显示和隐藏对话框。区别在于通过第二种方式创建的对话框会继承 Activity 的属性,如获得 Activity 的 menu 事件等。

创建 AlertDialog 的主要步骤如下。
(1)创建 Andorid 项目。
(2)获得 AlertDialog 的静态内部类 Builder 对象,由该类创建对话框。
(3)通过 Builder 对象设置对话框的标题、按钮及按钮将要响应的事件。
(4)调用 Builder 对象的 create()方法创建对话框。
(5)调用 AlertDialog 的 show()方法显示对话框。

下面通过实例演示不同形式的 AlertDialog 的创建方法。

5.3.1 提示对话框

实现效果图如图 5.12 所示。

图 5.12 提示对话框

实现代码片段：

```
AlertDialog.Builder builder = new Builder(DialogDemoActivity.this);
    builder.setMessage("确认退出吗？");
    builder.setTitle("提示");

    builder.setPositiveButton("确认", new DialogInterface.OnClickListener() {
     @Override
     public void onClick(DialogInterface dialog, int which) {
      dialog.dismiss();
      DialogDemoActivity.this.finish();
     }
    });

    builder.setNegativeButton("取消", new DialogInterface.OnClickListener() {
     @Override
     public void onClick(DialogInterface dialog, int which) {
      dialog.dismiss();
     }
    });
    builder.create().show();
```

5.3.2 多选对话框

实现效果图如 5.13 所示。

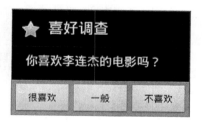

图 5.13　多选对话框

实现代码片段：

```
Dialog dialog = new AlertDialog.Builder(this).setIcon(
        android.R.drawable.btn_star).setTitle("喜好调查").setMessage(
        "你喜欢李连杰的电影吗？").setPositiveButton("很喜欢",
        new DialogInterface.OnClickListener() {

         @Override
         public void onClick(DialogInterface dialog, int which) {
          //TODO Auto-generated method stub
          Toast.makeText(DialogDemoActivity.this, "我很喜欢他的电影。",
```

```
                Toast.LENGTH_LONG).show();
            }
        }).setNegativeButton("不喜欢", new DialogInterface.OnClickListener() {

            @Override
            public void onClick(DialogInterface dialog, int which) {
                //TODO Auto-generated method stub
                Toast.makeText(DialogDemoActivity.this, "我不喜欢他的电影。", Toast
                .LENGTH_LONG)
                    .show();
            }
        }).setNeutralButton("一般", new DialogInterface.OnClickListener() {

            @Override
            public void onClick(DialogInterface dialog, int which) {
                //TODO Auto-generated method stub
                Toast.makeText(DialogDemoActivity.this, "谈不上喜欢不喜欢。", Toast.
                LENGTH_LONG)
                    .show();
            }
        }).create();

        dialog.show();
```

5.3.3 内容输入对话框

实现效果如图 5.14 所示。

图 5.14 内容输入对话框

实现代码片段:

```
new AlertDialog.Builder(this).setTitle("请输入").setIcon(
        android.R.drawable.ic_dialog_info).setView(
        new EditText(this)).setPositiveButton("确定", null)
        .setNegativeButton("取消", null).show();
```

5.3.4 单选对话框

实现效果图如图 5.15 所示。

第 5 章 事件处理及 Widget 组件　　75

图 5.15　单选对话框

实现代码片段：

```
new AlertDialog.Builder(this).setTitle("单选框").setIcon(
        android.R.drawable.ic_dialog_info).setSingleChoiceItems(
        new String[] { "Item1", "Item2" }, 0,
        new DialogInterface.OnClickListener() {
         public void onClick(DialogInterface dialog, int which) {
          dialog.dismiss();
         }
        }).setNegativeButton("取消", null).show();
```

5.3.5　复选对话框

实现效果图如图 5.16 所示。

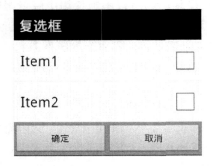

图 5.16　复选对话框

实现代码片段：

```
new AlertDialog.Builder(this).setTitle("复选框").setMultiChoiceItems(
            new String[] { "Item1", "Item2" }, null, null)
            .setPositiveButton("确定", null)
            .setNegativeButton("取消", null).show();
```

5.3.6　列表对话框

实现效果图如图 5.17 所示。

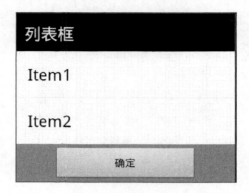

图 5.17 列表对话框

实现代码片段：

```
new AlertDialog.Builder(this).setTitle("列表框").setItems(
        new String[] { "Item1", "Item2" }, null).setNegativeButton(
        "确定", null).show();
```

5.4 消息提示

当程序有大量消息、图片需要向用户提示时，可以考虑使用前面介绍的对话框，但如果程序只有少量信息要向用户呈现，则可以考虑使用更轻量级的对话框，即 Android 提供的消息提示。本节主要讲解 Toast（吐司）显示提示信息框的功能及用法。

Toast 是一种非常方便的消息提示框，它会在程序界面上显示一个简单的提示信息，这个提示框用于向用户生成简单的提示信息。它具有两个特点。

- Toast 提示信息不会获得焦点。
- Toast 提示信息过一段时间会自动消失。

使用 Toast 来生成提示消息非常简单，只要按照如下步骤进行即可。

（1）调用 Toast 的构造器或 makeText 方法创建一个 Toast 对象。
（2）调用 Toast 的方法来设置该消息提示的对其方式、页边距、显示内容等。
（3）调用 Toast 的 show()方法将其显示出来。

Toast 的功能和用法都比较简单，大部分时候它只能显示简单的文本提示，如果程序需要显示图片、列表之类的复杂提示，一般用对话框完成比较适合。

下面实例讲解 Toast 的用法。

创建一个项目"toastDemo"，其布局非常简单，界面上只有两个文本信息不同的按钮，当单击其中一个按钮时，弹出一个 Toast，显示的内容为按钮上的文本，主程序代码如下。

```
package com.example.toastdemo;
import android.os.Bundle;
import android.app.Activity;
import android.view.Menu;
```

```java
import android.view.View;
import android.widget.Button;
import android.widget.Toast;
import android.view.View.OnClickListener;
public class ToastActivity extends Activity {
    @Override
    protected void onCreate(Bundle savedInstanceState) {
        super.onCreate(savedInstanceState);
        setContentView(R.layout.activity_toast);
        Button bt1=(Button)findViewById(R.id.button1);
        Button bt2=(Button)findViewById(R.id.button2);
        bt1.setOnClickListener(bt1lis);
        bt2.setOnClickListener(bt2lis);
    }
    OnClickListener bt1lis=new OnClickListener(){
        public void onClick(View v) {
            Toast toast=Toast.makeText(ToastActivity.this,"喜欢点击我！", Toast.LENGTH_SHORT);
            toast.show();

        }
    };
    OnClickListener bt2lis=new OnClickListener(){
        @Override
        public void onClick(View v) {
         Toast.makeText(ToastActivity.this,"讨厌点击我！", Toast.LENGTH_LONG).show();
        }

    };
    @Override
    public boolean onCreateOptionsMenu(Menu menu) {
        //Inflate the menu; this adds items to the action bar if it is present.
        getMenuInflater().inflate(R.menu.activity_toast, menu);
        return true;
    }
}
```

运行该程序，单击其中一个按钮，出现一个 Toast 提示框，一段时间后消失，效果如图 5.18 所示。

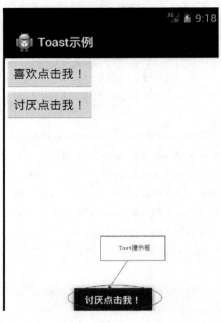

图 5.18　Toast 示例图

5.5　事件处理机制

无论是桌面应用程序还是手机应用程序，经常需要处理的就是用户动作——即需要为这种动作提供响应，这种为用户动作提供响应的机制就是事件处理。

Android 提供了强大的事件处理机制，包括两套处理机制：

- 基于监听的事件处理。
- 基于回调的事件处理。

基于监听的事件处理主要的做法是为组件绑定事件监听器，前面介绍组件时已经使用过此种方法。基于回调的事件处理，主要的做法就是重写组件特定的回调方法，或者重写 Activity 的回调方法。Android 为绝大部分组件都提供了事件相应的回调方法，开发者只需重写这些方法即可。

5.5.1　基于监听的事件处理

（1）基于监听的事件处理模型

基于监听的事件处理主要涉及以下三类对象。

- EventSource（事件源）：事件所发生的场所，通常是各组件，如按钮、窗口、菜单等。
- Event（事件）：通常是用户的某个操作，如单击、双击、长时间按下等。
- EventListener（事件监听器）：负责监听事件源所发生的事件，并对各种事件做出相应的响应。

基于监听的事件处理流程如图 5.19 所示。

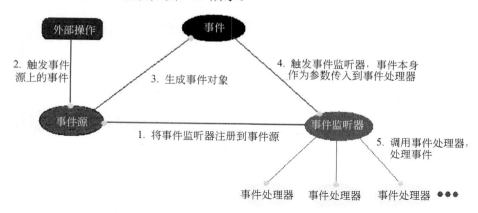

图 5.19 基于监听的事件处理流程

下面通过实例演示基于监听的事件处理模型。

创建工程，编写布局文件，代码如下。

```xml
<?xml version="1.0" encoding="utf-8"?>
<LinearLayout xmlns:android="http://schemas.android.com/apk/res/android"
    android:orientation="vertical"
    android:layout_width="fill_parent"
    android:layout_height="fill_parent"
    android:gravity="center_horizontal"
    >
<EditText
    android:id="@+id/txt"
    android:layout_width="fill_parent"
    android:layout_height="wrap_content"
    android:editable="false"
    android:cursorVisible="false"
    android:textSize="12pt"
    />
<!-- 定义一个按钮，该按钮将作为事件源 -->
<Button
    android:id="@+id/bn"
    android:layout_width="wrap_content"
    android:layout_height="wrap_content"
    android:text="单击我"
    />
</LinearLayout>
```

接着在程序中为按钮定义事件监听器，代码如下。

```
package org.crazyit.listener;
```

```
import android.app.Activity;
import android.os.Bundle;
import android.view.View;
import android.widget.Button;
import android.widget.EditText;
public class EventQs extends Activity
{
    public void onCreate(Bundle savedInstanceState)
    {
        super.onCreate(savedInstanceState);
        setContentView(R.layout.main);
        //获取应用程序中的 bn 按钮
        Button bn = (Button)findViewById(R.id.bn);
        //为按钮绑定事件监听器。
        bn.setOnClickListener(new MyClickListener());
    }
    //定义一个单击事件的监听器
    class MyClickListener implements View.OnClickListener
    {
        //实现监听器类必须实现的方法,该方法将会作为事件处理器
        public void onClick(View arg0)
        {
            EditText txt = (EditText)findViewById(R.id.txt);
            txt.setText("bn 按钮被单击了! ");
        }
    }
}
```

运行该项目,效果如图 5.20 所示。

图 5.20 基于监听器的事件处理

(2)事件和事件监听器

事件是在与 UI 交互式发生的,当单击一个按键时,可能就已经触发好几个事件,例如,我们点击数字键 "0",会涉及按下事件和一个弹起(松开)事件,在 android 中还可能涉及触摸屏事件,所以在 android 系统中,事件是作为常用的功能之一。

在 android 中,事件的发生是在监听器下进行的,android 系统可以响应按键事件和触摸屏事件,常见事件说明如下。

- onClick(View v)：一个普通的点击按钮事件。
- boolean onKeyMultiple(int keyCode,int repeatCount,KeyEvent event)：用于在多个事件连续时发生，用于按键重复，必须重载实现。
- boolean onKeyDown(int keyCode,KeyEvent event)：用于在按键进行按下时发生。
- boolean onKeyUp(int keyCode,KeyEvent event)：用于在按键进行释放时发生。
- onTouchEvent(MotionEvent event)：触摸屏事件，当在触摸屏上有动作时发生。
- boolean onKeyLongPress(int keyCode, KeyEvent event)：当长时间按键时发生。

事件监听器是对事件处理的方式，核心是事件处理的方法，也称为事件处理器。事件监听器，主要有以下几种实现方式。

（1）匿名内部类作为事件监听器类

```
Button button=( Button) findViewById(R.id.button);
button.setOnClickListener(new OnClickListener(){
    public void onClick(View v){
        System.out.println("匿名内部类作为事件监听器类");
    }
});
```

大部分时候，事件处理器都没有什么利用价值（可利用代码通常都被抽象成了业务逻辑方法），因此，大部分事件监听器只是临时使用一次，所以使用匿名内部类形式的事件监听器更合适，实际上，这种形式是目前是最广泛的事件监听器形式。上面的程序代码就是匿名内部类来创建事件监听器的。

对于使用匿名内部类作为监听器的形式来说，唯一缺点就是匿名内部类的语法不易掌握，如果读者 java 基础扎实，匿名内部类的语法掌握较好，通常建议使用匿名内部类作为监听器。

（2）内部类作为监听器

```
public class ButtonTest extends Activity
{
    public void onCreate(Bundle savedInstanceState)
    {
        super.onCreate(savedInstanceState);
        this.setContentView(R.layout.main);

        Button button=( Button) findViewById(R.id.button);
        MyButton listener=new MyButton();
        button.setOnClickListener(listener);
    }
    class MyButton implements OnClickListener
    {
        public void onClick(View v)
        {
          System.out.println("内部类作为事件监听器");
```

            }
        }
    }

将事件监听器类定义成当前类的内部类,其优势有以下两种:(1)使用内部类可以在当前类中复用监听器类,因为监听器类是外部类的内部类。(2)可以自由访问外部类的所有界面组件。这也是内部类的两个优势。上面代码就是以内部类的形式作为事件监听器。

(3) Activity 本身作为事件监听器

```
public class ButtonTest extends Activity implements OnClickListener
{
    public void onCreate(Bundle savedInstanceState)
    {
        super.onCreate(savedInstanceState);
        this.setContentView(R.layout.main);

        Button button=( Button) findViewById(R.id.button);
        MyButton listener=new MyButton();
        button.setOnClickListener(this);
    }
    public void onClick(View v)
    {
      System.out.println("Activity 本身作为事件监听器");
    }
}
```

这种形式使用 activity 本身作为监听器类,可以直接在 activity 类中定义事件处理器方法,这种形式非常简洁。但这种做法有两个缺点:① 这种形式可能造成程序结构混乱。Activity 的主要职责应该是完成界面初始化,但此时还需包含事件处理器方法,从而引起混乱。② 如果 activity 界面类需要实现监听器接口,让人感觉比较怪异。

上面的程序让 Activity 类实现了 OnClickListener 事件监听接口,从而可以在该 Activity 类中直接定义事件处理器方法:onClick(view v),当为某个组件添加该事件监听器对象时,直接使用 this 作为事件监听器对象即可。

(4) 外部类作为监听器

ButtonTest 类
```
public class ButtonTest extends Activity
{
    public void onCreate(Bundle savedInstanceState)
    {
        super.onCreate(savedInstanceState);
        this.setContentView(R.layout.main);

        Button button=( Button) findViewById(R.id.button);
        button.setOnClickListener(new MyButtonListener("外部类作为事件监听器"));
```

当用户单击 Button 按钮时，程序将触发 MyButtonListener 监听器。外部类 MyButtonListener 类代码如下。

```
class MyButtonListener implements OnClickListener
{
   Private String str;
   Public MyButtonListener(String str){
    super();
    this.str=str;
    }
   public void onClick(View v)
   {
     System.out.println(str);
   }
}
```

使用顶级类定义事件监听器类的形式比较少见，主要有两个原因：第一，事件监听器通常属于特定的界面，定义成外部类不利于提高程序的内聚性。第二，外部类形式的事件监听器不能自由访问创建界面类中的组件，编程不够简洁。

但如果某个事件监听器确实需要被多个界面所共享，而且主要是完成某种业务逻辑的实现，则可以考虑使用外部类的形式来定义事件监听器类。

（5）直接绑定到标签

Android 还有一种更简单的绑定事件监听器的的方式，直接在界面布局文件中为指定标签绑定事件处理方法。

对于很多 Android 标签而言，它们都支持如 onClick、onLongClick 等属性，这种属性的属性值就是一个形如 xxx(View source)的方法名。

比如，在布局文件中为 button 添加属性，代码如下。

```
<Button
    android:layout_width="wrap_content"
    android:layout_height="wrap_content"
    android:text="button"
    android:onClick="clickHandler"/>
```

为 Button 按钮绑定一个事件处理方法：clickHanlder，这意味着开发者需要在该界面布局对应的 Activity 中定义一个 void clickHanler(View source)方法，该方法将会负责处理该按钮上的单击事件。下面是该界面布局对应的 java 代码。

```
public class ButtonTest extends Activity
{
    public void onCreate(Bundle savedInstanceState)
    {
```

```
        super.onCreate(savedInstanceState);
        this.setContentView(R.layout.main);
    }
    Public void clickHanler(View source){
       System.out.println("直接绑定在标签上的按钮被点击了!!! ");
    }
}
```

5.5.2 基于回调机制的事件处理

回调机制实质就是将事件的处理绑定在组件上，由 GUI 组件自己处理事件，回调机制需要自定义 View 来实现，自定义 View 重写该 View 的事件处理方法即可。

Android 平台中，每个 View 都有自己的处理事件的回调方法，开发人员可以通过重写 View 中的这些回调方法来实现需要的响应事件。当某个事件没有被任何一个 View 处理时，便会调用 Activity 中相应的回调方法。Android 提供了以下回调方法供用户使用。

（1）onKeyDown

功能：该方法是接口 KeyEvent.Callback 中的抽象方法，所有的 View 全部实现了该接口并重写了该方法，该方法用来捕捉手机键盘被按下的事件。

声明：public boolean onKeyDown (int keyCode, KeyEvent event)

参数说明如下。

参数 keyCode，该参数为被按下的键值即键盘码，手机键盘中每个按钮都会有其单独的键盘码，应用程序都是通过键盘码才知道用户按下的是哪个键。

参数 event，该参数为按键事件的对象，其中包含了触发事件的详细信息，例如，事件的状态、事件的类型、事件发生的时间等。当用户按下按键时，系统会自动将事件封装成 KeyEvent 对象供应用程序使用。

返回值，该方法的返回值为一个 boolean 类型的变量，当返回 true 时，表示已经完整地处理了这个事件，并不希望其他的回调方法再次进行处理，而当返回 false 时，表示并没有完全处理完该事件，更希望其他回调方法继续对其进行处理，如 Activity 中的回调方法。

（2）onKeyUp

功能：onKeyUp 方法用来捕捉手机键盘按键抬起的事件。

声明：public boolean onKeyUp (int keyCode, KeyEvent event)

参数说明：同 onKeyDown。

（3）onTouchEvent

功能：该方法在 View 类中的定义，并且所有的 View 子类全部重写了该方法，应用程序可以通过该方法处理手机屏幕的触摸事件。

声明：public boolean onTouchEvent (MotionEvent event)

参数说明如下。

参数 event：参数 event 为手机屏幕触摸事件封装类的对象，其中，封装了该事件的所有信息，如触摸的位置、触摸的类型及触摸的时间等，该对象会在用户触摸手机屏幕时被创建。

返回值：该方法的返回值机理与键盘响应事件的相同，同样是当已经完整地处理了该事件且不希望其他回调方法再次处理时返回 true，否则返回 false。

详细说明如下。

该方法并不像之前介绍过的方法只处理一种事件，一般情况下以下三种情况的事件全部由 onTouchEvent 方法处理，只是三种情况中的动作值不同。

屏幕被按下：当屏幕被按下时，会自动调用该方法来处理事件，此时 MotionEvent.getAction()的值为 MotionEvent.ACTION_DOWN，如果在应用程序中需要处理屏幕被按下的事件，只需重新该回调方法，然后在方法中进行动作的判断即可。

屏幕被抬起：当触控笔离开屏幕时触发的事件，该事件同样需要 onTouchEvent 方法来捕捉，然后在方法中进行动作判断。当 MotionEvent.getAction()的值为 MotionEvent.ACTION_UP 时，表示是屏幕被抬起的事件。

在屏幕中拖动：该方法还负责处理触控笔在屏幕上滑动的事件，同样是调用 MotionEvent.getAction()方法来判断动作值是否为 MotionEvent.ACTION_MOVE 再进行处理。

（4）onTrackBallEvent

下面介绍手机中轨迹球的处理方法 onTrackBallEvent。所有的 View 同样全部实现了该方法。

声明：public boolean onTrackballEvent (MotionEvent event)

详细说明：该方法的使用方法与前面介绍过的各回调方法基本相同，可以在 Activity 中重写该方法，也可以在各 View 的实现类中重写。

参数 event：参数 event 为手机轨迹球事件封装类的对象，其中封装了触发事件的详细信息，同样包括事件的类型、触发时间等，一般情况下，该对象会在用户操控轨迹球时被创建。

返回值：该方法的返回值与前面介绍的各回调方法的返回值机制完全相同，因本书篇幅有限，不再赘述。

轨迹球与手机键盘的区别如下所示。

① 某些型号的手机设计出的轨迹球会比只有手机键盘时更美观，可增添用户对手机的整体印象。

② 轨迹球使用更为简单，例如，在某些游戏中使用轨迹球控制会更为合理。

③ 使用轨迹球会比键盘更为细化，即滚动轨迹球时，后台表示状态的数值会变得更细微、更精准。

提示：在模拟器运行状态下，可以通过 F6 键打开模拟器的轨迹球，然后便可以通过鼠标的移动来模拟轨迹球事件。

（5）onFocusChanged

功能：该方法是焦点改变的回调方法，某个控件重写了该方法后，当焦点发生变化时，会自动调用该方法来处理焦点改变的事件。

声明：protected void onFocusChanged (boolean gainFocus, int direction, Rect previouslyFocusedRect)

详细说明如下。

参数 gainFocus：表示触发该事件的 View 是否获得了焦点，当该控件获得焦点时，gainFocus 等于 true，否则等于 false。

参数 direction：表示焦点移动的方向，用数值表示，有兴趣的读者可以重写 View 中的该方法打印该参数进行观察。

参数 previouslyFocusedRect：表示在触发事件的 View 的坐标系中，前一个获得焦点的矩形区域，即表示焦点是从哪里来的。如果不可用，则为 null。

提示：焦点描述了按键事件（或者是屏幕事件等）的承受者，每次按键事件都发生在拥有焦点的 View 上。在应用程序中，可以对焦点进行控制，例如，从一个 View 移动另一个 View。下面列出一些与焦点有关的常用方法。

setFocusable 方法：设置 View 是否可以拥有焦点。
isFocusable 方法：监测此 View 是否可以拥有焦点。
setNextFocusDownId 方法：设置 View 的焦点向下移动后获得焦点 View 的 ID。
hasFocus 方法：返回了 View 的父控件是否获得了焦点。
requestFocus 方法：尝试让此 View 获得焦点。
isFocusableTouchMode 方法：设置 View 是否可以在触摸模式下获得焦点，在默认情况下是不可以获得的。

5.5.3 Handler

当应用程序启动时，Android 首先会开启一个主线程（也就是 UI 线程），主线程为管理界面中的 UI 控件，进行事件分发，比如说点击一个 Button，Android 会分发事件到 Button 上来响应操作。如果此时需要一个耗时的操作，例如，联网读取数据或者读取本地较大的一个文件时，不能把这些操作放在主线程中，如果放在主线程中，界面会出现假死现象，如果 5 秒钟还没有完成的话，会收到 Android 系统的一个错误提示"强制关闭"。这时候需要把这些耗时的操作，放在一个子线程中，因为子线程涉及 UI 更新，Android 主线程是线程不安全的，也就是说，更新 UI 只能在主线程中更新，在子线程中操作是危险的。Handler 就出现了，来解决这个复杂的问题，由于 Handler 运行在主线程中（UI 线程中），它与子线程可以通过 Message 对象来传递数据，Handler 就承担着接受子线程传过来的（子线程用 sedMessage()方法传弟）Message 对象及里面包含数据，把这些消息放入主线程队列中，配合主线程进行更新 UI。

Handler 可以分发 Message 对象和 Runnable 对象到主线程中，每个 Handler 实例都会绑定到创建他的线程中（一般是位于主线程），它有两个作用：① 安排消息或 Runnable 在某个主线程中某个地方执行；② 安排一个动作在不同的线程中执行。

下面的程序演示了 Handler 的使用。

```
public class MyHandlerActivity extends Activity {
    Button button;
    MyHandler myHandler;
    protected void onCreate(Bundle savedInstanceState) {
```

```java
        super.onCreate(savedInstanceState);
        setContentView(R.layout.handlertest);
        button = (Button) findViewById(R.id.button);
        myHandler = new MyHandler();
        //当创建一个新的 Handler 实例时，它会绑定到当前线程和消息的队列中，
        //开始分发数据
        //Handler 有两个作用，(1) 定时执行 Message 和 Runnalbe 对象 (2)：让一个
        //动作，在不同的线程中执行。它安排消息，用以下方法
        //post(Runnable)
        //postAtTime(Runnable,long)
        //postDelayed(Runnable,long)
        //sendEmptyMessage(int)
        //sendMessage(Message);
        //sendMessageAtTime(Message,long)
        //sendMessageDelayed(Message,long)
        //以上方法以 post 开头的允许你处理 Runnable 对象
        //sendMessage() 允许你处理 Message 对象(Message 里可以包含数据,)
        MyThread m = new MyThread();
        new Thread(m).start();
    }
    /**
     * 接受消息，处理消息，此 Handler 会与当前主线程一块运行
     * */
    class MyHandler extends Handler {
        public MyHandler() {
        }
        public MyHandler(Looper L) {
            super(L);
        }
        //子类必须重写此方法，接受数据
        @Override
        public void handleMessage(Message msg) {
            //TODO Auto-generated method stub
            Log.d("MyHandler", "handleMessage......");
            super.handleMessage(msg);
            //此处可以更新 UI
            Bundle b = msg.getData();
            String color = b.getString("color");
            MyHandlerActivity.this.button.append(color);
        }
    }
    class MyThread implements Runnable {
        public void run() {
            try {
                Thread.sleep(10000);
```

```
        } catch (InterruptedException e) {
          //TODO Auto-generated catch block
          e.printStackTrace();
        }
         Log.d("thread.......", "mThread........");
        Message msg = new Message();
        Bundle b = new Bundle();//存放数据
        b.putString("color", "我的");
        msg.setData(b);
         MyHandlerActivity.this.myHandler.sendMessage(msg);
                     //向 Handler 发送消息,更新 UI
      }
    }
```

5.6 本章小结

本章主要介绍了 Android 中常见组件的特性及使用方法，了解这些组件可以帮助快速构建 Android 应用。图形界面肯定需要与事件处理相结合，当开发了一个界面友好的应用后，用户在程序界面上执行操作时，程序必须能为这种操作提供响应，这种响应动作由事件处理来完成。Android 的事件处理机制主要有两种：基于回调的事件处理和基于监听的事件处理，开发者可以根据需要选择合适的事件处理机制。

第 6 章 Android 的门面——Activity

Acrivity 是 Android 的最基本组件之一，它就像是一个管理员。Activity 我们称之为"活动"。在应用程序中，一个活动（Activity）通常就是一个单独的屏幕，每一个活动都被实现为一个独立的类，并且从活动基类中继承而来，活动类将会显示由视图控件组成的用户接口，并对事件作出响应。所有应用的 Activity 都继承 android.app.Activity 类，该类是 Android 提供的基层类，其他的 Activity 继承该父类后，通过父类的方法来实现各种功能。本章深入介绍 Activity 的相关知识。

6.1 Activity 生命周期

当 Activity 处于 Android 应用中运行时，它的活动状态由 Android 及 Activity 栈的形式管理。当前活动的 Activity 位于栈顶。随着不同应用的运行，每个 Activity 都有可能从活动状态转入非活动状态，也可能从非活动状态转入活动状态。图 6.1 所示官方提供的 Activity 生命周期图。由图可以看到有三个关键的生命周期循环。

（1）一个 activity 完整的生命周期。自第一次调用 onCreate（Bundle）开始，直至调用 onDestroy()为止。activity 在 onCreate()中设置所有"全局"状态以完成初始化，而在 onDestroy()中释放所有系统资源。比如说，如果 activity 有一个线程在后台运行用来从网络上下载数据，它会以 onCreate()创建那个线程，而以 onDestroy()销毁那个线程。

（2）一个 activity 的可视生命周期。自 onStart()调用开始直到相应的 onStop()调用。在此期间，用户可以在屏幕上看到此 activity，尽管它也许并不是位于前台或正在与用户做交互。在这两个方法中，你可以管控用来向用户显示这个 activity 的资源。比如说，你可以在 onStart() 中注册一个 BroadcastReceiver 来监控会影响到你 UI 的改变，而在 onStop()中来取消注册，这时，用户是无法看到你程序显示的内容的。onStart()和 onStop()方法可以随着应用程序是否为用户可见而被多次调用。

（3）一个 activity 的前台生命周期。自 onResume() 调用起，至相应的 onPause()调用为止。在此期间，activity 位于前台最上面并与用户进行交互。activity 会经常在暂停和恢复之间进行状态转换——比如说，当设备转入休眠状态或有新 activity 启动时，将调用 onPause() 方法。当 activity 获得结果或接收到新的 intent 时会调用 onResume()方法。因此，在这两个方法中的代码应当是轻量级的。

由图 6.1 可以看出，Activity 在整个生命周期中大致会经过如下四个状态。
- 活动状态：当前 Activity 位于前台，用户可见，可以获得焦点。
- 暂停状态：其他 Activity 位于前台，该 Activity 依然可见，只是不能获得焦点。
- 停止状态：该 Acivity 不可见，失去焦点。
- 销毁状态：该 Activity 结束，或 Activity 所在的 Dalvik 进程被结束。

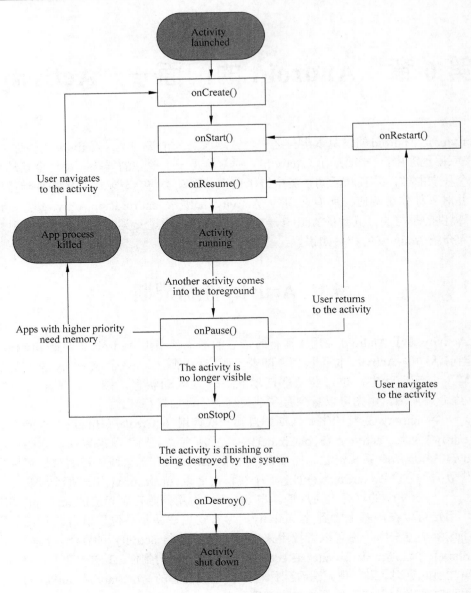

图 6.1 Activity 生命周期

Activity 的整个生命周期都定义在下面的接口方法中，所有方法都可以被重载。

```
public class Activity extends ApplicationContext {
    protected void onCreate(Bundle icicle);
    protected void onStart();
    protected void onRestart();
    protected void onResume();
    protected void onFreeze(Bundle outIcicle);
    protected void onPause();
    protected void onStop();
```

```
        protected void onDestroy();
}
```

下面对这些方法什么时候被调用,以及调用时做什么事情进行详细说明。

onCreate:当活动第一次启动时,触发该方法,可以在此时完成活动的初始化工作。

onStart:该方法的触发表示所属活动将被展现给用户。

onResume:当一个活动和用户发生交互时,触发该方法。

onPause:当一个正在前台运行的活动因为其他的活动需要由前台运行状态而转入后台运行时,触发该方法。这时需要将活动的状态持久化,比如,正在编辑的数据库记录等。

onStop:当一个活动不再需要展示给用户时,触发该方法。如果内存紧张,系统会直接结束这个活动,而不会触发 onStop 方法。所以保存状态信息应该在 onPause 时做,而不是 onStop 时做。活动如果没有在前台运行,都将被停止或者 Linux 管理进程为了给新的活动预留足够的存储空间而随时结束这些活动。因此,对于开发者来说,在设计应用程序时,必须时刻牢记这一原则。在一些情况下,onPause 方法或许是活动触发的最后方法,因此,开发者需要在这时候保存需要保存的信息。

onRestart:当处于停止状态的活动需要再次展现给用户时,触发该方法。

onDestroy:当活动销毁时,触发该方法。和 onStop 方法一样,如果内存紧张,系统会直接结束这个活动而不会触发该方法。

开发 Activity 时可以根据需要覆盖指定的方法。其中,最常见的就是覆盖 onCreate(Bundle saveStatus)方法。之前所有的示例都覆盖了此方法,该方法用于对该 Activity 执行初始化。此外,覆盖 onPause()方法也很常见:比如,用户正在玩一个游戏,此时有电话进来,那么我们需要将当前(游戏)暂停,并保存该游戏的进行状态,这就可以通过 onPause()方法来实现。

下面通过一个示例演示在一个 Activity 的生命周期中,这些方法何时被调用。

创建名为 activityLifeCycle 的工程,默认的 Activity 名字为 ActivityLifeCycle,其对应的界面非常简单,只有一个退出按钮,点击按钮可以退出程序,该 Activtiy 的代码如下。

```
package com.example.activitylifecycle;
import android.os.Bundle;
import android.app.Activity;
import android.util.Log;
import android.view.Menu;
import android.view.View;
import android.view.View.OnClickListener;
import android.widget.Button;
public class ActivityLifeCycle extends Activity {
    final String TAG="Hello Activity";
    protected void onCreate(Bundle savedInstanceState) {
        super.onCreate(savedInstanceState);
        setContentView(R.layout.activity_life_cycle);
        Log.d(TAG, "----onCreate----");
```

```java
        Button bt=(Button)findViewById(R.id.button1);
        bt.setOnClickListener(new OnClickListener(){
            public void onClick(View v) {
                ActivityLifeCycle.this.finish();
            }
        });
    }
    protected void onDestroy() {
        super.onDestroy();
        Log.d(TAG, "----onDestroy----");
    }
    protected void onPause() {
        super.onPause();
        Log.d(TAG, "----onPause----");
    }
    protected void onRestart() {
        super.onRestart();
        Log.d(TAG, "----onRestart----");
    }
    protected void onResume() {
        super.onResume();
        Log.d(TAG, "----onResume----");
    }
    protected void onStart() {
        super.onStart();
        Log.d(TAG, "----onStart----");
    }
    protected void onStop() {
        super.onStop();
        Log.d(TAG, "----onStop----");
    }
}
```

执行该程序，在 DDMS 的 Logcat 窗口将会看到如下信息，如图 6.2 所示。

Level	Time	PID	TID	Application	Tag	Text
D	12-25 02:11:30.019	580	580	com.example.activitylifecycle	Hello Activity	----onCreate----
D	12-25 02:11:30.019	580	580	com.example.activitylifecycle	Hello Activity	----onStart----
D	12-25 02:11:30.019	580	580	com.example.activitylifecycle	Hello Activity	----onResume----

图 6.2 启动 Activity 时调用的回调方法

当程序正在运行时，单击模拟器的 按钮，返回桌面，则当前的 Activity 变得不可见，失去焦点，但并未被销毁，只是处于暂停状态。此时，会看到 Logcat 中的输出信息如图 6.3 红色标记所示。

在模拟器列表里找到该应用程序，并再次启动，此时，可以在 Logcat 中看到如图 6.4 所示信息。

图 6.3 暂停 Activity 时的回调方法

图 6.4 重启 Activity 时的回调方法

在该程序界面点击退出按钮,该 Activity 将会结束自己,此时,可以在 Logcat 中看到如图 6.5 所示界面。

图 6.5 结束 Activity 时的回调方法

通过以上操作,相信读者对于 Activity 的生命周期状态及在不同状态之间切换时所回调的方法已经有了很清晰的认识了。

6.2 Activity 管理栈

之前提到开发者是无法控制 Activity 的状态的,那么,Activity 的状态是按照何种逻辑来运作的呢?这就需要用到 Activity 栈来对 Activity 进行管理。

一个程序一般由多个 Activity 组成,各 activities 之间关系很松散,它们之间没有直接的关联。必须有一个 activity 被指定为主 activity,它是程序启动时首先显示的界面。每个 activity 都可以随意启动其他的 activity。每当一个 activity 被启动,则前一个 activity 就被停止。一个程序中的所有启动的 activity 都被放在一个栈中,所以被停止的 activity 并没有销毁，而存在于栈中。新启动的 activity 先被存放于栈中，然后获得输入焦点。在当前活动的 Activity 上点返回键，它被从栈中取出，然后销毁，然后上一个 Activity 被恢复。每个 Activity 的状态是由它在 Activity 栈（是一个后进先出 LIFO，包含所有正在运行 Activity 的队列）中的位置决定的。当一个新的 Activity 启动时，当前活动的 Activity 将会移到 Activity 栈的顶部。如果用户使用后退按钮返回的话，或者前台的 Activity 结束，活动的 Activity 就会被移出栈消亡，而在栈上的上一个活动的 Activity 将会移上来并变为活动状态，如图 6.6 所示。

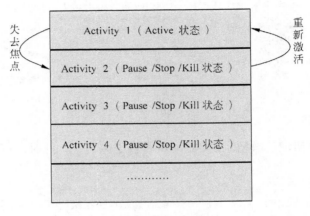

图 6.6　Activity 栈

一个应用程序的优先级是受最高优先级的 Activity 影响的。当决定某个应用程序是否要终结并释放资源时，Android 内存管理使用栈决定基于 Activity 的应用程序的优先级。

6.3　创建、配置和使用 Activity

6.3.1　创建 Activity

与开发 Web 应用时建立 Servlet 类类似，建立自己的 Activity 也需要继承 Activity 基类。当然，在不同的应用场景下，有时也要求继承 Activity 的子类。例如，如果应用程序界面只包括列表，则可以让应用程序继承 ListActivity；如果应用程序界面需要实现标签页效果，则可以让应用程序继承 TabActivity。

Activity 类间接或直接继承了 Context、ContextWrapper、ContextThemeWrapeper 等基类，它的直接子类有 AccountnAuthenticatorActivity、ActivityGroup、Aliasactivity、ExpandableListActivity、FragmentActivity、ListActivity、NativeActivity，间接子类有 LauncherActivity、PreferenceActivity、TabActivity。

当定义了一个 Activity 类，这个 Activity 类何时被实例化、它所包含的方法何时被调用，这些都不是由开发者决定的，而是由 Android 系统决定的。为了响应用户的操作，创建一个 Activity 需要实现一个或多个方法，其中，最常见的就是实现 onCreate(Bundle status)

方法，从前面内容可知，该方法将会在 Activity 启动时被回调，该方法调用 Activity 的 setContentView(View view)方法来显示要展现的 View。为了管理应用程序界面中的各组件，调用 Activity 的 findViewById(int id)方法来获取程序界面中的组件，接下来去修改各组件的属性和方法即可。这些步骤在之前的章节中都经常使用，本节将创建一个能处理用户常见动作的 Activity。

创建一个名为 activityDemo 的项目，为其创建一个默认的 Activity，名为 ActivityEvent，该 Activity 能对用户的动作做出反应，并以 Toast 显示提示信息，所以，此 Activity 需要重写处理事件的相关方法，代码如下。

```java
package com.example.activitydemo;
import android.os.Bundle;
import android.app.Activity;
import android.view.KeyEvent;
import android.view.Menu;
import android.view.MotionEvent;
import android.widget.Toast;
public class ActivityEvent extends Activity {
    protected void onCreate(Bundle savedInstanceState) {
        super.onCreate(savedInstanceState);
        setContentView(R.layout.activity_event);
    }
    public boolean onKeyDown(int keyCode, KeyEvent event) {
        showInfo("按键按下");
        return super.onKeyDown(keyCode, event);
    }
    public boolean onTouchEvent(MotionEvent event) {
        float x=event.getX();
        float y=event.getY();
        showInfo("您点击的坐标为:("+x+":"+y+")");
        return super.onTouchEvent(event);
    }
    public boolean onKeyUp(int keyCode, KeyEvent event) {
        showInfo("按键弹起！");
        return super.onKeyUp(keyCode, event);
    }
    public void showInfo(String info){
        Toast.makeText(ActivityEvent.this,info, Toast.LENGTH_SHORT).show();
    }
}
```

执行该程序，当按下按键时，结果如图 6.7 所示。
当松开按键时，结果如图 6.8 所示。
当点击屏幕时，结果如图 6.9 所示。

图 6.7 按键按下时界面　　　图 6.8 松开按键时界面　　　图 6.9 点击屏幕时界面

6.3.2 配置 Activity

Android 应用要求所有应用程序组件（Activity、Service、ContentProvider、BroadcastReceiver）都必须显示进行配置。之前我们的项目中都只有一个默认的 Activity，只需要编写相关的代码就可以运行，并没有进行配置，这是因为如果一个应用只有一个 Activity，系统就默认为它是整个程序的入口，并且自动在配置文件中进行配置，如上例 activityDemo 中的配置文件对 ActivityEvent 的默认配置片段，如下所示。

```xml
<activity
    android:name="com.example.activitydemo.ActivityEvent"
    android:label="@string/app_name" >
    <intent-filter>
        <action android:name="android.intent.action.MAIN" />
        <category
            android:name="android.intent.category.LAUNCHER" />
    </intent-filter>
</activity>
```

其中

```xml
<intent-filter>
    <action android:name="android.intent.action.MAIN" />
    <category
        android:name="android.intent.category.LAUNCHER" />
</intent-filter>
```

指明了该 Activity 是应用程序的入口，即程序一旦启动就会执行该 Activity。

但在实际开发中,一个项目不可能只有一个 Activity,可能包含多个 Activity,那么,就需要对每个 Activity 进行配置,即为<application.../>元素添加<activity.../>子元素。由以上的配置片段可以看出,配置 Activity 时通常指定如下三个属性。

- name:指定该 Activity 的实现类。
- icon:指定该 Activity 对应的图标。
- label:指定该 Activity 的标签。

此外,配置 Activity 时通常还需要指定一个或多个<intent-filter.../>元素,该元素用于指定该 Activity 可响应的 Intent。关于 Intent 和 IntentFileter 的介绍,详见后续章节。

6.3.3 启动关闭 Activity

正如前面介绍的那样,一个应用程序通常包含多个 Activity,但是只有一个 Activity 会作为程序的入口,至于应用中的其他 Activity,通常都是由入口 Activity 启动,或由入口 Activity 启动的 Activity 启动。

Activity 启动其他 Activity 有如下两个方法。

- startActivity(Intent intent):启动其他 Activity。
- startActivityForResult(Intent intent,int requestCode):以指定的请求码(requestCode)启动 Activity,而且程序将会等到新启动的 Activity 的结果(通过重写 onActivityResult()方法来获取)。

启动 Activity 时可指定一个 requestCode 参数,该参数代表了启动 Activity 的请求码。这个请求码的值由开发者根据业务自行设定,用于标识请求来源。

上面两个方法都用到了 Intent 类型的参数,Intent 是 Android 应用中各组件之间通信的重要方式,一个 Activity 通过 Intent 来表达自己的"意图"——即想要启动哪个组件,被启动的组件既可以是 Activity 组件,也可以是 Service 组件。由一个 Activity 通过 Intent 启动另一个 Activity 的典型方式,如图 6.10 所示。

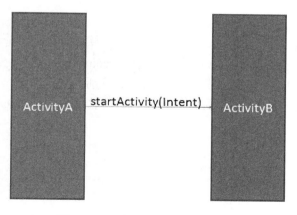

图 6.10 两个 Activity 之间的关系

Android 为关闭 Activity 提供了如下两个方法。

- finish():结束当前 Activity。

- finish(int requestCode)：结束以 startActivityForResult(Intent intent,int requestCode)方法启动的 Activity。

下面通过实例演示两个 Activty 之间的跳转，这两个 Activity 之间无需传递数据。

创建一个名为 multiActivity 的项目，默认的 Activity 名字为 FirstActivity，它的界面很简单，只有一个按钮，用来启动第二个 Activity，在此不给出布局文件，其对应的代码如下。

```java
package com.example.multiactivity;
import android.os.Bundle;
import android.app.Activity;
import android.content.Intent;
import android.view.Menu;
import android.view.View;
import android.view.View.OnClickListener;
import android.widget.Button;
public class FirstActivity extends Activity {
    protected void onCreate(Bundle savedInstanceState) {
        super.onCreate(savedInstanceState);
        setContentView(R.layout.activity_first);
        Button bt=(Button)findViewById(R.id.button1);
        bt.setOnClickListener(new OnClickListener(){
            public void onClick(View v) {
                //创建 Intent
                Intent intent=new Intent(FirstActivity.this,SecondActivity.class);
                //使用 Intent 启动另一个 Activity
                startActivity(intent);
            }
        });
    }
}
```

接着，在该项中再创建一个 Activty，名字为 SecondActivity，并为该 Activty 新建一个布局文件，该布局文件也很简单，只有两个按钮，分别用来返回第一个 Activity 和返回第一个 Activity 并结束自己，在此不再给出布局文件代码。编写 SecondeActivty，其代码如下。

```java
package com.example.multiactivity;
import android.app.Activity;
import android.content.Intent;
import android.os.Bundle;
import android.view.View;
import android.view.View.OnClickListener;
import android.widget.Button;
public class SecondActivity extends Activity {
    protected void onCreate(Bundle savedInstanceState) {
```

```
        super.onCreate(savedInstanceState);
        setContentView(R.layout.second);
        Button previous=(Button)findViewById(R.id.previous);
        Button close=(Button)findViewById(R.id.close);
        previous.setOnClickListener(new OnClickListener(){
            public void onClick(View v) {
                Intent intent=new Intent(SecondActivity.this,FirstActivity.
                class);
                startActivity(intent);
            }
        });
        close.setOnClickListener(new OnClickListener(){
            public void onClick(View v) {
                Intent intent=new Intent(SecondActivity.this,FirstActivity.
                class);
                startActivity(intent);
                //结束当前Activity
                finish();
            }
        });
    }
}
```

然后还要在配置文件中对 SecondActivity 进行配置，才能正常启动，对应的 AndroidManifest.xml 文件部分内容如下。

```
<application
    android:allowBackup="true"
    android:icon="@drawable/ic_launcher"
    android:label="@string/app_name"
    android:theme="@style/AppTheme" >
    <activity
        android:name="com.example.multiactivity.FirstActivity"
        android:label="@string/app_name" >
        <intent-filter>
            <action android:name="android.intent.action.MAIN" />
            <category
              android:name="android.intent.category.LAUNCHER" />
        </intent-filter>
    </activity>
    <activity
        android:name="com.example.multiactivity.SecondActivity"
        android:label="@string/app_name" >
    </activity>
</application>
```

可见，在此应用中有两个 Activity，第一个是默认的入口 Activity，程序一启动，自动执行该 Activity。运行该程序，效果如图 6.11 所示。

单击按钮，启动第二个 Activity，界面如图 6.12 所示。

图 6.11　启动 Activity 示例图　　　　图 6.12　第二个 Activity 启动后的界面

此时，单击两个按钮都会返回到图 6.11 所示界面，两者的区别在于单击第二个按钮会结束当前 Activity，即会调用回调方法 onDestroy()。

6.3.4　需要传递参数的 Activity 启动

前面的例子中，两个 Activity 进行跳转的过程中不需要传递任何参数，但是有时当一个 Activity 启动另一个 Activity 时，常常需要传递一些数据，这就像 Web 应用中从一个 Servlet 跳到另一个 Servlet 时，Web 应用习惯把需要交换的数据放入 requestScope、sessionScope 中。对于 Activity 而言，在 Activity 之间进行数据交换更简单：因为两个 Activity 之间本来就有一个桥梁：Intent，因此我们只要将需要交换的数据放入 Intent 即可。

Intent 提供了多个重载的方法来"携带"额外的数据，如下所示。

- putExtras(Bundle data)：向 Intent 中放入需要"携带"的数据。

Bundle 就是一个简单的数据携带包，该 Bundle 对象包含了多个方法来存入数据。

- putXxx(String key,Xxx data)：向 Bundle 放入 int、Long 等各种类型的数据。
- putSerializable(String key,Serializable data)：向 Bundle 中放入一个可序列化对象。

为了取出 Bundle 数据携带包中的数据，Bundle 提供了如下方法。

- getXxx(String key)：从 Bundle 取出 int、Long 等各种类型的数据。
- getSerializable(String key,Serializable data)：从 Bundle 取出一个可序列化对象。

下面通过一个实例演示两个 Activity 之间如何通过 Bundle 交换数据的。

本例包含两个 Activity，其中，LoginActivity 用于收集用户的登录信息，当用户单击"登录"按钮时，进入第二个 Activity——ResutActivity，此 Activity 将会获取 LoginActivity 中的数据，并显示出来。

LoginActivity 的布局文件内容如下。

```xml
<TableLayout xmlns:android="http://schemas.android.com/apk/res/android"
    android:layout_width="fill_parent"
    android:layout_height="fill_parent" >
    <TextView
        android:layout_width="fill_parent"
        android:layout_height="wrap_content"
        android:text="请输入您的登录信息"
        android:textSize="20sp" />
    <TableRow
        android:id="@+id/tableRow1"
        android:layout_width="fill_parent"
        android:layout_height="wrap_content"
        >
        <TextView
            android:textSize="16sp"
            android:layout_width="fill_parent"
            android:layout_height="wrap_content"
            android:text="用户名" />
        <EditText
            android:id="@+id/name"
            android:layout_width="fill_parent"
            android:layout_height="wrap_content"
            android:hint="请填写登录的用户名"
            android:selectAllOnFocus="true">
            <requestFocus />
        </EditText>
    </TableRow>
    <TableRow
        android:id="@+id/tableRow2"
        android:layout_width="fill_parent"
        android:layout_height="wrap_content" >
        <TextView
            android:textSize="16sp"
            android:layout_width="fill_parent"
            android:layout_height="wrap_content"
            android:text="请输入密码" />
        <EditText
            android:id="@+id/password"
            android:layout_width="fill_parent"
            android:layout_height="wrap_content"
            android:password="true"
            android:selectAllOnFocus="true">
            <requestFocus />
        </EditText>
    </TableRow>
```

```xml
<TableRow
    android:id="@+id/tableRow3"
    android:layout_width="fill_parent"
    android:layout_height="wrap_content" >
    <Button
        android:id="@+id/login"
        android:layout_width="wrap_content"
        android:layout_height="wrap_content"
        android:text="登录"
        android:textSize="16sp"/>
</TableRow>
</TableLayout>
```

此界面采用表格布局，表格包含两行，每一行又包含两个组件，第一行用来接收输入的用户名，第二行用来接收输入的密码。

编写 LoginActivity 的代码如下。

```java
import android.os.Bundle;
import android.app.Activity;
import android.content.Intent;
import android.view.Menu;
import android.view.View;
import android.view.View.OnClickListener;
import android.widget.Button;
import android.widget.EditText;
public class LoginActivity extends Activity {
    protected void onCreate(Bundle savedInstanceState) {
        super.onCreate(savedInstanceState);
        setContentView(R.layout.activity_login);
        Button bt=(Button)findViewById(R.id.login);
        bt.setOnClickListener(new OnClickListener(){
            public void onClick(View v) {
                EditText name=(EditText)findViewById(R.id.name);
                EditText password=(EditText)findViewById(R.id.password);
                //创建一个 Bundle 对象
                Bundle data=new Bundle();
                //向 Bundle 中绑定数据，以键值对的形式
                data.putString("name", name.getText().toString());
                data.putString("password", password.getText().toString());
                Intent intent=new Intent(LoginActivity.this,ResultActivity.class);
                //把 Bundle 绑定到 Intent 中
                intent.putExtras(data);
                startActivity(intent);
            }
```

```
        });
    }
    public boolean onCreateOptionsMenu(Menu menu) {
        // Inflate the menu; this adds items to the action bar if it is present.
        getMenuInflater().inflate(R.menu.activity_login, menu);
        return true;
    }
}
```

ResultActivity 对应的界面中只有两个 TextView，用来显示从 LoginActivity 中获取的用户名和密码，其布局文件内容如下。

```xml
<?xml version="1.0" encoding="utf-8"?>
<LinearLayout xmlns:android="http://schemas.android.com/apk/res/android"
    android:layout_width="match_parent"
    android:layout_height="match_parent"
    android:orientation="vertical" >
    <TextView
        android:id="@+id/nameShow"
        android:layout_width="fill_parent"
        android:layout_height="wrap_content"
        android:textSize="18sp"
        />
    <TextView
        android:id="@+id/passwordShow"
        android:layout_width="fill_parent"
        android:layout_height="wrap_content"
        android:textSize="18sp"
        />
</LinearLayout>
```

编写 ResutActivity 的代码，内容如下。

```java
package com.example.bundledemo;
import android.app.Activity;
import android.content.Intent;
import android.os.Bundle;
import android.widget.TextView;
public class ResultActivity extends Activity {
    protected void onCreate(Bundle savedInstanceState) {
        super.onCreate(savedInstanceState);
        setContentView(R.layout.result);
        TextView name=(TextView)findViewById(R.id.nameShow);
        TextView password=(TextView)findViewById(R.id.passwordShow);
        //获取 Intent
        Intent intent=getIntent();
```

```
        //获取 Intent 中绑定的 Bundle
        Bundle result=intent.getExtras();
        name.setText("您的用户名为："+result.getString("name"));
        password.setText("您的性别为："+result.getString("password"));
    }
}
```

运行此项目，结果如图 6.13 所示。

在此界面输入用户名和密码，单击"登录"按钮，则跳转到如图 6.14 所示界面。

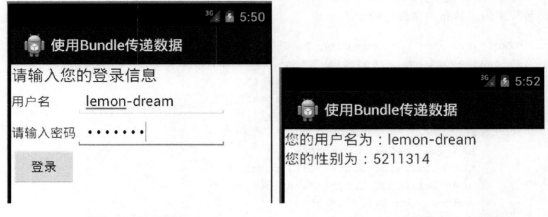

图 6.13　登录界面　　　　　　　　　　图 6.14　登录信息显示界面

6.3.5　启动其他 Activity 并返回结果

除了能够不带参数及带参数启动另一个 Activity，Activity 还提供了一个 startActivityForResult(Intent intent,int requestCode)方法来启动其他 Activity，并且期望从被启动的 Activity 中获取指定的结果。这种请求在实际应用中也是比较常见的，例如，应用程序的第一个界面需要用户进行选择——但需要选择的列表数据比较复杂，必须启动另一个 Activity 让用户选择。当用户在第二个 Activity 选择完成后，程序需要把选择的结果带回给第一个 Activity，并在此 Activity 中显示。这种情况，也是通过 Bundle 进行数据交换的。

为了获取被启动的 Activity 所返回的结果，当前 Activity 需要重写 onActivityResult(int requestCode,int resultCode,Intent intent)，其中，requestCode 代表请求码，而 resultCode 代表 Activity 返回的结果码，这个结果码也是由开发者根据业务自行设定的。

一个 Activity 中可能包含多个按钮，并调用多个 startActivityForResult()方法来打开多个不同的 Activity 处理不同的业务，当这些新 Activity 关闭后，系统都会调用前面 Activity 的 onActivityResult()方法。为了知道该方法是由哪个请求的结果触发的，可利用 requestCode 请求码；为了知道返回的数据来自于哪个新的 Activity，可利用 resultCode 结果码。

下面通过实例介绍如何启动 Activity 并获取被启动的 Activity 返回的结果。

创建一个名为 forResultDemo 的工程，默认的 Activity 为 ForResultActivity，其布局比较简单，界面上只有一个按钮和一个文本框。单击按钮可以进入下一个 Activity 即 SelectActivity，此 Activity 是 ExpandableListActivity 的子类，可提供一个可扩展的列表选项，无需布局文件，当选中其中一个选项时，结果返回给 ForResultActivity，并把其文本框内容更新为返回的信息。

编写 ForResultActivity 的代码如下。

```java
package com.example.forresultdemo;
import android.os.Bundle;
import android.app.Activity;
import android.content.Intent;
import android.view.Menu;
import android.view.View;
import android.view.View.OnClickListener;
import android.widget.Button;
import android.widget.EditText;
public class ForResultActivity extends Activity {
    Button bt;
    EditText special;
    protected void onCreate(Bundle savedInstanceState) {
        super.onCreate(savedInstanceState);
        setContentView(R.layout.activity_for_result);
        bt=(Button)findViewById(R.id.button1);
        special=(EditText)findViewById(R.id.editText1);
        bt.setOnClickListener(new OnClickListener(){
            public void onClick(View v) {
                Intent intent=new Intent(ForResultActivity.this,SelectActivity.class);
                startActivityForResult(intent, 0);
            }
        });
    }
    protected void onActivityResult(int requestCode, int resultCode, Intent data) {
        if(requestCode==0&&resultCode==0){
            Bundle info=data.getExtras();
            String resultSpecial=info.getString("special");
            special.setText(resultSpecial);
        }
    }
}
```

编写 SelectActivity 的代码如下。

```java
package com.example.forresultdemo;
```

```java
import android.app.Activity;
import android.app.ExpandableListActivity;
import android.content.Intent;
import android.os.Bundle;
import android.view.View;
import android.view.ViewGroup;
import android.widget.AbsListView;
import android.widget.BaseExpandableListAdapter;
import android.widget.ExpandableListAdapter;
import android.widget.ExpandableListView;
import android.widget.ExpandableListView.OnChildClickListener;
import android.widget.ImageView;
import android.widget.LinearLayout;
import android.widget.TextView;
public class SelectActivity extends ExpandableListActivity {
    private String[] colleges=new String[]{"计算机与通信工程学院","电器信息工程学院","艺术设计学院"};
    private String[][] specials=new String[][]{
        {"计算机","3G","物联网"},
        {"电器工程","自动化","电子信息工程"},
        {"工业设计","艺术设计","动画设计"}
    };
    protected void onCreate(Bundle savedInstanceState) {
        super.onCreate(savedInstanceState);
        ExpandableListAdapter adapter=new BaseExpandableListAdapter() {
            public boolean isChildSelectable(int groupPosition, int childPosition) {
                return true;
            }
            public boolean hasStableIds() {
                return true;
            }

            public View getGroupView(int groupPosition, boolean isExpanded,
                    View convertView, ViewGroup parent) {
                LinearLayout ll=new LinearLayout(SelectActivity.this);
                ll.setOrientation(0);
                ImageView logo=new ImageView(SelectActivity.this);
                ll.addView(logo);
                TextView tv=getTextView();
                tv.setText(getGroup(groupPosition).toString());
                ll.addView(tv);
                return ll;
            }
            public long getGroupId(int groupPosition) {
```

```java
            return groupPosition;
        }
        public int getGroupCount() {
            return colleges.length;
        }
        public Object getGroup(int groupPosition) {
            return colleges[groupPosition];
        }
        public int getChildrenCount(int groupPosition) {
            return specials[groupPosition].length;
        }
        private TextView getTextView(){
            AbsListView.LayoutParams lp=new AbsListView.LayoutParams(
                    ViewGroup.LayoutParams.FILL_PARENT,64);
            TextView textView=new TextView(SelectActivity.this);
            textView.setLayoutParams(lp);
            textView.setPadding(36, 0, 0, 0);
            textView.setTextSize(20);
            return textView;
        }
        public View getChildView(int groupPosition, int childPosition,
                boolean isLastChild, View convertView, ViewGroup parent){
            TextView textView=getTextView();
            textView.setText(getChild(groupPosition,childPosition).
            toString());
            return textView;
        }

        public long getChildId(int groupPosition, int childPosition){
            return childPosition;
        }

        public Object getChild(int groupPosition, int childPosition){
            return specials[groupPosition][childPosition];
        }
};
setListAdapter(adapter);
getExpandableListView().setOnChildClickListener(new
OnChildClickListener(){
    public boolean onChildClick(ExpandableListView parent, View v,
            int groupPosition, int childPosition, long id) {
        Intent intent=getIntent();
        Bundle data=new Bundle();
        data.putString("special", specials[groupPosition]
        [childPosition]);
```

```
                intent.putExtras(data);
                SelectActivity.this.setResult(0, intent);
                SelectActivity.this.finish();
                return false;
            }
        });
    }
}
```

运行此程序，界面如图 6.15 所示。

单击按钮，则跳转到如图 6.16 所示界面。

图 6.15　ForResultActivity 界面

图 6.16　专业列表

选择计算机与通信工程学院物联网专业，则跳转到如图 6.17 所示界面。

图 6.17　显示返回数据

6.4　启 动 模 式

在 android 的多 activity 开发中，activity 之间的跳转可能需要有多种方式，有时是普通的生成一个新实例,有时希望跳转到原来某个 activity 实例,而不是生成大量的重复的 activity。加载模式便是决定以哪种方式启动一个跳转到原来某个 Activity 实例。

在 android 里,有四种 activity 的启动模式,分别如下。

- standard:标准模式,一调用 startActivity()方法就会产生一个新的实例。
- singleTop:如果已经有一个实例位于 Activity 栈的顶部时,就不产生新的实例,而只是调用 Activity 中的 newInstance()方法。如果不位于栈顶,会产生一个新的实例。
- singleTask:会在一个新的 task 中产生这个实例,以后每次调用都会使用这个,不会去产生新的实例了。
- singleInstance:这个和 singleTask 基本相同,只有一个区别:在这个模式下的 Activity 实例所处的 task 中,只能有这个 activity 实例,不能有其他的实例。

这些启动模式可以在功能清单文件 AndroidManifest.xml 中通过 launchMode 属性值来设定。相关的代码中也有一些标志可以使用,比如,我们想只启用一个实例,则可以使用 Intent.FLAG_ACTIVITY_REORDER_TO_FRONT 标志,这个标志表示:如果这个 activity 已经启动了,就不产生新的 activity,而只是把这个 activity 实例加到栈顶来就可以了,例如,

```
Intent intent = new Intent(ReorderFour.this, ReorderTwo.class);
intent.addFlags(Intent.FLAG_ACTIVITY_REORDER_TO_FRONT);
    startActivity(intent);
```

Activity 的加载模式受启动 Activity 的 Intent 对象中设置的 Flag 和 manifest 文件中 Activity 的元素的特性值交互控制。

下面是影响加载模式的一些特性。

核心的 Intent Flag 如下。

- FLAG_ACTIVITY_NEW_TASK
- FLAG_ACTIVITY_CLEAR_TOP
- FLAG_ACTIVITY_RESET_TASK_IF_NEEDED
- FLAG_ACTIVITY_SINGLE_TOP

核心的特性如下。

- taskAffinity
- launchMode
- allowTaskReparenting
- clearTaskOnLaunch
- alwaysRetainTaskState
- finishOnTaskLaunch

上述四种启动模式又可以分为两类,standard 和 singleTop 属于一类,singleTask 和 singleInstance 属于另一类。

standard 和 singleTop 属性的 Activity 的实例可以属于任何任务(Task),并且可以位于 Activity 堆栈的任何位置。比较典型的一种情况是,一个任务的代码执行 startActivity(),如果传递的 Intent 对象没有包含 FLAG_ACTIVITY_NEW_TASK 属性,指定的 Activity

将被该任务调用，从而装入该任务的 Activity 堆栈中。 standard 和 singleTop 的区别在于：standard 模式的 Activity 在被调用时会创建一个新的实例，所有实例处理同一个 Intent 对象；但对于 singleTop 模式的 Activity，如果被调用的任务已经有一个这样的 Activity 在堆栈的顶端，那么不会有新的实例创建，任务会使用当前顶端的 Activity 实例来处理 Intent 对象，换句话说，如果被调用的任务包含一个不在堆栈顶端的 singleTop Activity，或者堆栈顶端为 singleTop 的 Activity 的任务不是当前被调用的任务，那么，仍然会有一个新的 Activity 对象被创建。

singleTask 和 singleInstance 模式的 Activity 仅可用于启动任务的情况，这种模式的 Activity 总是处在 Activity 堆栈的最底端，并且一个任务中只能被实例化一次。两者的区别在于：对于 singleInstance 模式的 Activity，任务的 Activity 堆栈中如果有这样的 Activity，那它将是堆栈中的唯一的 Activity，当前任务收到的 Intent 都由它处理，由它开启的其他 Activity 将在其他任务中被启动；对于 SingleTask 模式的 Activity，它在堆栈底端，其上方可以有其他 Activity 被创建，但是，如果发给该 Activity 的 Intent 对象到来时，该 Activity 不在堆栈顶端，那么该 Intent 对象将被丢弃，但是界面还是会切换到当前的 Activity。

在多 Activity 开发中，有可能是自己应用间的 activity 跳转，或者夹带其他应用的可复用 activity。可能会希望跳转到原来某个 activity 实例，而非产生多个重复的 activity。我们可借助 activity 四种启动模式来实现不同的需求。

- standard 默认模式 ——来了 intent，每次都创建新的实例。
- singleTop ——来了 intent, 每次都创建新的实例，仅一个例外：当栈顶的 activity 恰恰就是该 activity 的实例（即需要创建的实例）时，不再创建新实例，这解决了栈顶复用问题，想一想，你按两次 back 键，退出的都是同一个 activity，这感觉肯定不爽。
- singleTask ——来了 intent 后，检查栈中是否存在该 activity 的实例，如果存在就把 intent 发送给它，否则就创建一个新的该 activity 的实例，放入一个新的 task 栈的栈底。肯定位于一个 task 的栈底，而且栈中只能有它一个该 activity 实例，但允许其他 activity 加入该栈。解决了在一个 task 中共享一个 activity。
- singleInstance——肯定位于一个 task 的栈底，并且是该栈唯一的 activity，解决了多个 task 共享一个 activity。

6.5 本章小结

本章详细介绍了 Android 四大组件之一——Activity。Activity 就相当于 Android 应用程序的门面，一个 Activity 通常对应于手机的一屏，它负责把组件按照指定的布局呈现给用户，所以普通用户接触最多的就是 Activity。学习本章的重点就是在理解 Activity 生命周期的基础上掌握如何开发 Activity、如何配置 Activity。不仅如此，由于 Android 应用通常包含多个 Activity，因此，读者还需要掌握 Activity 之间的跳转，包括如何利用 Bundle 在不同的 Activity 之间传递数据。

第 7 章　Android 的邮递员——Intent

在一个 Android 应用中，主要由四种组件组成，分别为 Activity、Broadcast、Service、ContentProvider，而这些组件（ContentProvider 除外）之间的通讯中，主要是由 Intent 协助完成。系统会根据此 Intent 中的描述，到 AndroidManifest.xml 中找到满足此 Intent 要求的组件。

7.1　Intent 概述

Android 中提供 Intent 机制来协助应用间的交互与通讯，Intent 负责对应用中一次操作的动作、动作涉及数据、附加数据进行描述，Android 则根据此 Intent 的描述，负责找到对应的组件，将 Intent 传递给调用的组件，并完成组件的调用。Intent 不仅可用于应用程序之间，也可用于应用程序内部的 Activity/Service 之间的交互。因此，Intent 在这里起着一个媒体中介的作用，专门提供组件互相调用的相关信息，实现调用者与被调用者之间的解耦。

表 7.1 显示使用 Intent 启动不同组件的方法。

表 7.1　使用 Intent 启动不同组件的方法

组件类型	启动方法
Activity	Context.startActivity(Intent intent)
	Activity.startActivityForResult()
Service	Context.startService(Intent service)
	Context.bindService(Intent service, ServiceConnection conn, int flags)
BroadcastReceiver	Context.sendBroadcast(Intent intent)
	Context.sendOrderedBroadcast(Intent intent, String receiverPermission)
	Context.sendStickyBroadcast(Intent intent)

7.1.1　Intent 属性

（1）Action，也就是要执行的动作。使用一个字符串对所将执行动作的描述，为了方便引用，Intent 类中定义了一些标准的动作，如表 7.2 所示。

表 7.2　Action 常量

常量	目标组件	动作
ACTION_CALL	Activity	启动一个电话
ACTION_EDIT	Activity	显示数据编辑界面.

续表

常量	目标组件	动作
ACTION_MAIN	Activity	启动项目的初始界面
ACTION_SYNC	Activity	同步服务器和移动设备的数据
ACTION_BATTERY_LOW	Broadcast Receiver	警告电池电量低
ACTION_HEADSET_PLUG	Broadcast Receiver	耳机插入/拔掉设备
ACTION_SCREEN_ON	Broadcast Receiver	屏幕已打开
ACTION_TIMEZONE_CHANGED	Broadcast Receiver	时区的设置已经改变

当然，也可以自定义动作（自定义的动作在使用时，需要加上包名作为前缀，如"com.example.project.SHOW_COLOR"），并可定义相应的 Activity 来处理我们的自定义动作。

（2）Data，也就是执行动作要操作的数据。

Android 中采用指向数据的一个 URI 来表示，如在联系人应用中，一个指向某联系人的 URI 可能为 content://contacts/1。对于不同的动作，其 URI 数据的类型是不同的（可以设置 type 属性指定特定类型数据），如 ACTION_EDIT 指定 Data 为文件 URI，打电话为 tel:URI，访问网络为 http:URI，而由 content provider 提供的数据则为 content:URIs。

（3）Type，显式指定 Intent 的数据类型（MIME: Multipurpose Internet Mail Extensions，多用途互联网邮件扩展）。一般 Intent 的数据类型能够根据数据本身进行判定，但是通过设置这个属性，可以强制采用显式指定的类型而不再进行推导。

MIME 类型有两种形式。

单个记录的格式：vnd.android.cursor.item/vnd.yourcompanyname.content type，如 content://com.example.transportationprovider/trains/122（一条列车信息的 uri）的 MIME 类型是 vnd.android.cursor.item/vnd.example.rail；

多个记录的格式：vnd.android.cursor.dir/vnd.yourcompanyname.Content type，如 content://com.example.transportationprovider/trains（所有列车信息）的 MIME 类型是 vnd.android.cursor.dir/vnd.example.rail。

（4）Category，一个字符串，包含了关于处理该 intent 的组件的种类的信息。一个 intent 对象可以有任意个 category。intent 类定义了许多 category 常数，如表 7.3 所示。

表 7.3 Category 常量

常量	作用
CATEGORY_DEFAULT	默认的 Category
CATEGORY_BROWSABLE	该 Activity 能被浏览器安全调用
CATEGORY_HOME	设置该 Activity 随系统启动而运行
CATEGORY_LAUNCHER	Intent 的接受者应该在 Launcher 中作为顶级应用出现
CATEGORY_PREFERENCE	该 Activity 是参数面板

（5）Component，指定 Intent 的目标组件的类名称。通常 Android 会根据 Intent 中包含的其他属性的信息，如 Action、Data/Type、Category 进行查找，最终找到一个与之匹配的目标组件。但是，如果 Component 这个属性有指定的话，将直接使用它指定的组件，而不再执行上述查找过程。指定了这个属性以后，Intent 的其他所有属性都是可

选的。

（6）Extras（附加信息），是其他所有附加信息的集合。使用 Extras 可以为组件提供扩展信息，比如，如果要执行"发送电子邮件"这个动作，可以将电子邮件的标题、正文等保存在 Extras 里，传给电子邮件发送组件。

7.1.2　Intent 解析

理解 Intent 的关键之一是理解清楚 Intent 的两种基本用法：一种是显式的 Intent，即在构造 Intent 对象时就指定接收者；另一种是隐式的 Intent，即 Intent 的发送者在构造 Intent 对象时，并不知道也不关心接收者是谁，有利于降低发送者和接收者之间的耦合。

对于显式 Intent，Android 不需要去做解析，因为目标组件已经很明确，Android 需要解析的是那些隐式 Intent，通过解析，将 Intent 映射给可以处理此 Intent 的 Activity、Service 或 Broadcast Receiver。

Intent 解析机制主要是通过查找已注册在 AndroidManifest.xml 中的所有<intent -filter>及其中定义的 Intent，最终找到匹配的 Component。在这个解析过程中，Android 是通过 Intent 的 action、type、category 这三个属性来进行判断的，判断方法如下。

- 如果 Intent 指明定了 action，则目标组件的<intent-filter>的 action 列表中就必须包含有这个 action，否则不能匹配。
- 如果 Intent 没有提供 type，系统将从 data 中得到数据类型。和 action 一样，目标组件的数据类型列表中必须包含 Intent 的数据类型，否则不能匹配。
- 如果 Intent 中的数据不是 content：类型的 URI，而且 Intent 也没有明确指定它的 type，将根据 Intent 中数据的 scheme（如 http：或者 mailto：）进行匹配。同上，Intent 的 scheme 必须出现在目标组件的 scheme 列表中。
- 如果 Intent 指定了一个或多个 category，这些类别必须全部出现在组建的类别列表中，比如，Intent 中包含了两个类别：LAUNCHER_CATEGORY 和 ALTERNATIVE_CATEGORY，解析得到的目标组件必须至少包含这两个类别。

7.2　Intent Filter

活动、服务、广播接收者为了告知系统能够处理哪些隐式 intent，它们可以有一个或多个 intent 过滤器。每个过滤器描述组件的一种能力，即乐意接收的一组 intent。实际上，它筛掉不想要的 intents，也仅仅是不想要的隐式 intents。一个显式 intent 总是能够传递到它的目标组件，不管它包含什么；不考虑过滤器。但是一个隐式 intent，仅当它能够通过组件的过滤器之一才能够传递给它。

一个组件能够做的每一项工作都有独立的过滤器。例如，记事本中的 NoteEditer 活动有两个过滤器，一个是启动一个指定的记录，用户可以查看和编辑；另一个是启动一个新的、空的记录，用户能够填充并保存。

一个 intent 过滤器是一个 IntentFilter 类的实例。因为 Android 系统在启动一个组件之前必须知道它的能力，但是 intent 过滤器通常不在 Java 代码中设置，而是在应用程序的清单文件 AndroidManifest.xml 中以<intent-filter>元素设置。但有一个例外，广播接收者的过滤器通过调用 Context.registerReceiver()动态地注册，它直接创建一个 IntentFilter 对象。

一个过滤器有对应于 Intent 对象的动作、数据、种类的字段。过滤器要检测隐式 intent 的所有这三个字段，其中任何一个失败，Android 系统都不会传递 intent 给组件。然而，因为一个组件可以有多个 intent 过滤器，一个 intent 通不过组件的过滤器检测，其他的过滤器可能通过检测。

7.2.1 动作检测

清单文件中的<intent-filter>元素以<action>子元素列出动作，例如，

```
<intent-filter…>
    <action android:name="com.example.project.SHOW_CURRENT" />
    <action android:name="com.example.project.SHOW_RECENT" />
    <action android:name="com.example.project.SHOW_PENDING" />
    …
</intent-filter>
```

像例子所展示，虽然一个 Intent 对象仅是单个动作，但是一个过滤器可以列出不止一个，这个列表不能够为空，一个过滤器必须至少包含一个<action>子元素，否则它将阻塞所有的 intents。

要通过检测，Intent 对象中指定的动作必须匹配过滤器的动作列表中的一个。如果对象或过滤器没有指定一个动作，结果将如下。

如果过滤器没有指定动作，没有一个 Intent 将匹配，所有的 intent 将检测失败，即没有 intent 能够通过过滤器。

如果 Intent 对象没有指定动作，将自动通过检查。

7.2.2 种类检测

类似的，清单文件中的<intent-filter>元素以<category>子元素列出种类，例如，

```
<intent-filter … >
    <category android:name="android.intent.category.DEFAULT" />
    <category android:name="android.intent.category.BROWSABLE" />
    …
</intent-filter>
```

注意本文前面两个表格列举的动作和种类常量不在清单文件中使用，而是使用全字符串值。例如，例子中所示的"android.intent.category.BROWSABLE"字符串对应于本文前面提到的 BROWSABLE 常量。类似的，"android.intent.action.EDIT"字符串对应于 ACTION_EDIT 常量。

对于一个intent要通过种类检测，intent对象中的每个种类必须匹配过滤器中的一个。即过滤器能够列出额外的种类，但是intent对象中的种类都必须能够在过滤器中找到，只有一个种类在过滤器列表中没有，就算种类检测失败！

因此，原则上如果一个Intent对象中没有种类（即种类字段为空），应该总是通过种类测试，而不管过滤器中有什么种类。但是有个例外，Android对待所有传递给Context.startActivity()的隐式 intent 好像它们至少包含"android.intent.category.DEFAULT"（对应CATEGORY_DEFAULT常量）。因此，活动想要接收隐式intent必须要在intent过滤器中包含"android.intent.category.DEFAULT"。

注意："android.intent.action.MAIN"和"android.intent.category.LAUNCHER"设置，它们分别标记活动开始新的任务和带到启动列表界面。它们可以包含"android.intent.category.DEFAULT"到种类列表，也可以不包含。

7.2.3 数据检测

类似的，清单文件中的<intent-filter>元素以<data>子元素列出数据，例如，

```
<intent-filter …>
    <data android:mimeType="video/mpeg" android:scheme="http" . . . />
    <data android:mimeType="audio/mpeg" android:scheme="http" . . . />
    …
</intent-filter>
```

每个<data>元素指定一个URI和数据类型（MIME类型）。它有四个属性scheme、host、port、path对应于URI的每个部分：

```
scheme://host:port/path
```

例如，下面的URI：

```
content://com.example.project:200/folder/subfolder/etc
```

scheme是content，host是"com.example.project"，port是200，path是"folder/subfolder/etc"。host和port一起构成URI的凭据（authority），如果host没有指定，port也被忽略，这四个属性都是可选的，但它们之间并不都是完全独立的。要让authority有意义，scheme必须也要指定。要让path有意义，scheme和authority也都必须要指定。

当比较intent对象和过滤器的URI时，仅仅比较过滤器中出现的URI属性。例如，如果一个过滤器仅指定了scheme，所有有此scheme的URIs都匹配过滤器；如果一个过滤器指定了scheme和authority，但没有指定path，所有匹配scheme和authority的URIs都通过检测，而不管它们的path；如果四个属性都指定了，要都匹配才能算是匹配。然而，过滤器中的path可以包含通配符来要求匹配path中的一部分。

<data>元素的type属性指定数据的MIME类型。Intent对象和过滤器都可以用"*"通配符匹配子类型字段，例如"text/*"，"audio/*"表示任何子类型。

数据检测既要检测 URI，也要检测数据类型，规则如下。

一个 Intent 对象既不包含 URI，也不包含数据类型：仅当过滤器也不指定任何 URIs 和数据类型时，才不能通过检测；否则都能通过。

一个 Intent 对象包含 URI，但不包含数据类型：仅当过滤器也不指定数据类型，同时它们的 URI 匹配，才能通过检测。例如，mailto:和 tel:都不指定实际数据。

一个 Intent 对象包含数据类型，但不包含 URI：仅当过滤也只包含数据类型且与 Intent 相同，才通过检测。

一个 Intent 对象既包含 URI，也包含数据类型（或数据类型能够从 URI 推断出）：数据类型部分，只有与过滤器中之一匹配才算通过；URI 部分，它的 URI 要出现在过滤器中，或者它有 content:或 file: URI，又或者过滤器没有指定 URI。换句话说，如果它的过滤器仅列出了数据类型，组件假定支持 content:和 file:。

如果一个 Intent 能够通过不止一个活动或服务的过滤器，用户可能会被问哪个组件被激活。如果没有目标找到，会产生一个异常。

7.2.4　通用情况

上面最后一条规则表明组件能够从文件或内容提供者获取本地数据。因此，它们的过滤器仅列出数据类型且不必明确指出 content:和 file:scheme 的名字。这是一种典型的情况，一个<data>元素如下。

```
<data android:mimeType="image/*" />
```

告诉 Android 这个组件能够从内容提供者获取 image 数据并显示它。因为大部分可用数据由内容提供者（content provider）分发，过滤器指定一个数据类型但没有指定 URI 或许最通用。

另一种通用配置是过滤器指定一个 scheme 和一个数据类型。例如，一个<data>元素如下。

```
<data android:scheme="http" android:type="video/*" />
```

告诉 Android 这个组件能够从网络获取视频数据并显示它。那么当用户单击一个 Web 页面上的 link 时浏览器应用程序会做什么呢？它首先会试图显示数据（如果 link 是一个 HTML 页面，就能显示）。如果不能显示数据，将把一个隐式 Intent 加到 scheme 和数据类型，去启动一个能够做此工作的活动。如果没有接收者，它将请求下载管理者下载数据。这将在内容提供者的控制下完成，因此一个潜在的大活动池（他们的过滤器仅有数据类型）能够响应。

大部分应用程序能启动新的活动，而不引用任何特别的数据。活动有指定 android.intent.action.MAIN 的动作的过滤器，能够启动应用程序。如果它们出现在应用程序启动列表中，它们也指定 android.intent.category.LAUNCHER 种类：

```
<intent-filter . . . >
    <action android:name="code android.intent.action.MAIN" />
```

```
<category android:name="code android.intent.category.LAUNCHER" />
</intent-filter>
```

7.2.5 使用 intent 匹配

Intents 对照着 Intent 过滤器匹配,不仅去发现一个目标组件去激活,而且去发现设备上组件的其他信息。例如,Android 系统填充应用程序启动列表,最高层屏幕显示用户能够启动的应用程序:是通过查找所有的包含指定了 android.intent.action.MAIN 的动作和 android.intent.category.LAUNCHER 种类的过滤器的活动,然后在启动列表中显示这些活动的图标和标签。类似的,它通过查找有 android.intent.category. HOME 过滤器的活动发掘主菜单。

我们的应用程序也可以类似使用这种 Intent 匹配方式。PackageManager 有一组 query…()方法返回能够接收特定 intent 的所有组件,一组 resolve…()方法决定最适合的组件响应 intent。例如,queryIntentActivities()返回一组能够给执行指定的 intent 参数的所有活动,类似的 queryIntentServices()返回一组服务。这两个方法都不激活组件,它们仅列出所有能够响应的组件。对应广播接收者也有类似的方法 queryBroadcastReceivers()。

7.3 Intent 的调用

前面介绍的 Intent 属性和解析机制有些抽象,下面通过几个例子来加强大家对 Intent 的理解,以及如何在 Intent 调用的同时传递数据。

7.3.1 显式调用

显式 Intent 直接用组件的名称定义目标组件,这种方式很直接。但是由于开发人员往往并不清楚别的应程序的组件名称。因此,显式 Intent 更多用于在应用程序内部传递消息。

```
ComponentName comp = new ComponentName(MainActivity.this, OtherActivity.class);
Intent intent = new Intent();
intent.setComponent(comp);
startActivity(intent);
```

上面四行代码用于创建 ComponentName 对象,并将该对象设置成 Intent 对象的 Component 属性,这样,应用程序即可根据该 Intent 的"意图"去启动指定组件。这样写有点复杂,Intent 提供了一个构造函数,可以方便地指定要启动的组件。以下代码和上面的代码等价。

```
Intent intent = new Intent(MainActivity.this,OtherActivity.class);
startActivity(intent);
```

如果知道其他应用程序的包名和类名，可以采用如下方式进行调用：

```
Intent intent = new Intent();
intent.setClassName("com.example.test","com.example.test.OtherActivity");
startActivity(intent);
```

这个方法的前提条件是被调用的类已经安装在运行的模拟器或手机上。

7.3.2 隐式调用

如果 Intent 机制仅仅提供上面的显式 Intent 用法的话，这种相对复杂的机制似乎意义并不是很大。确实，Intent 机制更重要的作用在于下面这种隐式的 Intent，即 Intent 的发送者不指定接收者，很可能不知道也不关心接收者是谁，而由 Android 框架去寻找最匹配的接收者。

（1）最简单的隐式 Intent

我们先从最简单的例子开始。下面的 ImplicitIntent 程序用来启动 Android 自带的打电话功能的 Dialer 程序，程序运行的截图如图 7.1 所示。

图 7.1　ImplicitIntentTest 程序运行截图

ImplicitIntent 程序只包含一个 java 源文件 ImplicitIntent.java，代码如下所示。

```
package cn.wang.implicitintent;
import android.os.Bundle;
import android.app.Activity;
import android.content.Intent;
import android.view.Menu;
import android.view.View;
import android.view.View.OnClickListener;
import android.widget.Button;
```

```java
public class MainActivity extends Activity {
    //@Override
    protected void onCreate(Bundle savedInstanceState) {
        super.onCreate(savedInstanceState);
        setContentView(R.layout.activity_main);
        Button butDial = (Button) findViewById(R.id.butDial);
        butDial.setOnClickListener(new OnClickListener() {
            //@Override
            public void onClick(View v) {
                Intent intent = new Intent(Intent.ACTION_DIAL);
                startActivity(intent);
            }
        });
    }

    //   @Override
    public boolean onCreateOptionsMenu(Menu menu) {
        //Inflate the menu; this adds items to the action bar if it is present.
        getMenuInflater().inflate(R.menu.main, menu);
        return true;
    }
}
```

该程序在 Intent 的使用上，与上节中的使用方式有很大的不同，即根本不指定接收者，初始化 Intent 对象时，只是传入参数，设定 Action 为 Intent.ACTION_DIAL：

```
Intent intent = new Intent(Intent.ACTION_DIAL);
startActivity(intent);
```

这里使用的构造函数的原型如下。

```
Intent(String action);
```

有关 Action 的作用后文将有详细说明，这里读者可暂时将它理解为描述这个 Intent 的一种方式，这种使用方式看上去比较奇怪，Intent 的发送者只是指定了 Action 为 Intent.ACTION_DIAL，那么怎么找到接收者呢？来看下面的例子。

（2）增加一个接收者

事实上接收者如果希望能够接收某些 Intent，需要像上节例子中一样，通过在 AndroidManifest.xml 中增加 Activity 的声明，并设置对应的 Intent Filter 和 Action，才能被 Android 的应用程序框架所匹配。为了证明这一点，我们新建一个 Activity：MyDialActivity，修改 AndroidManifest.xml 文件，将 MyDialActivity 的声明部分改为：

```xml
<activity
    android:name="cn.wang.implicitintent.MyDialActivity"
    android:label="@string/title_activity_my_dial" >
    <intent-filter>
```

```
        <action android:name="android.intent.action.DIAL" />
        <category android:name="android.intent.category.DEFAULT"/>
    </intent-filter>
</activity>
```

然后再尝试运行 ImplicitIntentTest 程序，运行的截图如图 7.2 所示。

图 7.2 修改后的运行截图

这个截图中的第一幅表示可以选择 Dialer 或者 MyDialActivity 程序来完成 Intent.ACTION_DIAL，也就是说，针对 Intent.ACTION_DIAL，Android 框架找到了两个符合条件的 Activity，因此，它将这两个 Activity 分别列出，供用户选择。当选择"我的拨号"，单击"仅此一次"，就打开图 7.2 的第二幅。

回过头来看我们是怎么做到这一点的。仅仅在 AndroidManifest.xml 文件中增加了下面的两行：

```
<action android:name="android.intent.action.DIAL" />
<category android:name="android.intent.category.DEFAULT" />
```

这两行修改了原来的 Intent Filter，这样这个 Activity 才能够接收到我们发送的 Intent。我们通过这个改动及其作用，可以进一步理解隐式 Intent，Intent Filter 及 Action, Category 等概念——Intent 发送者设定 Action 来说明将要进行的动作，而 Intent 的接收者在 AndroidManifest.xml 文件中通过设定 Intent Filter 来声明自己能接收哪些 Intent。

（3）增加一个数据

修改 AndroidManifest.xml 文件，将 MyDialActivity 的声明部分改为：

```
<activity
    android:name="cn.wang.implicitintent.MyDialActivity"
    android:label="@string/title_activity_my_dial" >
    <intent-filter>
```

```xml
            <action android:name="android.intent.action.DIAL" />
            <category android:name="android.intent.category.DEFAULT"/>
        </intent-filter>
        <intent-filter>
            <action android:name="android.intent.action.MyDIAL" />
            <category android:name="android.intent.category.DEFAULT"/>
            <data android:scheme="wang" />
        </intent-filter>
        <intent-filter>
            <action android:name="android.intent.action.MyDIAL" />
            <category android:name="android.intent.category.DEFAULT"/>
            <data android:scheme="wang" android:host="www.study.com" android:port=
            "6666" />
        </intent-filter>
</activity>
```

在 MainActivity 中，增加两个按钮。为第一个按钮的单击事件增加如下代码：

```
Intent intent1 = new Intent("android.intent.action.MyDIAL");
intent1.setData(Uri.parse("wang:123456"));
startActivity(intent1);
```

为第二个按钮的单击事件增加如下代码：

```
Intent intent2 = new Intent("android.intent.action.MyDIAL");
intent2.setData(Uri.parse("wang://www.study.com:6666/123456"));
startActivity(intent2);
```

通过这两个按钮，可以直接打开 **MyDialActivity**，如图 7.2 右图所示。

7.3.3 在 Intent 中传递数据

Intent 除了定位目标组件外，另外一个职责就是传递数据信息。Intent 之间传递数据一般有两种常用的方法：一种是通过 data 属性，另一种是通过 extra 属性。data 属性是一种 URL，它可以指向 HTTP、FTP 等网络地址，也可以指向 ContentProvider 提供的资源。通过调用 Intent 的 setData 方法放入数据，使用 getData 方法取出数据。

例如，需要启动 Android 内置的浏览器，使用下面的代码可将网址通过 data 属性传递给它。打开的界面如图 7.3 所示。

```
Intent intent = new Intent(Intent.ACTION_VIEW);
intent.setData(Uri.parse("http://www.baidu.com"));
startActivity(intent);
```

由于 data 属性只能传递数据的 URL 地址，如果需要传

图 7.3 打开的浏览器界面

递一下数据对象,则需要使用 extra 属性。Intent 提供了多个重载的方法来"携带"额外的数据,如下。

putExtras(Bundle extras):向 Intent 中放入需要传递的数据。

putExtra(String name, Xxx value):向 Intent 中放入 Xxx 类型的数据。

putParcelableArrayListExtra(String name, ArrayList<? extends Parcelable> value):向 Intent 中放入 ArrayList 数据。

putIntegerArrayListExtra(String name, ArrayList<Integer> value):向 Intent 中放入 ArrayList 数据。

putStringArrayListExtra(String name, ArrayList<String> value):向 Intent 中放入 ArrayList 数据。

Bundle 是专门用来在 Android 的应用组件之间传递数据的一种对象,本质上是一个 Map 对象,可以将各种基本类型的数据保存在 Bundle 类中打包传输。

putXxx(String key, Xxx value):向 Bundle 中放入 int、long 等各种类型的数据。

putSerializable(String key, Serializable value):向 Bundle 中放入一个可系列化的对象。

putParcelable(String key, Parcelable value):向 Bundle 中放入一个可系列化的对象。

为了取出 Intent 中携带的数据,Intent 中提供了如下方法:

getXxxExtra(String name, Xxx defaultValue):从 Intent 中直接取出 Xxx 类型的数据。

也可以先使用 getExtras()方法获取一个 Bundle 对象,然后使用 Bundle 的如下方法来获取数据的值。

getXxx(String key):从 Bundle 中取出 Xxx 类型的数据。

getXxx(String key, Xxx defaultValue):从 Bundle 中取出 Xxx 类型的数据,如果取不到则使用 defaultValue。

下面通过一个示例来介绍两个 Activity 之间如何使用 Bundle 交换数据。

首先,新建一个新的 Android 项目:IntentSimpleData,该项目包含一个 Activity:MainActivity,然后创建一个 Activity:ReceiveDataActivity,用来接收 MainActivity 发送的数据。在 MainActivity 的布局文件中增加三个按钮,修改后的布局文件如下。

```xml
<LinearLayout xmlns:android="http://schemas.android.com/apk/res/android"
    xmlns:tools="http://schemas.android.com/tools"
    android:layout_width="match_parent"
    android:layout_height="match_parent"
    android:orientation="vertical"
    tools:context=".MainActivity" >

    <Button
        android:id="@+id/button1"
        android:layout_width="wrap_content"
        android:layout_height="wrap_content"
        android:text="Intent 直接传递"
        android:onClick="sendData" />

    <Button
```

```
        android:id="@+id/button2"
        android:layout_width="wrap_content"
        android:layout_height="wrap_content"
        android:text="使用 Bundle 传递"
        android:onClick="sendData" />

    <Button
        android:id="@+id/button3"
        android:layout_width="wrap_content"
        android:layout_height="wrap_content"
        android:text="传递 List"
        android:onClick="sendData" />

</LinearLayout>
```

在 MainActivity 类中增加三个按钮的单击事件处理，修改后的 MainActivity.java 文件如下。

```
package com.example.intentsimpledata;

import java.util.ArrayList;
import android.os.Bundle;
import android.view.View;
import android.app.Activity;
import android.content.Intent;

public class MainActivity extends Activity {
    private Intent intent = null;
    @Override
    protected void onCreate(Bundle savedInstanceState) {
        super.onCreate(savedInstanceState);
        setContentView(R.layout.activity_main);
    }

    public void sendData(View view){
        switch(view.getId()){
        case R.id.button1:
            intent = new Intent(this,ReceiveDataActivity.class);
            intent.putExtra("name", "Mike");
            intent.putExtra("gender", "男");
            intent.putExtra("age", 20);
            startActivity(intent);
            break;
        case R.id.button2:
            intent = new Intent(this,ReceiveDataActivity.class);
            Bundle bundle = new Bundle();
```

```java
                bundle.putString("name", "Mike");
                bundle.putString("gender", "男");
                bundle.putInt("age", 20);
                intent.putExtras(bundle);
                startActivity(intent);
                break;
            case R.id.button3:
                ArrayList<String> list = new ArrayList<String>();
                list.add("string1");
                list.add("string2");
                list.add("string3");
                intent = new Intent(this,ReceiveDataActivity.class);
                intent.putStringArrayListExtra("list", list);
                startActivity(intent);
                break;
        }
    }
}
```

然后，在 ReceiveDataActivity 类的 onCreate 方法中增加接收 intent 携带的参数信息。修改后的 ReceiveDataActivity.java 文件如下。

```java
package com.example.intentsimpledata;
import java.util.ArrayList;
import android.os.Bundle;
import android.app.Activity;
import android.content.Intent;
import android.widget.TextView;

public class ReceiveDataActivity extends Activity {

    //@Override
    protected void onCreate(Bundle savedInstanceState) {
        super.onCreate(savedInstanceState);
        setContentView(R.layout.activity_receive_data);

        TextView tv = (TextView) findViewById(R.id.textView1);
        Intent intent = getIntent();
        Bundle bundle = intent.getExtras();
        //使用 Intent 中的方法
        //String name = intent.getStringExtra("name");
        //String gender = intent.getStringExtra("gender");
        //int age = intent.getIntExtra("age",0);
        //使用 Bundle 的方法
        String name = bundle.getString("name", "");
        String gender = bundle.getString("gender", "");
```

```
            int age = bundle.getInt("age",0);
            String str = name+"\n"+gender+"\n"+age+"\n";
            //使用List的方法
            ArrayList<String> list = bundle.getStringArrayList("list");
            if(list!=null)
                str = str+list.toString();
            tv.setText(str);
    }
}
```

因为ReceiveDataActivity的布局文件没有修改，在此就不再给出。运行应用程序，单击前两个按钮，ReceiveDataActivity接收数据后的界面如图7.4（左图）所示；单击第三个按钮，传递一个ArrayList，接收数据后的界面如图7.4（右图）所示。

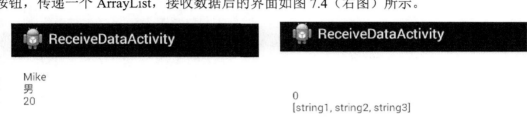

图7.4　使用Intent传递简单数据

7.3.4　在Intent中传递复杂对象

Android的Intent之间传递对象有两种方法，一种是Bundle.putSerializable(Key,Object)；另一种是Bundle.putParcelable(Key,Object)。方法中的Object要满足一定的条件，前者实现了Serializable接口，而后者实现了Parcelable接口。

Serializable和Parcelable这两种接口功能类似，但为什么Android不用内置的Java序列化机制，而偏偏要搞一套新东西呢？这是因为Android设计团队认为Java中的序列化太慢，难以满足Android的进程间通信需求，所以他们构建了Parcelable解决方案。Parcelable要求显式地序列化类的成员，但最终序列化对象的速度将快很多。在Android运行环境中推荐使用Parcelable接口，它不但可以利用Intent传递，还可以在远程方法调用中使用。

实现Parcelable接口需要实现三个方法。

（1）writeToParcel方法：该方法将类的数据写入外部提供的Parcel中。

声明：writeToParcel (Parcel dest, int flags)。

（2）describeContents方法：返回内容描述信息的资源ID，直接返回0就可以。

（3）静态的Parcelable.Creator<T>接口，本接口有两个方法，

createFromParcel(Parcel in)：实现从in中创建出类的实例的功能。

newArray(int size)：创建一个类型为T，长度为size的数组，returnnew T[size]即可。

下面，通过一个示例演示如何使用Serializable和Parcelable接口传递对象。

首先，创建一个新的 Android 项目：IntentObjectDemo，包含一个 Activity：MainActivity。然后创建两个类：SerializableUser 和 ParcelableUser，分别实现 Serializable 和 Parcelable 接口，这两个类的源码如下。

```
package com.example.intentobjectdemo;

import java.io.Serializable;

public class SerializableUser implements Serializable{
    private String userName;
    private String password;
    public SerializableUser(){

    }
    public SerializableUser(String userName,String password){
        this.userName = userName;
        this.password = password;
    }
    public String getUserName() {
        return userName;
    }
    public void setUserName(String userName) {
        this.userName = userName;
    }
    public String getPassword() {
        return password;
    }
    public void setPassword(String password) {
        this.password = password;
    }
}
package com.example.intentobjectdemo;

import android.os.Parcel;
import android.os.Parcelable;

public class ParcelableUser implements Parcelable {
    private String userName;
    private String password;
    public ParcelableUser(){

    }
    public ParcelableUser(String userName,String password){
        this.userName = userName;
        this.password = password;
```

```java
    }
    public String getUserName() {
        return userName;
    }
    public void setUserName(String userName) {
        this.userName = userName;
    }
    public String getPassword() {
        return password;
    }
    public void setPassword(String password) {
        this.password = password;
    }
    public static final Parcelable.Creator<ParcelableUser> CREATOR = new
    Creator<ParcelableUser>(){

        @Override
        public ParcelableUser createFromParcel(Parcel source) {
            ParcelableUser parcelableUser = new ParcelableUser();
            parcelableUser.userName = source.readString();
            parcelableUser.password = source.readString();
            return parcelableUser;
        }

        @Override
        public ParcelableUser[] newArray(int size) {
            return new ParcelableUser[size];
        }

    };
    @Override
    public int describeContents() {
        return 0;
    }

    @Override
    public void writeToParcel(Parcel dest, int flags) {
        dest.writeString(userName);
        dest.writeString(password);
    }
}
```

在 MainActivity 的布局文件中增加两个按钮,修改后的布局文件如下。

```
<LinearLayout xmlns:android="http://schemas.android.com/apk/res/android"
    xmlns:tools="http://schemas.android.com/tools"
```

```xml
    android:layout_width="match_parent"
    android:layout_height="match_parent"
    android:orientation="vertical"
    tools:context=".MainActivity" >

    <Button
        android:id="@+id/button1"
        android:layout_width="wrap_content"
        android:layout_height="wrap_content"
        android:text="传递Serializable对象"
        android:onClick="sendData" />

    <Button
        android:id="@+id/button2"
        android:layout_width="wrap_content"
        android:layout_height="wrap_content"
        android:text="传递Parcelable对象"
        android:onClick="sendData" />

</LinearLayout>
```

在MainActivity类中增加两个按钮的单击事件处理,分别发送SerializableUser和ParcelableUser对象。修改后的MainActivity.java文件如下。

```java
package com.example.intentobjectdemo;

import android.os.Bundle;
import android.view.View;
import android.app.Activity;
import android.content.Intent;

public class MainActivity extends Activity {

    @Override
    protected void onCreate(Bundle savedInstanceState) {
        super.onCreate(savedInstanceState);
        setContentView(R.layout.activity_main);
    }

    public void sendData(View view){
        switch(view.getId()){
        case R.id.button1:
            SerializableUser sUser = new SerializableUser("John", "123456");
            Intent intent = new Intent(this,ReceiveObjectActivity.class);
            Bundle bundle = new Bundle();
            bundle.putInt("type", 1);
```

```java
                bundle.putSerializable("serial", sUser);
                intent.putExtras(bundle);
                startActivity(intent);
                break;
            case R.id.button2:
                ParcelableUser pUser = new ParcelableUser("Mike", "123456");
                Intent intent1 = new Intent(this,ReceiveObjectActivity.class);
                Bundle bundle1 = new Bundle();
                bundle1.putInt("type", 2);
                bundle1.putParcelable("parcel", pUser);
                intent1.putExtras(bundle1);
                startActivity(intent1);
                break;
        }
    }
}
```

在当前项目中增加一个 Activity：ReceiveObjectActivity，用于传递过来的对象。修改该类的 onCreate 方法，修改后的代码如下。

```java
package com.example.intentobjectdemo;

import android.os.Bundle;
import android.widget.TextView;
import android.app.Activity;

public class ReceiveObjectActivity extends Activity {

    @Override
    protected void onCreate(Bundle savedInstanceState) {
        super.onCreate(savedInstanceState);
        setContentView(R.layout.activity_receive_object);

        TextView tv = (TextView)findViewById(R.id.textView1);

        Bundle bundle = getIntent().getExtras();
        int type = bundle.getInt("type");
        if(type==1){
            SerializableUser serializableUser = (SerializableUser) getIntent().
                getSerializableExtra("serial");
tv.setText(serializableUser.getUserName()+"\n"+serializableUser.getPassword());
        }else{
            ParcelableUser parcelableUser = (ParcelableUser) getIntent().
                getParcelableExtra("parcel");
```

```
            tv.setText(parcelableUser.getUserName()+"\n"+parcelableUser.getPassword());
        }
    }
}
```

由于 ReceiveObjectActivity 的布局文件没有修改，在此就不再给出。运行应用程序，单击第一个按钮，ReceiveObjectActivity 接收数据后的界面如图 7.5（左图）所示；单击第二个按钮，接收数据后的界面如图 7.5（右图）所示。

图 7.5 使用 Intent 传递对象

7.3.5 实现 Activity 之间的协同

如果想在 Activity 中得到新打开 Activity 关闭后返回的数据，需要使用系统提供的 startActivityForResult(Intent intent, int requestCode)方法打开新的 Activity，新的 Activity 关闭后会向前面的 Activity 传回数据，为了得到传回的数据，必须在前面的 Activity 中重写 onActivityResult(int requestCode, int resultCode, Intent data)方法。

使用 startActivityForResult(Intent intent, int requestCode)方法打开新的 Activity，我们需要为 startActivityForResult()方法传入一个请求码（第二个参数）。请求码的值是根据业务需要由自己设定，用于标识请求来源。例如：一个 Activity 有两个按钮，点击这两个按钮都会打开同一个 Activity，不管是哪个按钮打开新 Activity，当这个新 Activity 关闭后，系统都会调用前面 Activity 的 onActivityResult(int requestCode, int resultCode, Intent data)方法。在 onActivityResult()方法如果需要知道新 Activity 是由那个按钮打开的，并且要做出相应的业务处理，只要使用第一个参数 requestCode 即可区分开。

新 Activity 关闭前需要向前面的 Activity 返回数据，需要使用系统提供的 setResult(int resultCode, Intent data)方法。那么，这个结果码（resultCode）有什么作用呢？

在一个 Activity 中，可能会使用 startActivityForResult()方法打开多个不同的 Activity 处理不同的业务，当这些新 Activity 关闭后，系统都会调用前面 Activity 的 onActivityResult(int requestCode, int resultCode, Intent data)方法。为了知道返回的数据来自于哪个新 Activity，只需要它们设置的 resultCode 不同即可区分。

下面，通过一个实例演示如何使用 startActivityForResult 方法获取另一个 Activity 的返回值。

首先，新建一个 Android 项目：ForResultDemo，该项目包含一个 Activity：MainActivity，新建另外一个Activity：OtherActivity。在MainActivity的布局文件增加一个按钮，用来打开OtherActivity，修改后的布局文件如下。

```xml
<RelativeLayout xmlns:android="http://schemas.android.com/apk/res/android"
    xmlns:tools="http://schemas.android.com/tools"
    android:layout_width="match_parent"
    android:layout_height="match_parent"
    android:paddingBottom="@dimen/activity_vertical_margin"
    android:paddingLeft="@dimen/activity_horizontal_margin"
    android:paddingRight="@dimen/activity_horizontal_margin"
    android:paddingTop="@dimen/activity_vertical_margin"
    tools:context=".MainActivity" >

    <TextView
        android:id="@+id/textView1"
        android:layout_width="wrap_content"
        android:layout_height="wrap_content"
        android:text="@string/hello_world" />

    <Button
        android:id="@+id/button1"
        android:layout_width="wrap_content"
        android:layout_height="wrap_content"
        android:layout_alignLeft="@+id/textView1"
        android:layout_below="@+id/textView1"
        android:layout_marginTop="35dp"
        android:text="Open"
        android:onClick="forResult" />

</RelativeLayout>
```

在 MainActivity 类中增加按钮的事件处理和 onActivityResult 事件处理，修改后的MainActivity.java 文件如下。

```java
package cn.wang.forresultdemo;

import android.os.Bundle;
import android.view.View;
import android.widget.TextView;
import android.app.Activity;
import android.content.Intent;

public class MainActivity extends Activity {
    private TextView tView = null;
```

```java
    @Override
    protected void onCreate(Bundle savedInstanceState) {
        super.onCreate(savedInstanceState);
        setContentView(R.layout.activity_main);

        tView = (TextView) findViewById(R.id.textView1);
    }

    public void forResult(View view){
        Intent intent = new Intent(this,OtherActivity.class);
        intent.putExtra("name", "Tom");
        startActivityForResult(intent, 100);
    }

    @Override
    protected void onActivityResult(int requestCode, int resultCode, Intent data) {
        if(requestCode==100 && resultCode == RESULT_OK){
            String string = data.getStringExtra("result");
            tView.setText(string);
        }
        super.onActivityResult(requestCode, resultCode, data);
    }
}
```

在 OtherActivity 的布局文件中增加一个按钮，用于返回前一个 Activity，修改后的布局文件如下。

```xml
<RelativeLayout xmlns:android="http://schemas.android.com/apk/res/android"
    xmlns:tools="http://schemas.android.com/tools"
    android:layout_width="match_parent"
    android:layout_height="match_parent"
    android:paddingBottom="@dimen/activity_vertical_margin"
    android:paddingLeft="@dimen/activity_horizontal_margin"
    android:paddingRight="@dimen/activity_horizontal_margin"
    android:paddingTop="@dimen/activity_vertical_margin"
    tools:context=".OtherActivity" >

    <TextView
        android:id="@+id/textView1"
        android:layout_width="wrap_content"
        android:layout_height="wrap_content"
        android:text="@string/hello_world" />

    <Button
```

```xml
            android:id="@+id/button1"
            android:layout_width="wrap_content"
            android:layout_height="wrap_content"
            android:layout_alignLeft="@+id/textView1"
            android:layout_below="@+id/textView1"
            android:layout_marginLeft="23dp"
            android:layout_marginTop="47dp"
            android:text="Return"
            android:onClick="fanhui" />

</RelativeLayout>
```

在 OtherActivity 类中，为按钮添加事件处理，并在 onCreate 方法获取传递过来的参数。修改后的 OtherActivity.java 文件如下。

```java
package cn.wang.forresultdemo;

import android.os.Bundle;
import android.view.View;
import android.widget.TextView;
import android.app.Activity;
import android.content.Intent;

public class OtherActivity extends Activity {
    private TextView tView;
    private String param;
    @Override
    protected void onCreate(Bundle savedInstanceState) {
        super.onCreate(savedInstanceState);
        setContentView(R.layout.activity_other);
        
        tView = (TextView) findViewById(R.id.textView1);
        param = getIntent().getStringExtra("name");
        tView.setText(param);
    }
    
    public void fanhui(View view){
        Intent intent = getIntent();
        intent.putExtra("result", "Hi "+param);
        setResult(RESULT_OK, intent);
        finish();
    }
}
```

运行应用程序，在 MainActivity 的主界面中单击"Open"按钮，打开 OtherActivity，可以看到如图 7.6（左图）所示的界面，单击"Return"按钮，在图 7.6（右图）所示中可以看到返回值已经获取到了。

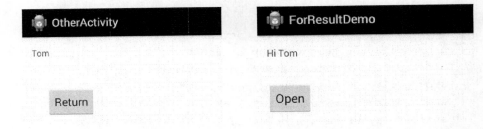

图 7.6 使用 startActivityForResult 获取结果

7.4 常用 Intent 组件的使用

用 Intent 调用系统中经常被用到的组件。

1. 调用拨号程序

```
//给移动客服 10086 拨打电话
Uri uri = Uri.parse("tel:10086");
Intent intent = new Intent(Intent.ACTION_DIAL, uri);
startActivity(intent);
```

2. 发送短信或彩信

```
//给 10086 发送内容为"Hello"的短信
Uri uri = Uri.parse("smsto:10086");
Intent intent = new Intent(Intent.ACTION_SENDTO, uri);
intent.putExtra("sms_body", "Hello");
startActivity(intent);
//发送彩信（相当于发送带附件的短信）
Intent intent = new Intent(Intent.ACTION_SEND);
intent.putExtra("sms_body", "Hello");
Uri uri = Uri.parse("content://media/external/images/media/23");
intent.putExtra(Intent.EXTRA_STREAM, uri);
intent.setType("image/png");
startActivity(intent);
```

3. 通过浏览器打开网页

```
//打开百度主页
Uri uri = Uri.parse("http://www.baidu.com");
Intent intent  = new Intent(Intent.ACTION_VIEW, uri);
startActivity(intent);
```

4. 发送电子邮件

```
someone@domain.com 发邮件
uri = Uri.parse("mailto:someone@domain.com");
```

```
Intent intent = new Intent(Intent.ACTION_SENDTO, uri);
startActivity(intent);
//给someone@domain.com发邮件发送内容为"Hello"的邮件
Intent intent = new Intent(Intent.ACTION_SEND);
intent.putExtra(Intent.EXTRA_EMAIL, "someone@domain.com");
intent.putExtra(Intent.EXTRA_SUBJECT, "Subject");
intent.putExtra(Intent.EXTRA_TEXT, "Hello");
intent.setType("text/plain");
startActivity(intent);
//给多人发邮件
Intent intent=new Intent(Intent.ACTION_SEND);
String[] tos = {"1@abc.com", "2@abc.com"};  //收件人
String[] ccs = {"3@abc.com", "4@abc.com"};  //抄送
String[] bccs = {"5@abc.com", "6@abc.com"};//密送
intent.putExtra(Intent.EXTRA_EMAIL, tos);
intent.putExtra(Intent.EXTRA_CC, ccs);
intent.putExtra(Intent.EXTRA_BCC, bccs);
intent.putExtra(Intent.EXTRA_SUBJECT, "Subject");
intent.putExtra(Intent.EXTRA_TEXT, "Hello");
intent.setType("message/rfc822");
startActivity(intent);
```

5．显示地图与路径规划

```
//打开Google地图中国北京位置（北纬39.9，东经116.3）
Uri uri = Uri.parse("geo:39.9,116.3");
Intent intent = new Intent(Intent.ACTION_VIEW, uri);
startActivity(intent);
//路径规划：从北京某地（北纬39.9，东经116.3）到上海某地（北纬31.2，东经121.4）
Uri uri = Uri.parse("http://maps.google.com/maps?f=d&saddr=39.9 116.3&daddr=31.2 121.4");
Intent intent = new Intent(Intent.ACTION_VIEW, uri);
startActivity(intent);
```

6．播放多媒体

```
Intent intent = new Intent(Intent.ACTION_VIEW);
Uri uri = Uri.parse("file:///sdcard/foo.mp3");
intent.setDataAndType(uri, "audio/mp3");
startActivity(intent);

Uri uri = Uri.withAppendedPath(MediaStore.Audio.Media.INTERNAL_CONTENT_URI, "1");
Intent intent = new Intent(Intent.ACTION_VIEW, uri);
startActivity(intent);
```

7. 拍照

```
//打开拍照程序
Intent intent = new Intent(MediaStore.ACTION_IMAGE_CAPTURE);
startActivityForResult(intent, 0);
//取出照片数据
Bundle extras = intent.getExtras();
Bitmap bitmap = (Bitmap) extras.get("data");
```

8. 获取并剪切图片

```
//获取并剪切图片
Intent intent = new Intent(Intent.ACTION_GET_CONTENT);
intent.setType("image/*");
intent.putExtra("crop", "true");           //开启剪切
intent.putExtra("aspectX", 1);             //剪切的宽高比为1:2
intent.putExtra("aspectY", 2);
intent.putExtra("outputX", 20);            //保存图片的宽和高
intent.putExtra("outputY", 40);
intent.putExtra("output", Uri.fromFile(new File("/mnt/sdcard/temp")));
                                            //保存路径
intent.putExtra("outputFormat", "JPEG");   //返回格式
startActivityForResult(intent, 0);
                                            //剪切特定图片
Intent intent = new Intent("com.android.camera.action.CROP");
intent.setClassName("com.android.camera", "com.android.camera.CropImage");
intent.setData(Uri.fromFile(new File("/mnt/sdcard/temp")));
intent.putExtra("outputX", 1);             //剪切的宽高比为1:2
intent.putExtra("outputY", 2);
intent.putExtra("aspectX", 20);            //保存图片的宽和高
intent.putExtra("aspectY", 40);
intent.putExtra("scale", true);
intent.putExtra("noFaceDetection", true);
intent.putExtra("output", Uri.parse("file:///mnt/sdcard/temp"));
startActivityForResult(intent, 0);
```

9. 打开 Google Market

```
//打开Google Market 直接进入该程序的详细页面
Uri uri = Uri.parse("market://details?id=" + "com.demo.app");
Intent intent = new Intent(Intent.ACTION_VIEW, uri);
startActivity(intent);
```

10. 安装和卸载程序

```
Uri uri = Uri.fromParts("package", "com.demo.app", null);
Intent intent = new Intent(Intent.ACTION_DELETE, uri);
```

```
startActivity(intent);
```

11. 进入设置界面

```
//进入无线网络设置界面
Intent intent = new Intent(android.provider.Settings.ACTION_WIRELESS
_SETTINGS);
startActivityForResult(intent, 0);
```

7.5 本章小结

本章主要介绍了 Android 系统中的 Intent 的功能和用法，Android 使用 Intent 封装了应用程序的启动"意图"，但这种"意图"并未直接与任何程序组件耦合，通过这种方式即可很好地提高系统的可扩展性和可维护性。学习本章需要重点掌握 Intent 的 Component、Action、Category、Data、Type 各属性的功能和用法，并掌握如何在 Manifest.xml 文件中配置。

第 8 章　Android 的隐形管理员——Service

服务（Service）是 Android 系统中 4 个应用程序组件之一。服务主要用于两个目的：后台运行和跨进程访问。通过启动一个服务，可以在不显示界面的前提下在后台运行指定的任务，这样可以不影响用户做其他事情。通过 AIDL 服务可以实现不同进程之间的通信，这也是服务的重要用途之一。

8.1　Service 概述

Android 中服务是运行在后台的东西，级别与 Activity 差不多。既然说，Service 是运行在后台的服务，那么它就是不可见的，没有界面的东西。你可以启动一个服务 Service 来播放音乐，或者记录你地理信息位置的改变，或者启动一个服务来运行并一直监听某种动作。

Service 和其他组件一样，都是运行在主线程中，因此，不能用它来做耗时的请求或动作。你可以在服务中开一个线程，在线程中做耗时动作。

服务一般分为两种：

（1）本地服务，Local Service 用于应用程序内部。在 Service 可以调用。Context.startService()启动，调用 Context.stopService()结束。在内部可以调用 Service.stopSelf()或 Service.stopSelfResult()来自己停止。无论调用了多少次 startService()，都只需调用一次 stopService()来停止。

（2）远程服务，Remote Service 用于 android 系统内部的应用程序之间。可以定义接口并把接口暴露出来，以便其他应用进行操作。客户端建立到服务对象的连接，并通过那个连接来调用服务。调用 Context.bindService()方法建立连接，并启动，以调用 Context.unbindService()关闭连接。多个客户端可以绑定至同一个服务。如果服务此时还没有加载，bindService()会先加载它。提供给可被其他应用复用，比如，定义一个天气预报服务，提供与其他应用调用即可。

8.2　Service 的生命周期

Service 生命周期可以从两种启动 Service 的模式开始讲起，分别是 context.startService()和 context.bindService()。

（1）startService 的启动模式下的生命周期。当我们首次使用 startService 启动一个服务时，系统会实例化一个 Service 实例，依次调用其 onCreate 和 onStartCommand 方法，然后进入运行状态，此后，如果再使用 startService 启动服务时，不再创建新的服务对象，系统会自动找到刚才创建的 Service 实例，调用其 onStart 方法；如果我们想要停掉一个服务，可使用 stopService 方法，此时 onDestroy 方法会被调用，注意，不管前面使用了多少次 startService，只需一次 stopService，即可停掉服务。

（2）bindService 启动模式下的生命周期。在这种模式下，当调用者首次使用 bindService 绑定一个服务时，系统会实例化一个 Service 实例，并一次调用其 onCreate 方法和 onBind 方法，然后调用者就可以和服务进行交互了，此后，如果再次使用 bindService 绑定服务，系统不会创建新的 Service 实例，也不会再调用 onBind 方法；如果我们需要解除与这个服务的绑定，可使用 unbindService 方法，此时 onUnbind 方法和 onDestroy 方法会被调用。

两种模式有以下几点不同之处：startService 模式下调用者与服务无必然联系，即使调用者结束了自己的生命周期，只要没有使用 stopService 方法停止这个服务，服务仍会运行；通常情况下，bindService 模式下服务是与调用者生死与共的，在绑定结束之后，一旦调用者被销毁，服务也就立即终止，就像江湖上的一句话：不求同生，但愿同死。

值得一提的是，以前我们在使用 startService 启动服务时都是习惯重写 onStart 方法，在 Android 2.0 时系统引进了 onStartCommand 方法取代 onStart 方法，为了兼容以前的程序，在 onStartCommand 方法中其实调用了 onStart 方法，不过我们最好是重写 onStartCommand 方法。

以上两种模式的流程如图 8.1 所示。

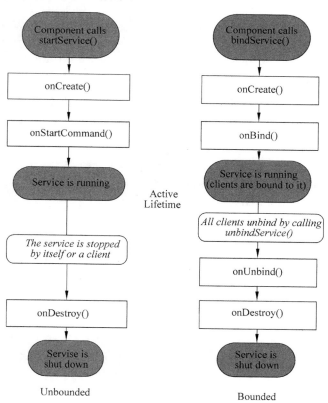

图 8.1　Service 生命周期

下面我们就结合实例来演示一下这两种模式的生命周期过程。

8.2.1 startService 启动服务

新建一个名为 ServiceTest 的项目，然后创建一个 MyService 的服务类，代码如下。

```java
package cn.wang.servicetest;

import android.app.Service;
import android.content.Intent;
import android.os.IBinder;
import android.util.Log;

public class MyService extends Service {
    private static final String TAG = "MyService";

    @Override
    public void onCreate() {
        super.onCreate();
        Log.i(TAG, "onCreate called.");
    }

    @Override
    public int onStartCommand(Intent intent, int flags, int startId) {
        Log.i(TAG, "onStartCommand called.");
        return super.onStartCommand(intent, flags, startId);
    }

    @Override
    public void onStart(Intent intent, int startId) {
        super.onStart(intent, startId);
        Log.i(TAG, "onStart called. ");
    }

    @Override
    public IBinder onBind(Intent intent) {
        Log.i(TAG, "onBind called. ");
        return null;
    }

    @Override
    public boolean onUnbind(Intent intent) {
        Log.i(TAG, "onUnbind called. ");
        return super.onUnbind(intent);
    }
```

```
    @Override
    public void onDestroy() {
        super.onDestroy();
        Log.i(TAG, "onDestroy called. ");
    }
}
```

然后在 AndroidManifest.xml 中配置服务信息，不然这个服务就不会生效，配置如下。

```xml
<service android:name=".MyService">
    <intent-filter>
        <action android:name="android.intent.action.MyService" />
        <category android:name="android.intent.category.DEFAULT" />
    </intent-filter>
</service>
```

如果服务只是在本应用中使用，大可以去掉<intent-filter>属性。

服务搭建完成之后，我们就来关注一下调用者 MainActivity，它只有两个按钮，一个是启动服务，另一个是停止服务，该服务的单击事件如下：

```java
//启动服务
public void startService(View view) {
    Intent intent = new Intent(this, MyService.class);
    startService(intent);
}

//停止服务
public void stopService(View view) {
    Intent intent = new Intent(this, MyService.class);
    stopService(intent);
}
```

接下来我们就先单击一次启动按钮，看看都发生了些什么。日志打印结果如图 8.2 所示。

再单击一次，我们会发现结果略有不同，如图 8.3 所示。

Tag	Text
MyService	onCreate called.
MyService	onStartCommand called.
MyService	onStart called.

图 8.2　LogCat 中日志输出

Tag	Text
MyService	onStartCommand called.
MyService	onStart called.

图 8.3　第二次启动服务日志输出

看到第二次单击时 onCreate 方法就不再被调用了，而是直接调用了 onStartCommand 方法（onStartCommand 中又调用了 onStart 方法）。我们选择"设置|应用|正在运行"就会发现刚刚启动的服务 ServiceTest，如图 8.4 所示。

然后单击停止按钮，试图停止服务，会发现 onDestroy 方法被调用了，此时服务就停止运行了，日志输出如图 8.5 所示。如果再次查看"应用 | 正在运行"，就会发现 ServiceTest 这个服务已全无踪迹。

图 8.4　查看已运行的服务　　　　图 8.5　停止服务日志输出

8.2.2　bindSerivce 启动服务

bindSerivce 的函数原型如下。

`bindSerivce(Intent service,ServiceConnection conn,int flags)`

参数说明如下。

service：通过该参数也就是 Intent 我们可以启动指定的 Service；

conn：该参数是一个 ServiceConnection 对象，这个对象用于监听访问者与 Service 之间的连接情况，当访问者与 Service 连接成功时将回调 ServiceConnection 对象的 onServiceConnected(ComponentName name,IBinder service)方法；如果断开将回调 onServiceDisconnected(CompontName name)方法；

flags：指定绑定时是否自动创建 Service。

由于 onServiceConnected 需要传入一个 IBinder 接口类型的参数，而前面服务中的 onBind 方法返回值为 null，这样是不行的，要想实现绑定操作，必须返回一个实现了 IBinder 接口类型的实例，该接口描述了与远程对象进行交互的抽象协议，有了它我们才能与服务进行交互。修改 onBind 的代码如下。

```
@Override
public IBinder onBind(Intent intent) {
    Log.i(TAG, "onBind called.");
    return new Binder(){};
}
```

在 MainActivity 的布局文件中增加两个按钮，增加如下代码，使其可以以 bindService 的方式启动一个服务，代码如下。

`private ServiceConnection conn = new ServiceConnection() {`

```
        @Override
        public void onServiceConnected(ComponentName name, IBinder service) {
            Log.i("MainActivity", "onServiceConnected called.");
        }

        @Override
        public void onServiceDisconnected(ComponentName name) {
            Log.i("MainActivity", "onServiceDisconnected called.");
        }
    };
    //绑定服务
    public void bind(View view) {
        Intent intent = new Intent(this, MyService.class);
        bindService(intent, conn, Context.BIND_AUTO_CREATE);
    }
    //解除绑定
    public void unbind(View view) {
        unbindService(conn);
    }
```

在使用 bindService 绑定服务时，我们需要一个 ServiceConnection 代表与服务的连接，它只有两个方法，onServiceConnected 和 onServiceDisconnected，前者是操作者在连接一个服务成功时被调用，而后者是在服务崩溃或被杀死导致的连接中断时被调用，而如果我们自己解除绑定时则不会被调用。

先单击一下绑定按钮，LogCat 的输出如图 8.6 所示。onServiceConnected 方法被调用了，看来绑定连接已经成功了，onCreate 方法和 onBind 方法被调用了，此时服务已进入运行阶段，如果再次单击绑定按钮，onCreate 和 onBinder 并不会再次被调用，这个过程中它们仅被调用一次。

Tag	Text
MyService	onCreate called.
MyService	onBind called.
MainActivity	onServiceConnected called.

图 8.6　启动绑定服务输出

然后单击解除绑定按钮，LogCat 的输出如图 8.7 所示。可以看到 onUnbind 方法和 onDestroy 方法被调用了，此时 MyService 已被销毁，整个生命周期结束。另一方面，当退出 MainActivity 时，服务也会随之而结束，从这一点上看，MyService 可以说是誓死追随着 MainActivity。

注意： 在连接中断状态再去做解除绑定操作会引起一个异常，如图 8.8 所示，在 MainActivity 销毁之前没有进行解除绑定也会导致后台出现异常信息，为了确保不会出现此类情况，需要对 MainActivity 作如下修改。

Tag	Text
MyService	onUnbind called.
MyService	onDestroy called.

图 8.7　解除绑定服务输出

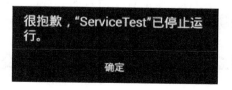

图 8.8　解除绑定引发异常

```java
private boolean binded = false;
private ServiceConnection conn = new ServiceConnection() {

    @Override
    public void onServiceConnected(ComponentName name, IBinder service) {
        Log.i("MainActivity", "onServiceConnected called.");
        binded = true;
    }

    @Override
    public void onServiceDisconnected(ComponentName name) {
        Log.i("MainActivity", "onServiceDisconnected called.");
    }
};
//绑定服务
public void bind(View view) {
    Intent intent = new Intent(this, MyService.class);
    bindService(intent, conn, Context.BIND_AUTO_CREATE);
}
//解除绑定
public void unbind(View view) {
    unbindService();
}
@Override
protected void onDestroy() {
    unbindService();
    super.onDestroy();
}

private void unbindService() {
    if(binded){
        unbindService(conn);
        binded = false;
    }
}
```

8.3 Service 的使用方法

Service 是运行在后台的服务，一般分为两种：本地服务和远程服务，下面主要从这两个方面介绍 Service 的常用的使用方法。

8.3.1 编写不需和 Activity 交互的本地服务

本地服务编写比较简单，和 8.2.1 小节使用的示例基本相同。这里写了一个计数服务的类，每秒钟为计数器加一。在服务类的内部，还创建了一个线程，用于实现后台执行上述业务逻辑。 只需把 MyService 类修改如下内容即可。

```java
public class MyService extends Service {
    private boolean threadDisable;
    private int count;

    @Override
    public IBinder onBind(Intent intent) {
        return null;
    }

    @Override
    public void onCreate() {
        super.onCreate();
        new Thread(new Runnable() {
            @Override
            public void run() {
                while (!threadDisable) {
                    try {
                        Thread.sleep(1000);
                    } catch (InterruptedException e) {
                    }
                    count++;
                    Log.v("CountService","Count is " + count);
                }
            }
        }).start();
    }

    @Override
    public void onDestroy() {
        super.onDestroy();
        this.threadDisable= true;
        Log.v("CountService","on destroy");
    }

    public int getCount() {
        return count;
    }
}
```

启动服务后，可通过日志查看到后台线程打印的计数内容。

8.3.2　编写本地服务和 Activity 交互

上面的示例是通过 startService 和 stopService 启动关闭服务的。适用于服务和 Activity

之间没有调用交互的情况。如果之间需要传递参数或者方法调用。需要使用 bind 和 unbind 方法。

首先，创建一个新的项目：LocalServiceDemo，新建的 Activity 命名为 LocalService-DemoActivity。服务类需要增加接口，如 ICountService，另外，服务类需要有一个内部类，这样可以方便访问外部类的封装数据，这个内部类需要继承 Binder 类并实现 ICountService 接口。还有，就是要实现 Service 的 onBind 方法，不能只传回一个 null 了。

新建立的接口 ICountService 源码如下。

```java
package cn.wang.localservicedemo;
public interface ICountService {
    public abstract int getCount();
}
```

然后建立服务类 CountService，源码如下。

```java
package cn.wang.localservicedemo;
import android.app.Service;
import android.content.Intent;
import android.os.Binder;
import android.os.IBinder;
import android.util.Log;
public class CountService extends Service implements ICountService {
    private boolean threadDisable;
    private int count;

    private ServiceBinder serviceBinder=new ServiceBinder();

    public class ServiceBinder extends Binder implements ICountService{
        @Override
        public int getCount(){
            return count;
        }
    }
    @Override
    public IBinder onBind(Intent intent) {
        return serviceBinder;
    }

    @Override
    public void onCreate() {
        super.onCreate();
        new Thread(new Runnable() {

            @Override
            public void run() {
```

```java
                while (!threadDisable) {
                    try {
                        Thread.sleep(1000);
                    } catch (InterruptedException e) {
                    }
                    count++;
                    Log.v("CountService","Count is " + count);
                }
            }
        }).start();
    }

    @Override
    public void onDestroy() {
        super.onDestroy();
        this.threadDisable= true;
        Log.v("CountService","on destroy");
    }

    public int getCount() {
        return count;
    }
}
```

修改 AndroidManifest.xml 文件，注册 CountService，源码如下。

```xml
<?xml version="1.0" encoding="utf-8"?>
<manifest xmlns:android="http://schemas.android.com/apk/res/android"
    package="cn.wang.localservicedemo"
    android:versionCode="1"
    android:versionName="1.0" >

    <uses-sdk
        android:minSdkVersion="8"
        android:targetSdkVersion="18" />

    <application
        android:allowBackup="true"
        android:icon="@drawable/ic_launcher"
        android:label="@string/app_name"
        android:theme="@style/AppTheme" >
        <activity
            android:name="cn.wang.localservicedemo.LocalServiceDemoActivity"
            android:label="@string/app_name" >
            <intent-filter>
                <action android:name="android.intent.action.MAIN" />
```

```xml
            <category android:name="android.intent.category.LAUNCHER" />
        </intent-filter>
    </activity>
    <service android:name="CountService">
        <intent-filter >
            <action android:name="cn.wang.CountService"/>
        </intent-filter>
    </service>
</application>
</manifest>
```

Acitity 代码需要通过 bindService 和 unbindService 启动关闭服务，另外，需要通过 ServiceConnection 的内部类实现来连接 Service 和 Activity。LocalServiceDemoActivity.java 的源码如下。

```java
package cn.wang.localservicedemo;

import android.os.Bundle;
import android.os.IBinder;
import android.util.Log;
import android.app.Activity;
import android.content.ComponentName;
import android.content.Intent;
import android.content.ServiceConnection;

public class LocalServiceDemoActivity extends Activity {
    private ICountService countService;

    private ServiceConnection serviceConnection = new ServiceConnection() {

        @Override
        public void onServiceDisconnected(ComponentName name) {
            countService = null;
        }

        @Override
        public void onServiceConnected(ComponentName name, IBinder service) {
            countService = (ICountService)service;
            Log.v("CountService", "on service connected,count is"+countService.
            getCount());
        }
    };
    @Override
    protected void onCreate(Bundle savedInstanceState) {
```

```
    super.onCreate(savedInstanceState);
    setContentView(R.layout.activity_local_service_demo);
    bindService(new Intent("cn.wang.CountService"),serviceConnection,
    BIND_AUTO_CREATE);
}

@Override
protected void onDestroy() {
    unbindService(serviceConnection);
    super.onDestroy();
}
}
```

运行建立的 LocalServiceDemo 项目，可以在 LogCat 看到如图 8.9 的输出结果，说明服务已经启动了。

Tag	Text
CountService	on service connected,count is0
CountService	Count is 1
CountService	Count is 2
CountService	Count is 3
CountService	Count is 4
CountService	Count is 5
CountService	Count is 6
CountService	Count is 7

图 8.9　本地服务与 Activity 交互

8.3.3　编写传递基本型数据的远程服务

上面的示例可以扩展为，让其他应用程序复用该服务。这样的服务叫远程（remote）服务，实际上是进程间通信（interprocess communication，简称 IPC）。Java 中是不支持跨进程内存共享的。因此，要传递对象，需要把对象解析成操作系统能够理解的数据格式，以达到跨界对象访问的目的。在 JavaEE 中，采用 RMI 通过序列化传递对象。在 Android 中，则采用 AIDL（Android Interface Definition Language：接口描述语言）方式实现。

AIDL 是一种接口定义语言，用于约束两个进程间的通讯规则，供编译器生成代码，实现 Android 设备上的两个进程间通信（IPC）。AIDL 的 IPC 机制和 EJB 所采用的 CORBA 很类似，进程之间的通信信息，首先会被转换成 AIDL 协议消息，然后发送给对方，对方收到 AIDL 协议消息后再转换成相应的对象。由于进程之间的通信信息需要双向转换，所以 android 采用代理类在背后实现了信息的双向转换，代理类由 android 编译器生成，对开发人员来说是透明的。

1．创建 AIDL 文件

使用 AIDL 来定义远程服务的接口，而不是上述那样简单的 Java 接口，扩展名为 aidl

而不是 Java。可用上面的 ICountService 改动而成 ICountSerivde.aidl，Eclipse 会自动生成相关的 Java 文件。为了和 8.3.2 小节的例子区分开，新建一个项目：AIDLDemo，然后新建一个 ICountSerivde.aidl 文件，将上例中的 ICountService.java 中的代码复制过来，并修改包名，删除掉权限修饰符。ICountSerivde.aidl 的源码如下。

```
package cn.wang.aidldemo;

interface ICountService{
    int getCount();
}
```

当完成 AIDL 文件创建后，Eclipse 会自动在项目的 gen 目录中同步生成 ICountSerivde.java 接口文件。接口文件中生成一个 Stub 的抽象类，里面包括 aidl 定义的方法，还包括一些其他辅助方法。值得关注的是 asInterface(IBinder iBinder)，它返回接口类型的实例，对于远程服务调用，远程服务返回给客户端的对象为代理对象，客户端在 onServiceConnected(ComponentName name, IBinder service)方法引用该对象时不能直接强转成接口类型的实例，而应该使用 asInterface(IBinder iBinder)进行类型转换。

编写 AIDL 文件时，需要注意下面几点。
（1）接口名和 AIDL 文件名相同。
（2）接口和方法前不用加访问权限修饰符 public、private、protected 等，也不能用 final、static。
（3）AIDL 默认支持的类型包话 Java 基本类型（int、long、boolean 等）和（String、List、Map、CharSequence），使用这些类型时不需要 import 声明。对于 List 和 Map 中的元素类型必须是 AIDL 支持的类型。如果使用自定义类型作为参数或返回值，自定义类型必须实现 Parcelable 接口。
（4）自定义类型和 AIDL 生成的其他接口类型在 AIDL 描述文件中，应该显式 import，即便在该类和定义的包在同一个包中。
（5）在 AIDL 文件中所有非 Java 基本类型参数必须加上 in、out、inout 标记，以指明参数是输入参数、输出参数还是输入输出参数。
（6）Java 原始类型默认的标记为 in，不能为其他标记。

2．建立服务类

定义好 AIDL 接口之后，接下来就可定义一个服务（Service）类，该 Service 的 onBind()方法返回的 IBinder 对象应该是 ICountSerivde.Stub 的子类的实例。至于其他部分，则与开发本地 Service 完全一样。

新建一个 CountService 类，其源码如下。

```
package cn.wang.aidldemo;

import android.app.Service;
import android.content.Intent;
import android.os.IBinder;
```

```java
import android.os.RemoteException;
import android.util.Log;

public class CountService extends Service{
    private boolean threadDisable;
    private int count;

    private ICountService.Stub serviceBinder = new ICountService.Stub() {

        @Override
        public int getCount() throws RemoteException {
            return count;
        }
    };

    @Override
    public IBinder onBind(Intent intent) {
        return serviceBinder;
    }

    @Override
    public void onCreate() {
        super.onCreate();
        new Thread(new Runnable(){

            @Override
            public void run() {
                while(!threadDisable){
                    try{
                        Thread.sleep(1000);
                    }catch(Exception e){

                    }
                    count++;
                    Log.v("CountService", "Count is "+count);
                }
            }

        }).start();
    }

    @Override
    public void onDestroy() {
        super.onDestroy();
        threadDisable = true;
```

```
        Log.v("CountService", "on destroy");
    }
}
```

3. 注册服务

注册 CountService 和上面的示例类似,注册后的配置文件 AndroidManifest.xml 如下。

```xml
<?xml version="1.0" encoding="utf-8"?>
<manifest xmlns:android="http://schemas.android.com/apk/res/android"
    package="cn.wang.aidldemo"
    android:versionCode="1"
    android:versionName="1.0" >

    <uses-sdk
        android:minSdkVersion="8"
        android:targetSdkVersion="18" />

    <application
        android:allowBackup="true"
        android:icon="@drawable/ic_launcher"
        android:label="@string/app_name"
        android:theme="@style/AppTheme" >
        <activity
            android:name="cn.wang.aidldemo.AIDLDemoActivity"
            android:label="@string/app_name" >
            <intent-filter>
                <action android:name="android.intent.action.MAIN" />

                <category android:name="android.intent.category.LAUNCHER" />
            </intent-filter>
        </activity>
        <service android:name="CountService">
            <intent-filter >
                <action android:name="cn.wang.aidlCountService"/>
            </intent-filter>
        </service>
    </application>
</manifest>
```

4. 访问 CountService

在 Activity 中使用服务的差别不大,只需要对 ServiceConnection 中的调用远程服务的方法时,要捕获异常。如果是另外的应用程序使用远程服务,需要做的是复制上面的 aidl 文件和相应的包复制到应用程序中,下面通过一个示例来演示一下。

首先，新建一个新的 Android 项目：AIDLClientDemo，并将 AIDLDemo 项目中的 aidl 文件复制过来，该项目的结构如图 8.10 所示。

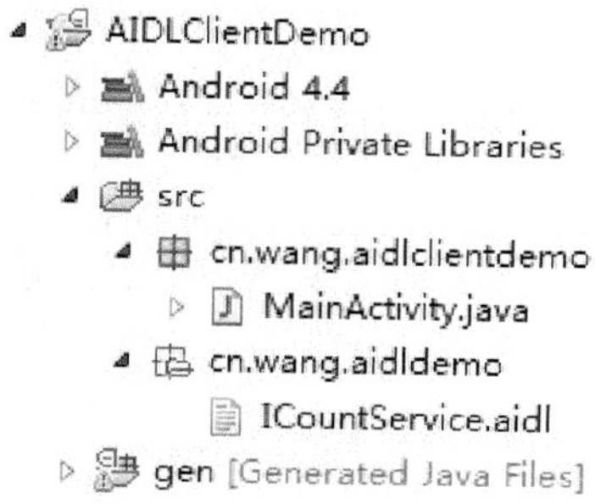

```
▲ AIDLClientDemo
  ▷ ■ Android 4.4
  ▷ ■ Android Private Libraries
  ▲ ⊞ src
    ▲ ⊞ cn.wang.aidlclientdemo
      ▷ J MainActivity.java
    ▲ ⊞ cn.wang.aidldemo
         ICountService.aidl
  ▷ ⊞ gen [Generated Java Files]
```

图 8.10 AIDLClientDemo 项目结构

修改 MainActivity.java，源码如下。

```java
package cn.wang.aidlclientdemo;

import cn.wang.aidldemo.ICountService;
import android.os.Bundle;
import android.os.IBinder;
import android.os.RemoteException;
import android.app.Activity;
import android.content.ComponentName;
import android.content.Intent;
import android.content.ServiceConnection;
import android.util.Log;

public class MainActivity extends Activity {
    private ICountService countService;
    
    private ServiceConnection serviceConnection = new ServiceConnection() {
        
        @Override
        public void onServiceDisconnected(ComponentName name) {
            countService = null;
        }
        
        @Override
        public void onServiceConnected(ComponentName name, IBinder service) {
            countService = ICountService.Stub.asInterface(service);
            try {
                Log.v("CountService", "on service connected,count is "+
                countService.getCount());
            } catch (RemoteException e) {
                e.printStackTrace();
```

```java
            }
        }
    };
    @Override
    protected void onCreate(Bundle savedInstanceState) {
        super.onCreate(savedInstanceState);
        setContentView(R.layout.activity_main);
        bindService(new Intent("cn.wang.aidlCountService"),serviceConnection,
        BIND_AUTO_CREATE);
    }

    @Override
    protected void onDestroy() {
        unbindService(serviceConnection);
        super.onDestroy();
    }
}
```

由上面的代码可以看出，和 8.3.2 小节中的示例基本相同，只是在 onServiceConnected 方法中的代码略有不同，countService 需要通过 ICountService.Stub.asInterface(service)进行赋值，另外一个就是需要处理 getCount()抛出的异常：RemoteException。

首先运行一次 AIDLDemo 项目，然后运行 AIDLClientDemo 项目，并将之关闭，就可以看到如图 8.11 所示的输出，说明可以在不同的应用程序中使用远程服务的方式和自己定义的服务交互了。

Tag	Text
CountService	on service connected,count is 0
CountService	Count is 1
CountService	Count is 2
CountService	Count is 3
CountService	on destroy
CountService	Count is 4

图 8.11 使用远程服务

8.3.4 编写传递复杂数据类型的远程服务

远程服务往往不只是传递 Java 基本数据类型，这时需要注意 AIDL 的一些限制和规定：
- AIDL 支持 Java 原始数据类型。
- AIDL 支持 String 和 CharSequence。
- 如果需要在 aidl 中使用其他 AIDL 接口类型，需要 import，即使是在相同包结构下。
- AIDL 允许传递实现 Parcelable 接口的类，需要 import。
- AIDL 支持集合接口类型 List 和 Map，但是有一些限制，元素必须是基本型或者上述三种情况，不需要 import 集合接口类，但是需要元素涉及的类型 import。
- 非基本数据类型，也不是 String 和 CharSequence 类型的，需要有方向指示，包括 in、out 和 inout，in 表示由客户端设置，out 表示由服务端设置，inout 是两者均可设置。

也就是说，在不同的进程间传递一个类对象，该类必须实现 Parcelable 接口。Parcelable 接口会告诉 Android 运行时，在封送（marshalling）和解封送（unmarshalling）过程中如何实现序列化和反序列化对象。我们很容易联想到 java.io.Serializable 接口，可能会有疑问，两种接口功能确实类似，但为什么 Android 不用内置的 Java 序列化机制，而偏偏要搞一套新东西呢？这是因为 Android 设计团队认为 Java 中的序列化太慢，难以满足 Android 的进程间通信需求，所以他们构建了 Parcelable 解决方案。Parcelable 要求显式地序列化类的成员，但最终序列化对象的速度将快很多。另外注意，Android 提供了两种机制来将数据传递给另一个进程，第一种是使用 Intent 将数据束（Bundle）传递给活动，第二种也就是 Parcelable 传递给服务。这两种机制不可互换，不要混淆。也就是说，Parcelable 无法传递给活动，只能用作 AIDL 定义的一部分。

那么，如何创建这样的类呢？必须满足如下要求。
（1）实现 Parcelable 接口。
（2）实现 writeToParcel 方法，该方法会将对象的属性值写入 Parcel。
（3）添加一个静态属性 CREATOR，该属性需要实现 android.os.Parcelable.Creator<T> 接口。
（4）创建一个声明该类的 AIDL 文件。

下面，通过一个示例来演示如何在进程间传递复制数据类型。

首先，新建一个 Android 项目：PersonAidlService，并在 src 下新建一个包：cn.wang.personaidl，然后在该包中创建一个 Person 类，实现 Parcelable 接口。Person.java 的源码如下。

```java
package cn.wang.personaidl;

import android.os.Parcel;
import android.os.Parcelable;

public class Person implements Parcelable{
    private String name;
    private int age;

    public static final Parcelable.Creator<Person> CREATOR = new Parcelable.Creator<Person>() {

        @Override
        public Person createFromParcel(Parcel source) {
            return new Person(source);
```

```java
        }

        @Override
        public Person[] newArray(int size) {
            return new Person[size];
        }
    };

    public Person(){

    }

    private Person(Parcel source){
        readFromParcel(source);
    }

    @Override
    public int describeContents() {
        return 0;
    }

    @Override
    public void writeToParcel(Parcel dest, int flags) {
        dest.writeString(name);
        dest.writeInt(age);
    }

    public void readFromParcel(Parcel source){
        name = source.readString();
        age = source.readInt();
    }

    public String getName(){
        return name;
    }

    public void setName(String name){
        this.name = name;
    }

    public int getAge(){
        return age;
    }

    public void setAge(int age){
```

```
            this.age = age;
        }
    }
```

Person.aidl 文件很简单,就是定义了一个 Parcelable 类,告诉系统需要序列化和反序列化的类型。注意,readFromParcel 方法是从 Parcel 中读取数据,为了避免出错,应该和 writeToParcel 方法的写入顺序保持一致。

然后,需要在同一包下建立一个与包含复杂类型的 Person.java 文件匹配的 Person.aidl 文件,代码如下。

```
package cn.wang.personaidl;
parcelable Person;
```

接下来,需要创建一个 IGreetService.aidl 文件,以接收类型为 Person 的输入参数,以便客户端可以将 Person 传递给服务。

```
package cn.wang.personaidl;
import cn.wang.personaidl.Person;
interface IGreetService{
    String greet(in Person person);
}
```

注意:我们需要在参数上加入方向指示符 in,代表参数由客户端设置,我们还需要为 Person 提供一个 import 语句(虽然说在同一个包下)。

此时,在 eclipse 插件 ADT 的帮助下,AIDL 编译器会自动编译生成一个 IGreetService.java 文件,请读者自行查看。

接下来,在 cn.wang.personaidlservice 包中创建 Service 类:PersonService。PersonService.java 的源码如下。

```
package cn.wang.personaidlservice;

import cn.wang.personaidl.IGreetService;
import cn.wang.personaidl.Person;
import android.app.Service;
import android.content.Intent;
import android.os.IBinder;
import android.os.RemoteException;

public class PersonService extends Service {

    IGreetService.Stub stub = new IGreetService.Stub() {

        @Override
        public String greet(Person person) throws RemoteException {
            String strRet = "Hello, "+person.getName()+", your age is
```

```
            "+person.getAge();
            return strRet;
        }
    };
    @Override
    public IBinder onBind(Intent intent) {
        return stub;
    }
}
```

最后，在 AndroidManifest.xml 中配置该服务，AndroidManifest.xml 的源码如下。

```xml
<?xml version="1.0" encoding="utf-8"?>
<manifest xmlns:android="http://schemas.android.com/apk/res/android"
    package="cn.wang.personaidlservice"
    android:versionCode="1"
    android:versionName="1.0" >

    <uses-sdk
        android:minSdkVersion="8"
        android:targetSdkVersion="18" />

    <application
        android:allowBackup="true"
        android:icon="@drawable/ic_launcher"
        android:label="@string/app_name"
        android:theme="@style/AppTheme" >
        <activity
            android:name="cn.wang.personaidlservice.MainActivity"
            android:label="@string/app_name" >
            <intent-filter>
                <action android:name="android.intent.action.MAIN" />

                <category android:name="android.intent.category.LAUNCHER" />
            </intent-filter>
        </activity>
        <service android:name=".PersonService">
            <intent-filter >
                <action android:name="cn.wang.PersonService"/>
            </intent-filter>
        </service>
    </application>
</manifest>
```

这样，服务端就完成了，服务端的结构如图 8.12 所示。

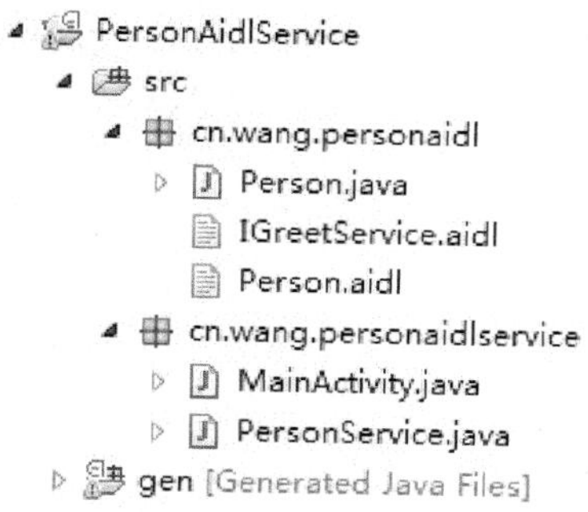

```
    ▲ 🗁 PersonAidlService
        ▲ 📁 src
            ▲ ⊞ cn.wang.personaidl
                ▷ 🇯 Person.java
                    📄 IGreetService.aidl
                    📄 Person.aidl
            ▲ ⊞ cn.wang.personaidlservice
                ▷ 🇯 MainActivity.java
                ▷ 🇯 PersonService.java
        ▷ 🎁 gen [Generated Java Files]
```

图 8.12 PersonAidlService 结构

下面，创建一个新的 Android 工程：PersonAidlClient，去访问刚创建的 PersonService。首先，将 PersonAidlService 工程里的 cn.wang.personaidl 包复制到 PersonAidlClient 中，然后在 MainActivity 的布局文件中增加三个按钮，分别用来绑定服务、调用 greet 方法和解除绑定服务。activity_main.xml 文件如下。

```xml
<RelativeLayout xmlns:android="http://schemas.android.com/apk/res/android"
    xmlns:tools="http://schemas.android.com/tools"
    android:layout_width="match_parent"
    android:layout_height="match_parent"
    android:paddingBottom="@dimen/activity_vertical_margin"
    android:paddingLeft="@dimen/activity_horizontal_margin"
    android:paddingRight="@dimen/activity_horizontal_margin"
    android:paddingTop="@dimen/activity_vertical_margin"
    tools:context=".MainActivity" >

    <TextView
        android:id="@+id/textView1"
        android:layout_width="wrap_content"
        android:layout_height="wrap_content"
        android:text="@string/hello_world" />

    <Button
        android:id="@+id/btnBind"
        android:layout_width="wrap_content"
        android:layout_height="wrap_content"
        android:layout_below="@+id/textView1"
        android:layout_marginTop="31dp"
        android:text="Bind" />

    <Button
        android:id="@+id/btnHello"
        android:layout_width="wrap_content"
        android:layout_height="wrap_content"
        android:layout_alignBaseline="@+id/btnBind"
```

```xml
        android:layout_alignBottom="@+id/btnBind"
        android:layout_centerHorizontal="true"
        android:enabled="false"
        android:text="Hello" />

    <Button
        android:id="@+id/btnUnbind"
        android:layout_width="wrap_content"
        android:layout_height="wrap_content"
        android:layout_alignBaseline="@+id/btnHello"
        android:layout_alignBottom="@+id/btnHello"
        android:layout_marginLeft="18dp"
        android:layout_toRightOf="@+id/btnHello"
        android:enabled="false"
        android:text="Unbind" />
</RelativeLayout>
```

接着，修改 MainActivity.java，增加按钮的处理事件。MainActivity.java 的源码如下。

```java
package cn.wang.personaidlclient;
import cn.wang.personaidl.IGreetService;
import cn.wang.personaidl.Person;
import android.os.Bundle;
import android.os.IBinder;
import android.os.RemoteException;
import android.view.View;
import android.view.View.OnClickListener;
import android.widget.Button;
import android.widget.TextView;
import android.app.Activity;
import android.content.ComponentName;
import android.content.Intent;
import android.content.ServiceConnection;

public class MainActivity extends Activity {
    private Button btnBind,btnHello,btnUnbind;
    private TextView tv;
    private IGreetService greetService;
    private ServiceConnection conn = new ServiceConnection(){
        @Override
        public void onServiceConnected(ComponentName name, IBinder service) {
            greetService = IGreetService.Stub.asInterface(service);
        }

        @Override
        public void onServiceDisconnected(ComponentName name) {
```

```java
        }
    };
    @Override
    protected void onCreate(Bundle savedInstanceState) {
        super.onCreate(savedInstanceState);
        setContentView(R.layout.activity_main);

        tv = (TextView)findViewById(R.id.textView1);
        btnBind = (Button)findViewById(R.id.btnBind);
        btnBind.setOnClickListener(new OnClickListener() {

            @Override
            public void onClick(View v) {
                Intent intent = new Intent("cn.wang.PersonService");
                bindService(intent, conn, BIND_AUTO_CREATE);

                btnBind.setEnabled(false);
                btnHello.setEnabled(true);
                btnUnbind.setEnabled(true);
            }
        });
        btnHello = (Button)findViewById(R.id.btnHello);
        btnHello.setOnClickListener(new OnClickListener() {

            @Override
            public void onClick(View v) {
                Person p = new Person();
                p.setName("Mike");
                p.setAge(30);
                try {
                    String ret = greetService.greet(p);
                    tv.setText(ret);
                } catch (RemoteException e) {
                    e.printStackTrace();
                }
            }
        });
        btnUnbind = (Button)findViewById(R.id.btnUnbind);
        btnUnbind.setOnClickListener(new OnClickListener() {

            @Override
            public void onClick(View v) {
                unbindService(conn);
```

```
                    btnBind.setEnabled(true);
                    btnHello.setEnabled(false);
                    btnUnbind.setEnabled(false);
                }
            });
        }
    }
```

先运行一次 PersonAidlService 工程，然后运行 PersonAidlClient 工程，单击"Bind"按钮启动服务，然后单击"Hello"按钮，可以看到 TextView 里已经显示了 greet()返回的字符串（图 8.13），说明在进程间传递复杂数据类型已经成功了。

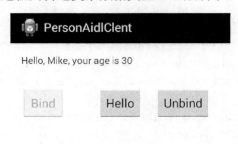

图 8.13　进程间传递复杂数据类型

8.4　IntentService

不管是何种 Service，它默认都是在应用程序的主线程（亦即 UI 线程）中运行的。所以，如果你的 Service 将要运行非常耗时或可能被阻塞的操作时，应用程序将会被挂起，甚至会出现 ANR 错误。为了避免这一问题，你应该在 Service 中重新启动一个新的线程来进行这些操作。可以通过以下两种方法来解决。

（1）直接在 Service 的 onStartCommand()方法中重启一个线程来执行，如

```
@Override
    public int onStartCommand(Intent intent, int flags, int startId) {
        MyServiceActivity.updateLog(TAG + " ----> onStartCommand()");
        new Thread(new Runnable() {
            @Override
            public void run() {
                // 此处进行耗时的操作，这里只是简单地让线程睡眠了 1s
                try {
                    Thread.sleep(1000);
                } catch (Exception e) {
                    e.printStackTrace();
                }
            }
```

```
    }).start();
    return START_STICKY;
}
```

（2）Android SDK 中为我们提供了一个现成的 Service 类来实现这个功能，它就是 IntentService，它主要负责以下几个方面。

- 生成一个默认的且与主线程互相独立的工作者线程来执行所有传送至 onStartCommand() 方法的 Intent。
- 生成一个工作队列来传送 Intent 对象给你的 onHandleIntent()方法，同一时刻只传送一个 Intent 对象，这样，就不必担心多线程的问题。
- 在所有的请求（Intent）都被执行完以后会自动停止服务，所以，不需要自己去调用 stopSelf()方法来停止该服务。
- 提供了一个 onBind()方法的默认实现，它返回 null。
- 提供了一个 onStartCommand()方法的默认实现，它将 Intent 先传送至工作队列，然后从工作队列中每次取出一个传送至 onHandleIntent()方法，在该方法中对 Intent 对相应的处理。

简单地说，IntentService 是继承于 Service 并处理异步请求的一个类，在 IntentService 内有一个工作线程来处理耗时操作，启动 IntentService 的方式和启动传统 Service 相同，同时，当任务执行完后，IntentService 会自动停止，而不需要我们去手动控制。另外，可以启动 IntentService 多次，而每一个耗时操作会以工作队列的方式在 IntentService 的 onHandleIntent 回调方法中执行，并且，每次只会执行一个工作线程，执行完第一个再执行第二个，以此类推。

那么，使用 IntentService 有什么好处呢？首先，省去了在 Service 中手动开线程的麻烦，第二，当操作完成时，不用手动停止 Service，第三，简单易用。

下面，通过一个示例来演示 IntentService 的使用方法。

首先新建一个 Android 项目：IntentServiceDemo。在项目里增加一个类：IntentServiceDemo，该类继承 IntentService。继承 IntentService 时，必须提供一个无参构造函数，且在该构造函数内，需要调用父类的构造函数。在该类中需要实现 onHandleIntent()，该方法中对 Intent 相应的处理，在此模拟了两个耗时的操作，根据不同的参数在 LogCat 做不同的输出，然后让线程等待 2 秒。同时，为了便于了解 IntentService 的生命周期，对其他方法也进行了重写。

IntentServiceDemo.java 文件的源码如下。

```
package cn.wang.intentservicedemo;

import android.app.IntentService;
import android.content.Intent;
import android.os.IBinder;
import android.util.Log;

public class IntentServiceDemo extends IntentService {
```

```java
private static final String TAG = "IntentServiceDemo";
public IntentServiceDemo() {
    super("IntentServiceDemo");
}

@Override
protected void onHandleIntent(Intent intent) {
    //Intent 是从 Activity 发过来的, 携带识别参数, 根据参数不同执行不同的任务
    String action = intent.getExtras().getString("param");
    long id = Thread.currentThread().getId();
    if (action.equals("oper1")) {
        Log.i(TAG,"Operation1 in thread:"+id);
    }else if (action.equals("oper2")) {
        Log.i(TAG,"Operation2 in thread:"+id);
    }

    try {
        Thread.sleep(2000);
    } catch (InterruptedException e) {
        e.printStackTrace();
    }
}

@Override
public void setIntentRedelivery(boolean enabled) {
    Log.i(TAG, "setIntentRedelivery");
    super.setIntentRedelivery(enabled);
}

@Override
public void onCreate() {
    long id = Thread.currentThread().getId();
    Log.i(TAG, "onCreate in thread:"+id);
    super.onCreate();
}

@Override
public void onStart(Intent intent, int startId) {
    Log.i(TAG, "onStart");
    super.onStart(intent, startId);
}

@Override
public int onStartCommand(Intent intent, int flags, int startId) {
    Log.i(TAG, "onStartCommand");
```

```java
        return super.onStartCommand(intent, flags, startId);
    }

    @Override
    public void onDestroy() {
        Log.i(TAG, "onDestroy");
        super.onDestroy();
    }

    @Override
    public IBinder onBind(Intent intent) {
        Log.i(TAG, "onBind");
        return super.onBind(intent);
    }

}
```

然后,需要在 AndroidManifest.xml 文件中注册刚创建的服务,AndroidManifest.xml 文件源码如下。

```xml
<?xml version="1.0" encoding="utf-8"?>
<manifest xmlns:android="http://schemas.android.com/apk/res/android"
    package="cn.wang.intentservicedemo"
    android:versionCode="1"
    android:versionName="1.0" >

    <uses-sdk
        android:minSdkVersion="8"
        android:targetSdkVersion="18" />

    <application
        android:allowBackup="true"
        android:icon="@drawable/ic_launcher"
        android:label="@string/app_name"
        android:theme="@style/AppTheme" >
        <activity
            android:name="cn.wang.intentservicedemo.MainActivity"
            android:label="@string/app_name" >
            <intent-filter>
                <action android:name="android.intent.action.MAIN" />

                <category android:name="android.intent.category.LAUNCHER" />
            </intent-filter>
        </activity>
        <service android:name=".IntentServiceDemo">
            <intent-filter >
```

```xml
            <action android:name="cn.wang.intentservice"/>
        </intent-filter>
    </service>
</application>
</manifest>
```

最后，修改 MainActivity.java 文件，在 onCreate()方法中通过两个 Intent 对象，携带不同的参数分别启动服务。MainActivity.java 文件的源码如下。

```java
package cn.wang.intentservicedemo;

import android.os.Bundle;
import android.util.Log;
import android.app.Activity;
import android.content.Intent;

public class MainActivity extends Activity {

    @Override
    protected void onCreate(Bundle savedInstanceState) {
        super.onCreate(savedInstanceState);
        setContentView(R.layout.activity_main);

        Intent intent = new Intent("cn.wang.intentservice");
        intent.putExtra("param", "oper1");
        startService(intent);

        Intent intent2 = new Intent("cn.wang.intentservice");
        intent2.putExtra("param", "oper2");
        startService(intent2);

        Log.i("MainActivity", "onCreate in thread:"+Thread.currentThread().
            getId());
    }
}
```

运行 IntentServiceDemo，在 LogCat 窗口可以看到如图 8.14 所示的输出。从图 8.14 中可以看到，启动了两个服务，onCreate 方法只执行了一次，而 onStartCommand 和 onStart 方法执行了两次，开启了两个线程，MainActivity 和 IntentServiceDemo 的 onCreate 的线程号都是 1，说明二者是在同一个线程中执行的；而在 onHandleIntent()方法中处理 Intent 则是在不同的线程中执行的。说明不管启动多少次，但 IntentService 的实例只有一个，这跟传统的 Service 是一样的。Operation1 也是先于 Operation2 打印，并且我让两个操作间停顿了 2 秒，最后是 onDestroy 销毁了 IntentService。这就是 IntentService，一个方便我们处理业务流程的类，它是一个 Service，但是比 Service 更智能。

```
Tag                    Text
MainActivity           onCreate in thread:1
IntentServiceDemo      onCreate in thread:1
IntentServiceDemo      onStartCommand
IntentServiceDemo      onStart
IntentServiceDemo      Operation1 in thread:79
IntentServiceDemo      onStartCommand
IntentServiceDemo      onStart
Choreographer          Skipped 34 frames!  The application may be doing too much
                       read.
IntentServiceDemo      Operation2 in thread:79
IntentServiceDemo      onDestroy
```

图 8.14 IntentService 调用输出

8.5 本章小结

作为 Android 四大组件之一的 Service，主要承担两项职能：长期运行的耗时工作和进程间的交互，因此，也对应两种启动模式：启动模式和绑定模式。学习 Service 需要重点掌握创建、配置 Service 组件，以及如何启动、停止 Service，同时需要区分清楚两种启动方式的区别。本章的重点、难点就是如何远程 AIDL Service 和调用远程 AIDL Service，需要多加练习。最后介绍了 Intent Service，需要掌握它的特点，以及和 Service 之间的区别，达到能够灵活运用的目的。

第 9 章　Android 的接收员 ——BroadcastReceiver

广播是一种广泛运用在应用程序之间传输信息的机制。而 BroadcastReceiver 是 Android 的四大组件之一，这个组件实际上是一个全局的监听器，用于监听系统全局的广播消息。BroadcastReceiver 对发送出来的广播进行过滤接收并进行响应或处理，可以方便地实现系统中不同组件之间的通信。

9.1　BroadcastReceiver 概述

BroadcastReceiver 也就是"广播接收者"的意思，顾名思义，它就是用来接收来自系统和应用中的广播。Broadcast 是一种广泛运用在应用程序之间传输信息的机制。而 BroadcastReceiver 是对发送出来的 Broadcast 进行过滤接受并响应的一类组件。

在 Android 系统中，广播体现在方方面面，例如，当开机完成后系统会产生一条广播，接收到这条广播就能实现开机启动服务的功能；当网络状态改变时系统会产生一条广播，接收到这条广播就能及时地做出提示和保存数据等操作；当电池电量改变时，系统会产生一条广播，接收到这条广播就能在电量低时告知用户及时保存进度，等等。

应用程序可以拥有任意数量的广播接收器以对所有它感兴趣的通知信息予以响应。所有的接收器均继承自 BroadcastReceiver 基类。广播接收器没有用户界面。然而，它们可以启动一个 activity 来响应收到的信息，或者用 NotificationManager 来通知用户。通知可以用多种方式吸引用户的注意力——闪动背灯、震动、播放声音等。一般来说，是在状态栏上放一个持久的图标，用户可以打开它并获取消息。

BroadcastReceiver 事件分类如下。
- 系统广播事件，比如 ACTION_BOOT_COMPLETED（系统启动完成后触发），ACTION_TIME_CHANGED（系统时间改变时触发），ACTION_BATTERY_LOW（电量低时触发）等等。
- 用户自定义的广播事件。

BroadcastReceiver 事件的编程流程如下。
- 注册广播事件：注册方式有两种，一种是静态注册，就是在 AndroidManifest.xml 文件中定义，注册的广播接收器必须要继承 BroadcastReceiver 类；另一种是动态注册，是在程序中使用 Context.registerReceiver 注册，注册的广播接收器相当于一个匿名类，两种方式都需要 IntentFilter。

- 发送广播事件：通过 Context.sendBroadcast 来发送，由 Intent 来传递注册时用到的 Action。
- 接收广播事件：当发送的广播被接收器监听后，会调用它的 onReceive()方法，并将包含消息的 Intent 对象传给它。onReceive 中代码的执行时间不要超过 10 秒，否则 Android 会弹出超时窗口。

9.2 广播消息

Android 中的广播机制设计得非常出色，很多事情原本需要开发者亲自操作，现在只需等待广播告知自己就可以了，大大减少了开发的工作量和开发周期。而作为应用开发者来说，就需要熟练掌握 Android 系统提供的一个开发利器，那就是 BroadcastReceiver。下面就对 BroadcastReceiver 逐一分析和演练，了解和掌握其各种功能和用法。

9.2.1 自定义 BroadcastReceiver

通过一个示例来演示自定义一个 BroadcastReceiver，并让这个 BroadcastReceiver 能够运行起来。

首先，创建一个名为 MyReceiverTest 的工程。要创建自己的 BroadcastReceiver 对象，需要继承 android.content.BroadcastReceiver，并实现其 onReceive 方法。下面创建一个名为 MyReceiver 广播接收者，代码如下。

```java
package cn.wang.myreceivertest;

import android.content.BroadcastReceiver;
import android.content.Context;
import android.content.Intent;
import android.util.Log;

public class MyReceiver extends BroadcastReceiver {

    @Override
    public void onReceive(Context context, Intent intent) {
        String msg = intent.getStringExtra("msg");
        Log.i("MyReceiver", msg);
    }
}
```

在创建完 BroadcastReceiver 后，还不能够使它进入工作状态，需要为其注册一个指定的广播地址。没有注册广播地址的 BroadcastReceiver 就像一个缺少选台按钮的收音机，虽然功能俱备，但也无法收到电台的信号。下面介绍如何为 BroadcastReceiver 注册广播地址。

（1）静态注册

静态注册是在 AndroidManifest.xml 文件中配置的，在 application 节点内增加如下代码来为 MyReceiver 注册一个广播地址。

```xml
<receiver android:name=".MyReceiver">
    <intent-filter>
        <action android:name="android.intent.action.MYRECEIVER"/>
        <category android:name="android.intent.category.DEFAULT" />
    </intent-filter>
</receiver>
```

配置了以上信息后，只要是 android.intent.action.MYRECEIVER 这个地址的广播，MyReceiver 都能够接收的到。注意，这种方式的注册是常驻型的，也就是说当应用关闭后，如果有广播信息传来，MyReceiver 也会被系统调用而自动运行。

在 MainActivity 的布局文件中增加一个按钮，该按钮的单击事件如下。

```java
//启动静态注册 Receiver
public void sendStatic(View view){
    Intent intent = new Intent("android.intent.action.MYRECEIVER");
    intent.putExtra("msg", "Hello BroadcastReceiver");
    sendBroadcast(intent);
}
```

启动工程，单击该按钮，即可发送一个广播，MyReceiver 接收到广播后，即在 LogCat 窗口输出接收到的信息，如图 9.1 所示。

Tag	Text
MyReceiver	Hello BroadcastReceiver

图 9.1　MyReceiver 输出

sendBroadcast 方法有如下两个重载。

```
sendBroadcast(Intent intent)
sendBroadcast(Intent intent, String receiverPermission)
```

刚刚使用的是第一种，没有对接收权限进行限制，如果使用第二种形式，并且指定接收权限，则只有在 AndroidManifest.xml 中使用标签<uses-permission>声明了拥有此权限的 BroadcastReceiver 才有可能接收到发送来的 Broadcast。同样，若注册 BroadcastReceiver 时指定了可接收的 Broadcast 权限，则只有在包内的 AndroidManifest.xml 中用标签<uses-permission>进行声明，拥有此权限的对象发出的 Broadcast 才能被这个 BroadcastReceiver 所接收。

如果将上面的 sendStatic 方法修改为

```java
public void sendStatic(View view){
    Intent intent = new Intent("android.intent.action.MYRECEIVER");
    intent.putExtra("msg", "Hello BroadcastReceiver");
```

```
sendBroadcast(intent,"cn.wang.permission");
}
```

那么，无论我们怎么单击该按钮，MyReceiver 都接收不到该广播。若想接收到，需要在 AndroidManifest.xml 文件中增加如下代码。

```
<permission android:name="cn.wang.permission"></permission>
<uses-permission android:name="cn.wang.permission"/>
```

说明：第一行注册一个 permission，第二行使用该 permission。

如果要限制广播的发送者必须具有某个权限，那么，注册广播接收者时增加 android:permission 属性，代码如下。

```
<receiver android:name=".MyReceiver" android:permission="cn.wang.test">
    <intent-filter>
        <action android:name="android.intent.action.MYRECEIVER"/>
        <category android:name="android.intent.category.DEFAULT" />
    </intent-filter>
</receiver>
```

在其他工程中，给该接收者发送广播，必须在 AndroidManifest.xml 文件中增加相应的权限，代码如下。

```
<permission android:name="cn.wang.test"></permission>
<uses-permission android:name="cn.wang.test"/>
```

（2）动态注册

动态注册需要在代码中动态地指定广播地址并注册，通常是在 Activity 或 Service 中注册一个广播，为了确保 Activity 或 Service 启动后已完成注册，可以将注册语句添加到 onStart 方法中。但是，在 Activity 或 Service 中注册了一个 BroadcastReceiver，当这个 Activity 或 Service 被销毁时如果没有解除注册，系统会报一个异常，提示我们是否忘记解除注册了。所以，需要在 onDestroy 方法中添加解除注册操作，具体代码如下。

```
MyReceiver receiver = new MyReceiver();
@Override
protected void onStart() {
    IntentFilter filter = new IntentFilter();
    filter.addAction("android.intent.action.RECEIVERTEST");

    registerReceiver(receiver, filter);
    super.onStart();
}

@Override
protected void onDestroy() {
    unregisterReceiver(receiver);
```

```
        super.onDestroy();
    }
    public void sendDyn(View view){
        Intent intent = new Intent("android.intent.action.RECEIVERTEST");
        intent.putExtra("msg", "Hi BroadcastReceiver");
        sendBroadcast(intent);
    }
```

启动工程，单击该按钮，即可发送一个广播，MyReceiver 接收到广播后，在 LogCat 窗口输出接收到的信息和图 9.1 类似。

上面的例子只是一个接收者来接收广播，如果有多个接收者都注册了相同的广播地址，又会是什么情况呢，能同时接收到同一条广播吗，相互之间会不会有干扰呢？这就涉及普通广播和有序广播的概念了。

9.2.2 普通广播

普通广播（Normal Broadcast） 对于多个接收者来说是完全异步的，通常每个接收者都无需等待即可以接收到广播，接收者相互之间不会有影响。对于这种广播，接收者无法终止广播，即无法阻止其他接收者的接收动作。

为了验证以上论断，在 MyReceiverTest 工程里新建两个 BroadcastReceiver，分别为 MyReceiver2 和 MyReceiver3，代码如下。

```
package cn.wang.myreceivertest;

import android.content.BroadcastReceiver;
import android.content.Context;
import android.content.Intent;
import android.util.Log;

public class MyReceiver2 extends BroadcastReceiver {

    @Override
    public void onReceive(Context context, Intent intent) {
        String msg = intent.getStringExtra("msg");
        Log.i("MyReceiver2", msg);
    }
}
public class MyReceiver3 extends BroadcastReceiver {

    @Override
    public void onReceive(Context context, Intent intent) {
        String msg = intent.getStringExtra("msg");
        Log.i("MyReceiver3", msg);
```

    }
}
```

然后在 AndroidManifest.xml 文件中注册这两个 BroadcastReceiver，代码如下。

```xml
<receiver android:name=".MyReceiver2">
    <intent-filter>
        <action android:name="android.intent.action.MYRECEIVER"/>
        <category android:name="android.intent.category.DEFAULT" />
    </intent-filter>
</receiver>
<receiver android:name=".MyReceiver3">
    <intent-filter>
        <action android:name="android.intent.action.MYRECEIVER"/>
        <category android:name="android.intent.category.DEFAULT" />
    </intent-filter>
</receiver>
```

然后再次单击发送按钮，发送一条广播，在 LogCat 窗口的输入如图 9.2 所示。

```
Tag             Text
MyReceiver      Hello BroadcastReceiver
MyReceiver2     Hello BroadcastReceiver
MyReceiver3     Hello BroadcastReceiver
```

图 9.2　普通广播

看来这三个接收者都接收到这条广播了，修改一下三个接收者，在 onReceive 方法的最后一行添加以下代码，试图终止广播：

```
abortBroadcast();
```

再次点击发送按钮，会发现三个接收者仍然都打印了自己的日志，表明接收者并不能终止广播。

在上面的例子中，发送广播使用的是 sendBroadcast(Intent intent)，但系统还提供了一个方法 sendStickyBroadcast(Intent intent)来发送广播，二者有什么区别呢？

sendStickyBroadcast 发出的广播会一直滞留（等待），以便有人注册这则广播消息后能尽快地收到这条广播。其他功能与 sendBroadcast 相同。但是使用 sendStickyBroadcast 发送广播需要获得 BROADCAST_STICKY permission，如果没有这个 permission 则会抛出异常。

下面通过示例演示一下 sendStickyBroadcast，便于更好地理解二者的区别。

（1）创建一个新的 Android 项目：BroadcastTest。

（2）在 BroadcastTest 新建一个 Activity：ReceiverActivity。在该 Activity 中定义了一个广播接收者 mReceiver，并动态为 mReceiver 注册了两个 Action，代码如下。

```
package cn.wang.broadcasttest;
```
```

```java
import android.os.Bundle;
import android.app.Activity;
import android.content.BroadcastReceiver;
import android.content.Context;
import android.content.Intent;
import android.content.IntentFilter;

public class ReceiverActivity extends Activity {
    private IntentFilter mIntentFilter;
    @Override
    protected void onCreate(Bundle savedInstanceState) {
        super.onCreate(savedInstanceState);
        setContentView(R.layout.activity_receiver);
        mIntentFilter = new IntentFilter();
        mIntentFilter.addAction("cn.android.my.action");
        mIntentFilter.addAction("cn.android.my.action.sticky");
    }

    private BroadcastReceiver mReceiver = new BroadcastReceiver() {

        @Override
        public void onReceive(Context context, Intent intent) {
            final String action = intent.getAction();
            System.out.println("action: "+action);
        }
    };

    @Override
    protected void onResume() {
        super.onResume();
        registerReceiver(mReceiver, mIntentFilter);
    }

    @Override
    protected void onPause() {
        super.onPause();
        unregisterReceiver(mReceiver);
    }
}
```

（3）在 MainActivity 的布局文件中添加三个按钮，并为这三个按钮添加 onClick 事件。增加按钮后的布局文件如下。

```xml
<LinearLayout xmlns:android="http://schemas.android.com/apk/res/android"
    xmlns:tools="http://schemas.android.com/tools"
    android:layout_width="match_parent"
```

```xml
    android:layout_height="match_parent"
    tools:context=".MainActivity"
    android:orientation="vertical" >

    <Button
        android:id="@+id/sendBroadcast"
        android:layout_width="wrap_content"
        android:layout_height="wrap_content"
        android:text="SendBroadcast" />

    <Button
        android:id="@+id/startActivity"
        android:layout_width="wrap_content"
        android:layout_height="wrap_content"
        android:text="StartActivity" />

    <Button
        android:id="@+id/sendStickyBroadcast"
        android:layout_width="wrap_content"
        android:layout_height="wrap_content"
        android:text="SendStickyBroadcast" />

</LinearLayout>
```

修改后的 MainActivity.java 文件如下。

```java
package cn.wang.broadcasttest;

import android.os.Bundle;
import android.app.Activity;
import android.content.Intent;
import android.view.View;
import android.view.View.OnClickListener;
import android.widget.Button;

public class MainActivity extends Activity {
    Button btnSend1;
    Button btnSend2;
    Button btnStart;

    @Override
    protected void onCreate(Bundle savedInstanceState) {
        super.onCreate(savedInstanceState);
        setContentView(R.layout.activity_main);
```

```java
        btnSend1 = (Button) findViewById(R.id.sendBroadcast);
        btnSend2 = (Button) findViewById(R.id.sendStickyBroadcast);
        btnStart = (Button) findViewById(R.id.startActivity);
        btnSend1.setOnClickListener(new OnClickListener() {

            @Override
            public void onClick(View v) {
                Intent intent = new Intent();
                intent.setAction("cn.android.my.action");
                intent.setFlags(1);
                sendBroadcast(intent);
            }

        });

        btnStart.setOnClickListener(new OnClickListener() {

            @Override
            public void onClick(View v) {
                Intent intent = new Intent(MainActivity.this,
                        ReceiverActivity.class);

                startActivity(intent);
            }
        });

        btnSend2.setOnClickListener(new OnClickListener() {

            @Override
            public void onClick(View v) {
                Intent intent = new Intent();
                intent.setAction("cn.android.my.action.sticky");
                intent.setFlags(2);
                sendStickyBroadcast(intent);
            }
        });
    }
}
```

（4）修改 AndroidManifest.xml 文件，增加 android.permission.BROADCAST_STICKY 权限，代码如下。

```
<uses-permission android:name="android.permission.BROADCAST_STICKY"/>
```

运行应用程序，主界面如图 9.3 所示。

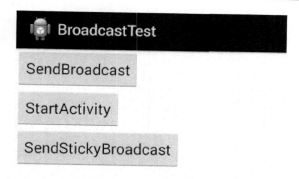

图 9.3 主界面

首先单击 Start Activity，从代码中可以看到 Receiver 已经注册，但 Log 无输出，这是因为没有广播发出自然就不会有人响应了。

按 Back 后退到图 9.3，分别单击 Send Broadcast 和 Send StickyBroadcast 按钮，随便单击几次，此时对应的 receiver 并没有注册（onPause 里 unregisterReceiver 了），所以是不会有人响应这两条广播的。然后单击 Start Activity，当打开新的 Activity 后对应的 Receiver 被注册，此时从日志中就能看出已经收到了 Send StickyBroadcast 发出的广播，但没有 Send Broadcast 发出的广播（如图 9.4 所示）。这就是 sendStickyBroadcast 的特别之处，它将发出的广播保存起来，一旦发现有人注册这条广播，则立即能接收到。

Tag	Text
System.out	action cn.android.my.action.sticky

图 9.4 Log 输出

从上面的日志信息可以看出 sendStickyBroadcast 只保留最后一条广播，并且一直保留下去，这样，即使已经处理了这条广播，但当再一次注册这条广播后依然可以收到它。

如果你只想处理一遍，可以在 onReceive 方法的最后添加以下语句：

`removeStickyBroadcast(intent);`

该方法将处理完的广播删除掉。

9.2.3 有序广播

有序广播（Ordered Broadcast）比较特殊，它每次只发送到优先级较高的接收者那里，然后由优先级高的接收者再传播到优先级低的接收者那里，优先级高的接收者有能力终止这个广播。

为了演示有序广播的流程，首先创建一个新的工程 MyOrderedReceiverTest，然后创建三个接收者，代码如下。

```
package cn.wang.myorderedreceivertest;;
```

```java
import android.content.BroadcastReceiver;
import android.content.Context;
import android.content.Intent;
import android.os.Bundle;
import android.util.Log;

public class MyReceiver extends BroadcastReceiver {

    @Override
    public void onReceive(Context context, Intent intent) {
        String msg = intent.getStringExtra("msg");
        Log.i("MyService", msg);

        Bundle bundle = new Bundle();
        bundle.putString("msg", msg + ", MyReceiver");
        setResultExtras(bundle);
    }
}
public class MyReceiver2 extends BroadcastReceiver {

    @Override
    public void onReceive(Context context, Intent intent) {
        String msg = intent.getStringExtra("msg");
        Log.i("MyService2", msg);

        msg = getResultExtras(true).getString("msg");
        Log.i("MyService2", msg);

        Bundle bundle = new Bundle();
        bundle.putString("msg", msg + ", MyReceiver2");
        setResultExtras(bundle);
    }
}
public class MyReceiver3 extends BroadcastReceiver {

    @Override
    public void onReceive(Context context, Intent intent) {
        String msg = intent.getStringExtra("msg");
        Log.i("MyService3", msg);

        msg = getResultExtras(true).getString("msg");
        Log.i("MyService3", msg);
    }
}
```

在 MyReceiver 和 MyReceiver2 中最后都使用了 setResultExtras 方法将一个 Bundle 对象设置为结果集对象，传递到下一个接收者那里，这样，优先级低的接收者可以用 getResultExtras 获取到最新的经过处理的信息集合。

接着，需要为三个接收者注册广播地址，注册的方法和前面讲的方法基本相同，但是这里需要体现广播接收者的优先级，需要在<intent-filter>里增加"android:priority"属性，这个属性的范围在–1000 到 1000，数值越大，优先级越高。在 AndroidMainfest.xml 文件中注册三个广播接收者的代码如下。

```xml
<receiver android:name=".MyReceiver">
    <intent-filter android:priority="1000">
        <action android:name="android.intent.action.MYORDEREDRECEIVER"/>
        <category android:name="android.intent.category.DEFAULT" />
    </intent-filter>
</receiver>
<receiver android:name=".MyReceiver2">
    <intent-filter android:priority="999">
        <action android:name="android.intent.action.MYORDEREDRECEIVER"/>
        <category android:name="android.intent.category.DEFAULT" />
    </intent-filter>
</receiver>
<receiver android:name=".MyReceiver3">
    <intent-filter android:priority="998">
        <action android:name="android.intent.action.MYORDEREDRECEIVER"/>
        <category android:name="android.intent.category.DEFAULT" />
    </intent-filter>
</receiver>
```

在 MainActivity 的布局文件中增加一个按钮，该按钮的单击事件如下。

```java
//启动静态注册 Receiver
public void sendStatic(View view){
    Intent intent = new Intent("android.intent.action.MYRECEIVER");
    intent.putExtra("msg", "Hello BroadcastReceiver");
    sendOrderedBroadcast(intent, null);
}
```

启动工程，单击该按钮，即可发送一个广播，MyReceiver 接收到广播后，即在 LogCat 窗口输出接收到的信息，如图 9.5 所示。

Tag	Text
MyReceiver	Hello BroadcastReceiver
MyReceiver2	Hello BroadcastReceiver
MyReceiver2	Hello BroadcastReceiver, MyReceiver
MyReceiver3	Hello BroadcastReceiver
MyReceiver3	Hello BroadcastReceiver, MyReceiver, MyReceiver2

图 9.5 有序广播

在图 9.3 中可以看到，MyReceiver2 和 MyReceiver3 分别输出了两次，第一次是使用

intent.getStringExtra("msg")获取 intent 中的数据，而第二次是使用 getResultExtras(true).getString("msg")，获取的是优先级高的接收者使用 setResultExtras 传递下来的数据。

修改一下接收者 MyReceiver2，在 onReceive 方法的最后一行添加以下代码，终止广播：

```
abortBroadcast();
```

再次单击发送按钮，在 LogCat 窗口输出接收到的信息，如图 9.6 所示。可以发现，MyReceiver3 没有接收到广播，说明在有序广播模式下，优先级高的接收者可以中断广播的传播。

Tag	Text
MyReceiver	Hello BroadcastReceiver
MyReceiver2	Hello BroadcastReceiver
MyReceiver2	Hello BroadcastReceiver, MyReceiver

图 9.6　广播传输中断

和普通广播一样，系统提供了另外一个方法 sendStickyOrderedBroadcast(intent, resultReceiver, scheduler, initialCode, initialData, initialExtras) 可以发送广播，这个方法具有有序广播的特性也有异步广播的特性。

9.3　处理系统广播消息

在广播消息中，有一类特殊的广播消息，它们特殊在只能由 Android 系统发出，这类广播称为系统广播。系统广播被用来通知一些重要的系统事件，如电池电量低、应用软件的安装卸载、SD 卡插入拔出、外部电源的插拔等，这些系统广播都被定义为 Action 常量，存放在 android.content.Intent 中，如表 9.1 所示。

表 9.1　常用的系统广播

Action 常量名	说明
Intent.ACTION_BATTERY_CHANGED	充电状态，或者电池的电量发生变化
Intent.ACTION_BATTERY_LOW	电池电量低
Intent.ACTION_BATTERY_OKAY	电池电量充足，即从电池电量低变化到饱满时会发出广播
Intent.ACTION_BOOT_COMPLETED	系统启动完成后，这个动作被广播一次
Intent.ACTION_CAMERA_BUTTON	按下照相时的拍照按键（硬件按键）时发出的广播
Intent.ACTION_DATE_CHANGED	设备日期发生改变时会发出此广播
Intent.ACTION_HEADSET_PLUG	在耳机口上插入耳机时发出的广播
Intent.ACTION_PACKAGE_ADDED	成功安装 APK 后发出的广播
Intent.ACTION_PACKAGE_REMOVED	成功删除某个 APK 后发出的广播
Intent.ACTION_POWER_CONNECTED	插上外部电源时发出的广播
Intent.ACTION_POWER_DISCONNECTED	已断开外部电源连接时发出的广播
Intent.ACTION_SCREEN_OFF	屏幕被关闭之后的广播
Intent.ACTION_SCREEN_ON	屏幕被打开之后的广播
Intent.ACTION_SHUTDOWN	关闭系统时发出的广播
Intent.ACTION_TIME_CHANGED	时间被设置时发出的广播

（1）开机启动服务

经常会有这样的应用场合，需要实现开机启动某个服务。要实现这个功能，我们就可以订阅系统"启动完成"这条广播，接收到这条广播后就可以启动自己的服务了。首先创建一个工程BootReveiverTest，接着新建两个类：BootCompleteReceiver 和 MyService，其中，MyService 是开机启动的服务，而 BootCompleteReceiver 在接收到 Intent.ACTION_BOOT_COMPLETED 广播后，启动 MyService，二者的具体实现如下。

```java
package cn.wang.bootreceivertest;

import android.app.Service;
import android.content.Intent;
import android.os.IBinder;
import android.util.Log;

public class MyService extends Service {

    @Override
    public IBinder onBind(Intent intent) {
        return null;
    }

    @Override
    public void onCreate() {
        super.onCreate();
        Log.i("MyService", "onCreate called.");
    }

    @Override
    public int onStartCommand(Intent intent, int flags, int startId) {
        Log.i("MyService", "onStartCommand called.");
        return super.onStartCommand(intent, flags, startId);
    }
}
```

```java
package cn.wang.bootreceivertest;

import android.content.BroadcastReceiver;
import android.content.Context;
import android.content.Intent;
import android.util.Log;

public class BootCompleteReceiver extends BroadcastReceiver {

    @Override
    public void onReceive(Context context, Intent intent) {
```

```
            Intent service = new Intent(context, MyService.class);
            context.startService(service);
            Log.i("BootCompleteReceiver", "Boot Complete. Starting MyService...");
        }
    }
```

然后需要在 AndroidManifest.xml 中注册服务和广播接收者，增加如下代码：

```xml
<!-- 需要开机启动的服务 -->
<service android:name=".MyService"/>

<!-- 开机广播接收者 -->
<receiver android:name=".BootCompleteReceiver">
    <intent-filter>
        <!-- 注册开机广播地址-->
        <action android:name="android.intent.action.BOOT_COMPLETED"/>
        <category android:name="android.intent.category.DEFAULT" />
    </intent-filter>
</receiver>
```

系统要求必须声明接收开机启动广播的权限，于是再声明使用下面的权限：

```xml
<uses-permission android:name="android.permission.RECEIVE_BOOT_COMPLETED" />
```

将应用运行在模拟器上，然后重启模拟器，在 LogCat 窗口输出接收到的信息，如图 9.7 所示。由输出信息可以看出，我们定义的 MyService 已经启动了，同样也可以通过"设置|应用|正在运行"查看 MyService 是否运行。

Tag	Text
BootComplete...	Boot Complete. Starting MyService...
MyService	onCreate called.
MyService	onStartCommand called.

图 9.7　开机启动服务

（2）网络状态变化

在某些场合，比如，用户浏览网络信息时，网络突然断开，我们要及时地提醒用户网络已断开。要实现这个功能，可以接收网络状态改变这样一条广播，当由连接状态变为断开状态时，系统就会发送一条广播，接收到之后，再通过网络的状态做出相应的操作。下面来实现这个功能：首先创建一个工程 NetStateReveiver，接着新建一个类：BootCompleteReceiver，实现如下。

```java
package cn.wang.netstatereceiver;

import android.content.BroadcastReceiver;
```

```java
import android.content.Context;
import android.content.Intent;
import android.net.ConnectivityManager;
import android.net.NetworkInfo;
import android.util.Log;
import android.widget.Toast;

public class NetReceiver extends BroadcastReceiver {
private static int type = -1;
    @Override
    public void onReceive(Context context, Intent intent) {
        Log.i("NetStateReceiver", "network state changed.");
        if (!isNetworkAvailable(context)) {
            Toast.makeText(context, "network disconnected!", Toast.LENGTH_LONG).show();
        }else{
         String string="";
         switch (type) {
            case ConnectivityManager.TYPE_MOBILE:
                string = "Mobile";
                break;
            case ConnectivityManager.TYPE_WIFI:
                string = "Wifi";
                break;
            case ConnectivityManager.TYPE_BLUETOOTH:
                string = "BlueTooth";
                break;
            default:
                break;
        }
         Toast.makeText(context, string, Toast.LENGTH_LONG).show();
        }
    }
    public static boolean isNetworkAvailable(Context context) {
        ConnectivityManager mgr = (ConnectivityManager) context.getSystemService(Context.CONNECTIVITY_SERVICE);
        NetworkInfo[] info = mgr.getAllNetworkInfo();
        if (info != null) {
            for (int i = 0; i < info.length; i++) {
                if (info[i].getState() == NetworkInfo.State.CONNECTED) {
                    type = info[i].getType();
                    return true;
                }
            }
        }
```

```
            return false;
    }
}
```

然后需要在 AndroidManifest.xml 中注册广播接收者,增加如下代码。

```xml
<receiver android:name="cn.wang.netstatereceiver.NetReceiver">
    <intent-filter>
        <action android:name="android.net.conn.CONNECTIVITY_CHANGE"/>
        <category android:name="android.intent.category.DEFAULT" />
    </intent-filter>
</receiver>
```

系统要求必须声明访问网络状态广播的权限,于是再声明使用下面的权限:

```xml
<uses-permission android:name="android.permission.ACCESS_NETWORK_STATE"/>
```

将应用运行在模拟器上,然后将模拟器设为"飞行模式",会看到 9.8 左图所示的提示,说明网络已经断开;如果取消"飞行模式",将看到 9.8 右图所示的提示,说明网络已经使用 Mobile 方式连接网络。

图 9.8 网络状态变化

(3) 拦截短信

Android 系统在接收到短信时会发送一条广播:android.provider.Telephony.SMS_RECEIVED,这个广播是以有序广播的形式发送的,所以可以监听这条信号,在传递给系统的接收程序时,我们将自定义的广播接收程序的优先级大于它,并且取消广播的传播,这样就可以实现拦截短信的功能了。

SMSReceiver 类继承了 BroadcastReceiver,读取接收到短信的内容,代码如下。

```java
package cn.wang.smsreceiverdemo;

import android.content.BroadcastReceiver;
import android.content.Context;
import android.content.Intent;
import android.os.Bundle;
import android.telephony.SmsMessage;
import android.widget.Toast;

public class SMSReceiver extends BroadcastReceiver {

    //当接收到短信时被触发
```

```java
@Override
public void onReceive(Context context, Intent intent)
{
    //如果是接收到短信
    if (intent.getAction().equals(
        "android.provider.Telephony.SMS_RECEIVED"))
    {
        //取消广播（这行代码将会让系统收不到短信）
        abortBroadcast();
        StringBuilder sb = new StringBuilder();
        //接收由 SMS 传过来的数据
        Bundle bundle = intent.getExtras();
        //判断是否有数据
        if (bundle != null)
        {
            // 通过 pdus 可以获得接收到的所有短信消息
            Object[] pdus = (Object[]) bundle.get("pdus");
            //构建短信对象 array,并依据收到的对象长度来创建 array 的大小
            SmsMessage[] messages = new SmsMessage[pdus.length];
            for (int i = 0; i < pdus.length; i++)
            {
                messages[i] = SmsMessage
                    .createFromPdu((byte[]) pdus[i]);
            }
            //将送来的短信合并自定义信息于 StringBuilder 当中
            for (SmsMessage message : messages)
            {
                sb.append("短信来源:");
                //获得接收短信的电话号码
                sb.append(message.getDisplayOriginatingAddress());
                sb.append("\n------短信内容------\n");
                //获得短信的内容
                sb.append(message.getDisplayMessageBody());
            }
        }
        Toast.makeText(context, sb.toString()
            , Toast.LENGTH_LONG).show();
    }
}
```

然后需要在 AndroidManifest.xml 中注册广播接收者 SMSReceiver，增加如下代码。

```xml
<receiver android:name="cn.wang.smsreceiverdemo.SMSReceiver" >
    <intent-filter android:priority="1000" >
        <action android:name="android.provider.Telephony.SMS_RECEIVED" />
        <category android:name="android.intent.category.DEFAULT" />
    </intent-filter>
</receiver>
```

在 AndroidManifest.xml 文件中添加以下权限：

`<uses-permission android:name="android.permission.RECEIVE_SMS" />`

将应用运行在模拟器上，使用另一个模拟器给该模拟器发送一条短信，SMSReceiver 拦截到短信后会使用 Toast 将内容显示出来，如图 9.9 所示。

图 9.9　拦截短信

9.4　BroadcastReceiver 的生命周期

一个广播接收者有一个回调方法：void onReceive(Context context, Intent intent)。当一个广播消息到达接收者时，Android 调用它的 onReceive() 方法并传递给它包含消息的 Intent 对象。广播接收者被认为仅当它执行这个方法时是活跃的。当 onReceive() 返回后，它是不活跃的。

有一个活跃的广播接收者的进程是受保护的，不会被杀死。但是当占用的别的内存进程需要时，系统可以在任何时候杀死仅有不活跃组件的进程。

如果 onReceive() 方法在 10 秒内没有执行完毕，Android 会认为该程序无响应。所以在 BroadcastReceiver 里不能做一些比较耗时的操作，否则会弹出"Application No Response"的对话框，即 Android 的 ANR。

这带来一个问题，当一个广播消息的响应是费时的，因此应该在独立的线程中做这些事，远离用户界面其他组件运行的主线程。如果 onReceive() 衍生线程然后返回，整个进程，包括新的线程，被判定为不活跃的（除非进程中的其他应用程序组件是活跃的），将使它处于被杀的危机。解决这个问题的方法是 onReceive() 启动一个服务，及时服务做这个工作，因此，系统知道进程中有活跃的工作在做。

9.5　本 章 小 结

BroadcastReceiver 是实现消息异步处理的组件。学习 BroadcastReceiver 需要掌握创建和配置 BroadcastReceiver 组件，还需要掌握在程序中发送 Broadcast 的方法。注意区分配置 BroadcastReceiver 组件两种方法的区别，发送 Broadcast 的不同方法的区别，以及常用系统广播的接收和处理。

第 10 章　Android 的数据存储

所有的应用程序都必然涉及数据的输入、输出，Android 应用也不例外，应用程序的参数设置、运行状态数据只有保存到外部存储器上，系统在关机之后数据才不会丢失。本章重点介绍如何在 Android 应用程序中对数据进行存储和读取。

10.1　数据存储概述

在 Android 中一共提供了四种数据存储的方式，但是由于存储的这些数据都是其应用程序的私有数据，所以如果需要在其他应用程序中使用这些数据，就要使用 Android 提供的 Content Providers。

Android 提供的五种数据存储方式如下。

SharedPreferences：它是一个轻量级的键值（key-value）存储机制，只可以存储基本数据类型，主要是针对系统配置信息的保存。

Files：Android 使用的是基于 Linux 的文件系统，程序开发人员可以建立和访问程序自身的私有文件，也可以访问保存在资源目录中的原始文件和 XML 文件，还可以在 SD 卡等外部存储设备中保存文件。

SQLite：Android 提供的标准数据库，支持 SQL 语句，可以用来存储大量的数据。

ContentProvider：主要用于在应用程序间的数据共享和交换。

NetWork：通过网络存储和获得数据。

本章主要介绍前面三种，ContentProvider 会在下一章介绍。

10.2　SharedPreferences

通常很多软件都会有配置文件，存放该程序运行中的各个属性值，由于这些信息数据量较小，而且数据的格式很简单，都是普通的字符串、标量类型的值等，通常不采用数据库的存储方式。在 Android 中，提供了 SharedPreferences 进行保存。

10.2.1　使用 SharedPreferences

SharedPreferences 是 Android 平台上一个轻量级的存储类，是基于 XML 文件来存储 key-value 键值对数据，通常用来存储一些简单的配置信息，其存储位置在/data/data/<包名>/shared_prefs 目录下。主要是保存一些常用的配置，如窗口状态，一般在 Activity 中重载窗口状态 onSaveInstance State 保存一般使用 SharedPreferences 完成，它提供了 Android 平台常规的 Long 长整形、Int 整形、String 字符串型的保存。

SharedPreferences 是一个接口，程序无法直接创建 SharedPreferences 实例，只能通过 Context 提供的 getSharedPreferences(String name,int mode)方法来获取 SharedPreferences 实例，该方法的第一个参数指定 xml 文件的名字，第二个参数支持如下几个值。

- Context.MODE_PRIVATE：指定该 SharedPreferences 数据只能被本应用程序读写。
- Context.MODE_WORLD_READABLE：指定该 SharedPreferences 数据能被其他应用程序读。
- Context.MODE_WORLD_WRITEABLE：指定该 SharedPreferences 数据能被其他应用程序写。

在 Activity 中提供了如下方法，可以创建一个 SharedPreferences，默认名为当前的 activity 的类名。

```
public SharedPreferences getPreferences(int mode)
```

也可以使用 PreferenceManager 中提供的 getDefaultSharedPreferences 来创建一个 SharedPreferences，默认名为<项目名>_preferences。

```
public static SharedPreferences getDefaultSharedPreferences(Context context)
```

如果查看它们的源码，就会发现其实还是调用了 Context.getSharedPreferences(String name,int mode)来创建 SharedPreferences。

SharedPreferences 接口主要负责读取应用程序的 Perferences 数据，它提供了如下常用的方法来访问 SharedPreferences 中的 key-value 对。

- boolean contains(String key)：判断 SharedPreferences 是否包含 key 的数据。
- Map<String,?> getAll()：获取 SharedPreferences 中全部的 key-value 对。
- xxx getXxx(String key, xxx defValue)：获取 SharedPreferences 中指定 key 的 value。如果该 key 不存在，返回默认值 defValue。其中 xxx 可以是 boolean、float、int、long、String。
- SharedPreferences.Editor edit()：返回一个 Editor 用于操作 SharedPreferences。

SharedPreferences 对象本身只能获取数据而不支持存储和修改，存储修改是通过 Editor 对象实现。Editor 提供了如下方法向 SharedPreferences 写入数据。

- SharedPreferences.Editor putXxx(String key,xxx value)：向 SharedPreferences 存入指定 key 对应的数据。其中，xxx 可以是 boolean、float、int、long、String。
- SharedPreferences.Editor clear()：清空 SharedPreferences 中所有数据。
- SharedPreferences.Editor remove(String key)：删除 SharedPreferences 中指定 key 对应的数据项。
- boolean commit()：编辑完成后，调用该方法提交修改。

实现 SharedPreferences 存储的步骤如下。

（1）根据 Context 获取 SharedPreferences 对象。
（2）利用 edit()方法获取 Editor 对象。
（3）通过 Editor 对象存储 key-value 键值对数据。
（4）通过 commit()方法提交数据。

下面是通过使用 SharedPreferences 存储数据的关键语句代码：

```
SharedPreferences sp = getContext().getSharedPreferences("test",Context.MODE_PRIVATE);
    //获取它的编辑对象
    Editor editor =sp.edit();
    //写入数据
    editor.putString("name", "jack");
    editor.putBoolean("married",true);
    editor.putInt("age", 89);
    //保存并提交
editor.commit();
```

生成的 SharedPreferences 文件名为 test.xml，保存在应用程序文件夹下的 shared_prefs 文件夹内，其位置如图 10.1 所示。

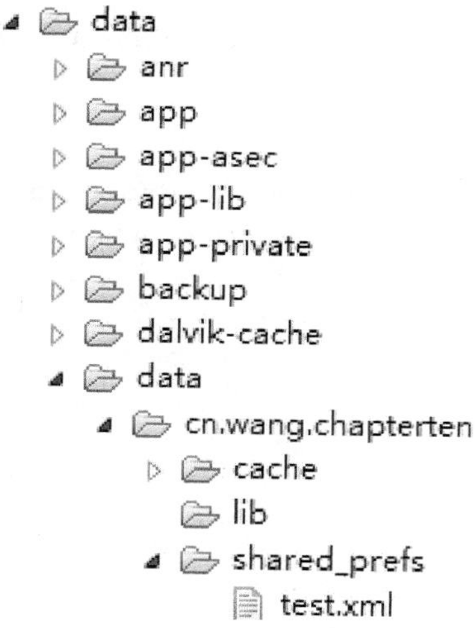

```
▲ 📂 data
   ▷ 📂 anr
   ▷ 📂 app
   ▷ 📂 app-asec
   ▷ 📂 app-lib
   ▷ 📂 app-private
   ▷ 📂 backup
   ▷ 📂 dalvik-cache
   ▲ 📂 data
      ▲ 📂 cn.wang.chapterten
         ▷ 📂 cache
           📂 lib
         ▲ 📂 shared_prefs
              📄 test.xml
```

图 10.1　SharedPreferences 存储文件的位置

查看 test.xml，其内容如下。

```xml
<?xml version='1.0' encoding='utf-8' standalone='yes' ?>
<map>
    <string name="name">jack</string>
    <boolean name="married" value="true" />
    <int name="age" value="89" />
</map>
```

若将上面程序中的 SharedPreferences sp = getContext().getSharedPreferences ("test", Context.MODE_PRIVATE);用 SharedPreferences sp = getPreferences (MODE_ PRIVATE);或 SharedPreferences sp = PreferenceManager. getDefaultSharedPreferences(this);替换，请读者自行查看生产的文件及内容。

下面以一个例子来说明如何使用 sharedpreferences 来存储数据，该例子是在 Activity

退出时保存界面的基本信息，当再次运行该程序时，就会读取上次保存的信息。界面中有一个账号 EditText，密码 EditText 和一个记住密码的 CheckBox，输入账号和密码，如果选中记住密码复选框，下次打开程序时，则会显示账号和密码；如果没有选中记住密码复选框，下次打开程序时，则只是显示账号。示例运行的界面如图 10.2、图 10.3 所示。

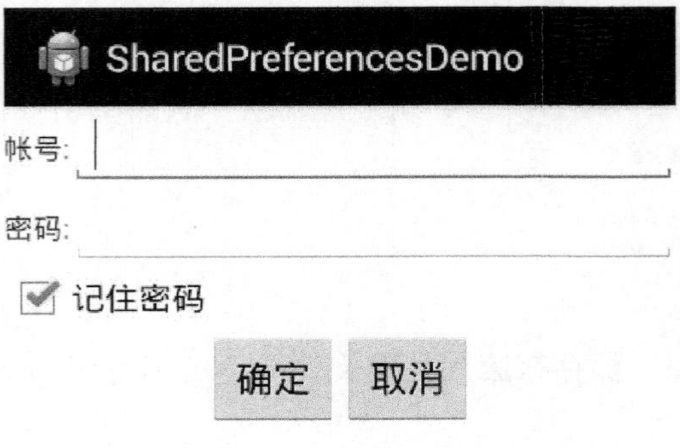

图 10.2　首次运行的登录界面 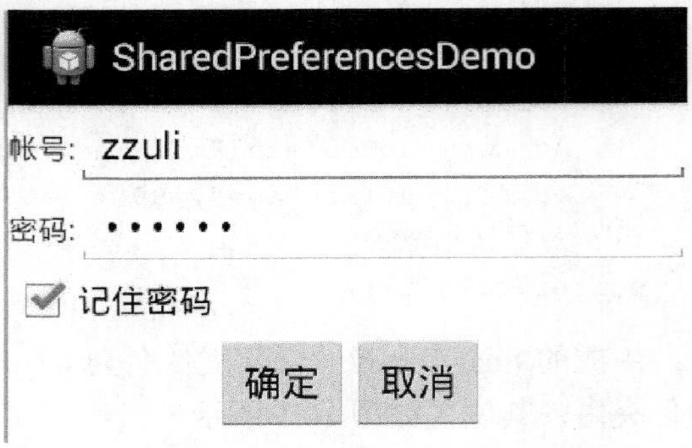　　　　图 10.3　再次运行的登录界面

res/layout/main.xml 中的代码如下。

```xml
<LinearLayout xmlns:android="http://schemas.android.com/apk/res/android"
    xmlns:tools="http://schemas.android.com/tools"
    android:layout_width="match_parent"
    android:layout_height="match_parent"
    android:orientation="vertical"
    tools:context=".MainActivity" >

    <LinearLayout
      android:orientation="horizontal"
      android:layout_width="fill_parent"
      android:layout_height="wrap_content">
      <TextView
        android:layout_width="wrap_content"
        android:layout_height="wrap_content"
        android:text="账号:"/>
      <EditText
        android:id="@+id/username"
        android:layout_height="wrap_content"
        android:layout_width="fill_parent"/>
    </LinearLayout>
     <LinearLayout
      android:orientation="horizontal"
      android:layout_width="fill_parent"
      android:layout_height="wrap_content">
      <TextView
        android:layout_width="wrap_content"
```

```xml
        android:layout_height="wrap_content"
        android:text="密码:"/>
    <EditText
        android:id="@+id/password"
        android:layout_height="wrap_content"
        android:layout_width="fill_parent"
        android:inputType="textPassword"/>
</LinearLayout>
<CheckBox
    android:id="@+id/ischecked"
    android:layout_width="fill_parent"
    android:layout_height="wrap_content"
    android:text="记住密码"/>
<LinearLayout
    android:orientation="horizontal"
    android:layout_width="wrap_content"
    android:layout_height="wrap_content"
    android:layout_gravity="center_horizontal">
    <Button
        android:layout_width="wrap_content"
        android:layout_height="wrap_content"
        android:text="确定"
        android:onClick="login"/>
    <Button
        android:layout_width="wrap_content"
        android:layout_height="wrap_content"
        android:text="取消"
        android:onClick="exit"/>
</LinearLayout>
</LinearLayout>
```

MainActivity 中的代码如下。

```java
package cn.wang.sharedpreferencesdemo;

import android.os.Bundle;
import android.app.Activity;
import android.content.SharedPreferences;
import android.view.View;
import android.widget.CheckBox;
import android.widget.CompoundButton;
import android.widget.EditText;

public class MainActivity extends Activity {
```

```java
        private final String PREFERENCES_NAME = "userinfo";
        private EditText username,password;
        private CheckBox cbRemember;

        private String userName,passWord;
        private Boolean isRemember = false;
        @Override
        protected void onCreate(Bundle savedInstanceState) {
            super.onCreate(savedInstanceState);
            setContentView(R.layout.activity_main);

            username = (EditText)findViewById(R.id.username);
            password = (EditText)findViewById(R.id.password);

            cbRemember = (CheckBox)findViewById(R.id.ischecked);
            cbRemember.setOnCheckedChangeListener(new          CompoundButton.OnCheckedChangeListener() {

                @Override
                public void onCheckedChanged(CompoundButton buttonView, boolean isChecked) {
                    isRemember = isChecked;
                }
            });
            //读取 userinfo.xml 中存储的数据,设置相应的属性
            SharedPreferences       preferences    =    getSharedPreferences(PREFERENCES_NAME, Activity.MODE_PRIVATE);
            username.setText(preferences.getString("UserName", null));
            cbRemember.setChecked(preferences.getBoolean("Remember", true));
            if(cbRemember.isChecked()){
                password.setText(preferences.getString("PassWord", null));
            }else{
                password.setText(null);
            }
        }
        //当 Activity 关闭时,将新的用户信息保存至 userinfo.xml
        @Override
        public void onStop() {
            super.onStop();
            SharedPreferences    agPreferences    =    getSharedPreferences(PREFERENCES_NAME, Activity.MODE_PRIVATE);
            SharedPreferences.Editor editor = agPreferences.edit();

            userName = username.getText().toString();
```

```
            passWord = password.getText().toString();
            editor.putString("UserName", userName);
            editor.putString("PassWord", passWord);
            editor.putBoolean("Remember", isRemember);
            editor.commit();
    }
    //单击"确定"按钮执行
    public void login(View view){
        //1.获取用户名和密码
        userName = username.getText().toString();
        passWord = password.getText().toString();
        //2.进行验证
        
        //3.验证通过，跳转至主界面
    }
    //单击"取消"按钮执行
    public void exit(View view){
        finish();
    }
}
```

10.2.2 PreferenceActivity

在开发应用程序的过程中有很大的机会需要用到参数设置功能，那么在 Android 应用中，如何实现参数设置界面及参数存储呢，根据刚刚学过的知识 很快一个念头闪过即 Activity + Preference 组合，前者用于界面构建，后者用于设置数据存放。虽然这是正确的；但比较繁琐。因为，每个设置选项都要建立与其对应的 Preference。

下面介绍 Android 中的一个特殊 Activity：PreferencesActivity。PreferenceActivity 是 android 提供的对系统信息和配置进行自动保存的 Activity，它通过 SharedPreference 方式将信息保存在 XML 文件当中，当然，也可以通过 SharedPreferences 来获取 PreferenceActivity 设置的值。使用 PreferenceActivity 不需要对 SharedPreference 进行操作，系统会自动对 Activity 的各种 View 上的改变进行保存。

下面，用一个实例来介绍如何使用 PreferencesActivity。

（1）创建 Android 项目 PreferenceActivityDemo，选择"New|Other..."菜单，打开图 10.4 所示窗口，选择"Android XMLFile"，添加一个 Android xml 文件。单击"Next"按钮，打开"New Android XML File"窗口（图 10.5 所示），改变资源类型（Resource Type）为 Preference，输入文件名：preference，将会在 res/xml 路径下新增一个文件 preference.xml，然后单击"Finish"按钮。

（2）Android 为我们提供两种编辑模式,可视化的结构设计及 XML 源码设计。Preference XML 文件中的 View 是有限的，只有下面几个。

- CheckBoxPreference：CheckBox 选择项，对应的值 ture 或 flase。

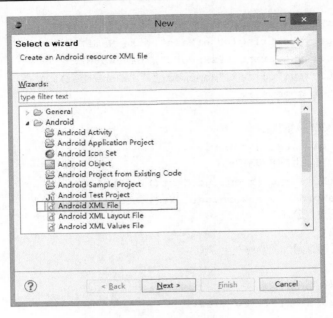

图 10.4 新建 Android Resource

图 10.5 新建 Android XML 文件

android:key：唯一标识。

android:title：显示标题（大字体显示）。

android:summary：副标题（小字体显示）。

android:defaultValue：默认值（true 或 false）。

- SwitchPreference：和 CheckBoxPreference 基本相同，只不过是显示的形式略有不同。

- EditTextPreference：输入编辑框，值为 String 类型，会弹出对话框供输入。

android:key：唯一标识。

android:title：显示标题（大字体显示）。

- ListPreference：列表选择，弹出对话框供选择。下拉框内显示的内容和具体的值需要在 res/values/array.xml 中设置两个 array 来表示。

android:key：唯一标识。

android:title：显示标题（大字体显示） android:dialogTitle：弹出对话框的标题。

android:entries：列表中显示的值。为一个数组，通过资源文件进行设置。

androide:entryValues：列表中实际保存的值，与 entries 对应，为一个数组，通过资源文件进行设置。

- MultiSelectListPreference：和 ListPreference 基不相同，可以多选。
- Preference：只进行文本显示，需要与其他进行组合使用。

android:key：唯一标识。

android:title：显示标题（大字体显示）。

android:summary：副标题（小字体显示）。

android:dependency：附属，即标识此元素附属于某一个元素（通常为 CheckBoxPreference），dependency 值为所附属元素的 key。

- PreferenceCategory：用于分组。

android:title：显示的标题。

android:key：唯一标识符。SharedPreferences 也将通过此 Key 值进行数据保存，也可以通过 key 值获取保存的信息。

- PreferenceScreen：PreferenceActivity 的根元素，设置页面，可嵌套形成二级设置页面，用 Title 参数设置标题。
- RingtonePreference：系统铃声选择。

android:title：设置标题 android:summary：设置说明。

android:dialogTitle：设置铃声选择框的标题。

下面，可以通过可视化界面进行结构设计或直接编辑 XML 源码，增加相应的组件，最终的 preference.xml 文件源码如下。

```
<?xml version="1.0" encoding="utf-8"?>
<PreferenceScreen xmlns:android="http://schemas.android. com/apk/ res/
android" >
    <PreferenceCategory android:title="无线和网络设置" >
        <CheckBoxPreference
            android:key="apply_fly"
            android:summary="禁用所有无线连接"
            android:title="飞行模式" >
        </CheckBoxPreference>
        <CheckBoxPreference
            android:key="apply_internet"
            android:summary="禁用通过 USB 共享 Internet 连接"
```

```xml
        android:title="Internet 共享" >
</CheckBoxPreference>
<CheckBoxPreference
    android:key="apply_wifi"
    android:summary="打开 Wi-Fi"
    android:title="Wi-Fi" >
</CheckBoxPreference>

<Preference
    android:dependency="apply_wifi"
    android:key="wifi_setting"
    android:summary="设置和管理无线接入点"
    android:title="Wi-Fi 设置" >
</Preference>

<SwitchPreference
    android:key="apply_bluetooth"
    android:summaryOn="启用蓝牙"
    android:summaryOff="关闭蓝牙"/>
<Preference
    android:dependency="apply_bluetooth"
    android:key="bluetooth_setting"
    android:summary="管理连接、设备名称和可检测性"
    android:title="蓝牙设置" >
</Preference>

<EditTextPreference
    android:key="number_edit"
    android:title="输入电话号码" >
</EditTextPreference>

<ListPreference
    android:dialogTitle="选择部门"
    android:entries="@array/department"
    android:entryValues="@array/department_value"
    android:key="depart_value"
    android:title="部门设置" >
</ListPreference>
<MultiSelectListPreference
    android:dialogTitle="你喜欢"
    android:entries="@array/entries_love"
    android:entryValues="@array/entriesvalue_love"
    android:key="MultiSelect"
    android:summary="你喜欢"
    android:title="爱好" />
```

```xml
        <RingtonePreference
            android:key="ring_key"
            android:ringtoneType="all"
            android:showDefault="true"
            android:showSilent="true"
            android:title="铃声" >
        </RingtonePreference>
    </PreferenceCategory>
</PreferenceScreen>
```

由于 ListPreference 和 MultiSelectListPreference 需要用到数组，所以，在 res/values 里新建一个 array.xml 文件，该文件定义了四个数组，源码如下。

```xml
<?xml version="1.0" encoding="utf-8"?>
<resources>
    <string-array name="entries_love">
        <item>旅游</item>
        <item>唱歌</item>
        <item>爬山</item>
    </string-array>
    <string-array name="entriesvalue_love">
        <item>1</item>
        <item>2</item>
        <item>3</item>
    </string-array>
    <string-array name="department">
        <item>计算机</item>
        <item>机电</item>
        <item>电气</item>
    </string-array>
    <string-array name="department_value">
        <item>1</item>
        <item>2</item>
        <item>3</item>
    </string-array>
</resources>
```

最后，修改 MainActivity.java 文件，让 MainActivity 继承 PreferenceActivity 类，并调用 addPreferencesFromResource(R.xml.preference) 将刚定义的 perference 加载进来。MainActivity.java 的源码如下。

```java
package cn.wang.preferenceactivitydemo;

import android.os.Bundle;
import android.preference.PreferenceActivity;

public class MainActivity extends PreferenceActivity {
```

```
    @Override
    protected void onCreate(Bundle savedInstanceState) {
        super.onCreate(savedInstanceState);
        addPreferencesFromResource(R.xml.preference);
    }
}
```

示例运行的主界面如图 10.6 所示，CheckBoxPreference、SwitchPreference 和 Android 系统设置界面完全相同，在此不再赘述。单击"输入电话号码"，弹出的 EditTextPreference 输入窗口如图 10.7 所示；单击"部门设置"，弹出的 ListPreference 选择窗口如图 10.8 所示；单击"爱好"，弹出 MultiSelectListPreference 选择窗口，如图 10.9 所示；单击"铃声"，弹出铃声选择窗口，如图 10.10 所示。在此所做的修改，均会自动保存，在下次打开时，会自动读取相应的值。

图 10.6　PreferenceActivityDemo 主界面

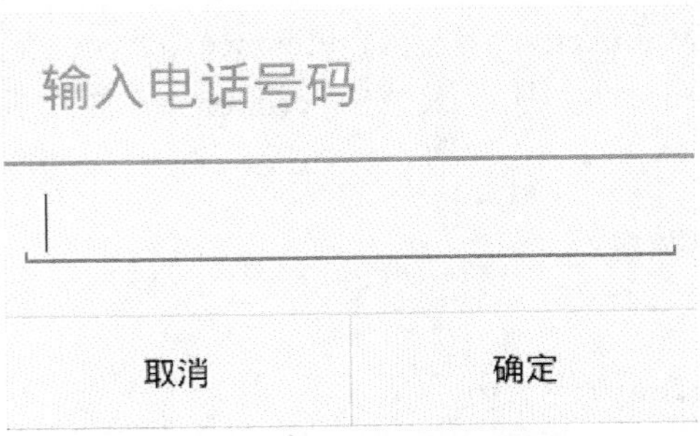

图 10.7　输入电话号码 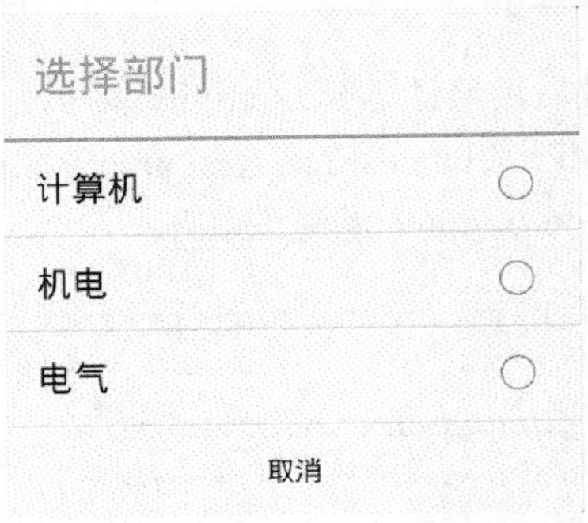　　　　　　　　　　10.8　选择部门

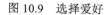

图 10.9　选择爱好　　　　　　　　　　图 10.10　选择铃声

10.3　文　　件

　　Shared Preferences 用来保存一些配置信息虽然很方便，但只能存储 boolean、float、integer、long、String 等基本类型的数据，并且保存的数据只能局限在 Android 应用内部访问。为了在更大的范围内交换复杂内容格式的消息，还要通过文件系统完成。

　　Android 系统下的文件，可以分为两类：一类是共享的文件，如存储在 SD 卡上的文件，这种文件任何 Android 应用都可以访问；另外一种是私有文件，即 Android 应用自己创建的文件。Android 的文件读写与 JavaSE 的文件读写相同，都是使用 IO 流。但是对于私有文件，只有具有操作权限的用户才能进行操作，故 Android 提供了一组特有的 API 来访问私有文件。

```
FileInputStream openFileInput(String name)
FileOutputStream openFileOutput(String name, int mode)
```

　　参数说明如下。

　　name：文件名，不能包含路径分隔符。

　　mode：操作模式，包括

　　　　　　Context.MODE_PRIVATE：新内容覆盖原内容。

　　　　　　Context.MODE_APPEND：新内容追加到原内容后。

　　　　　　Context.MODE_WORLD_READABLE：允许其他应用程序读取。

　　　　　　Context.MODE_WORLD_WRITEABLE：允许其他应用程序写入，会覆盖原数据。

　　　　　　可以使用+连接这些权限。

　　Context 对象还可以通过调用 fileList() 方法来获得私有文件目录下所有的文件名组成的字符串数组，调用 deleteFile(String name) 来删除文件名为 name 的文件。

　　Activity 还提供了 getCacheDir() 和 getFilesDir() 方法。

　　getCacheDir() 方法用于获取 /data/data/<package name>/cache 目录（一些临时文件可以放在缓存目录用完了就删了）。

getFilesDir()方法用于获取/data/data/<package name>/files 目录。

其他程序获取文件路径的方法如下。

（1）绝对路径：/data/data/packagename/files/filename；

（2）context：context.getFilesDir()+"/filename"；

（3）缓存目录：/data/data/packagename/Cache 或 getCacheDir()；

10.3.1 应用程序文件读写

下面通过示例来演示 Android 下的文件操作。

res/layout/main.xml 中的代码如下。

```xml
<LinearLayout xmlns:android="http://schemas.android.com/apk/res/android"
    xmlns:tools="http://schemas.android.com/tools"
    android:layout_width="match_parent"
    android:layout_height="match_parent"
    tools:context=".FileTestActivity"
    android:orientation="vertical" >
    <LinearLayout
        android:orientation="horizontal"
        android:layout_width="fill_parent"
        android:layout_height="wrap_content">
        <TextView
            android:layout_width="wrap_content"
            android:layout_height="wrap_content"
            android:text="文件名:"/>
        <EditText
            android:id="@+id/title"
            android:layout_height="wrap_content"
            android:layout_width="fill_parent"/>
    </LinearLayout>
    <TextView
        android:layout_width="wrap_content"
        android:layout_height="wrap_content"
        android:text="内容: " />
    <EditText
        android:id="@+id/content"
        android:layout_height="wrap_content"
        android:layout_width="fill_parent"
        android:layout_weight="1"
        android:gravity="top"/>
    <LinearLayout
        android:orientation="horizontal"
        android:layout_width="wrap_content"
        android:layout_height="wrap_content"
```

```xml
            android:layout_gravity="center_horizontal">
    <Button
        android:layout_width="wrap_content"
        android:layout_height="wrap_content"
        android:text="保存"
        android:onClick="saveFile" />
    <Button
        android:layout_width="wrap_content"
        android:layout_height="wrap_content"
        android:text="读取"
        android:onClick="readFile" />
    </LinearLayout>
</LinearLayout>
```

MainActivity 中的代码如下。

```java
package cn.wang.filetest;

import java.io.FileInputStream;
import java.io.FileNotFoundException;
import java.io.FileOutputStream;
import java.io.IOException;

import android.os.Bundle;
import android.app.Activity;
import android.view.View;
import android.widget.EditText;

public class FileTestActivity extends Activity {
    private EditText etTitle,etContent;

    @Override
    protected void onCreate(Bundle savedInstanceState) {
        super.onCreate(savedInstanceState);
        setContentView(R.layout.activity_file_test);

        etTitle = (EditText) findViewById(R.id.title);
        etContent = (EditText) findViewById(R.id.content);
    }

    public void saveFile(View view){
        String title = etTitle.getText().toString();
        String content = etContent.getText().toString();

        FileOutputStream fos = null;
        try {
```

```java
            fos = openFileOutput(title, MODE_PRIVATE);
            fos.write(content.getBytes());
        } catch (FileNotFoundException e) {
            e.printStackTrace();
        } catch (IOException e) {
            e.printStackTrace();
        }finally{
            if(fos!= null){
                try {
                    fos.close();
                } catch (IOException e) {
                    e.printStackTrace();
                }
            }
        }
    }

    public void readFile(View view){
        FileInputStream fis = null;
        String title = etTitle.getText().toString();

        try {
            fis = openFileInput(title);
            byte[] aa = new byte[1024];
            int len = fis.read(aa);
            String string = new String(aa,0,len);
            etContent.setText(string);
        } catch (FileNotFoundException e) {
            e.printStackTrace();
        } catch (IOException e) {
            e.printStackTrace();
        }finally{
            if(fis != null){
                try {
                    fis.close();
                } catch (IOException e) {
                    e.printStackTrace();
                }
            }
        }
    }
}
```

示例运行主界面如图 10.11 所示，然后编辑一个文件（如图 10.12 所示），单击"保存"按钮，切换到 DDMS 窗口的 File Explore 子窗口（如图 10.13 所示），在目录

data/data/cn.wang.filetest/files 下，可以看到刚刚创建的文件 hello。选择文件 hello，单击右上角的图标 ，可以将文件同步到计算机的文件系统中，便于查看文件的详细信息。

图 10.11　示例的运行界面

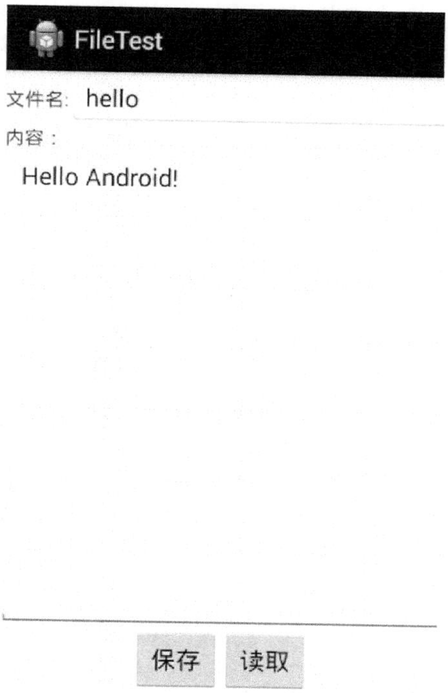

图 10.12　文件的读取与保存

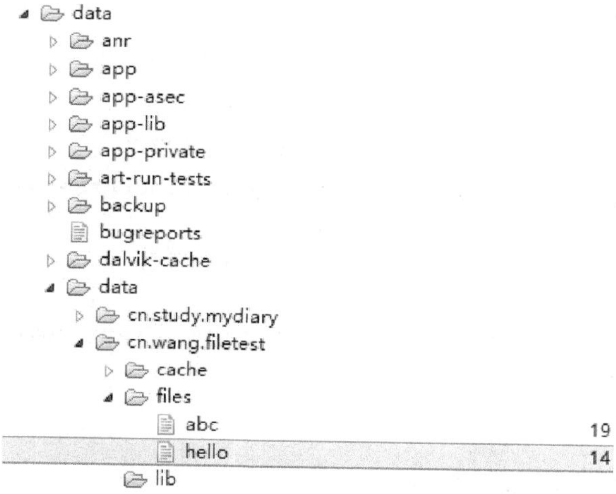

图 10.13　DDMS 的 File Explore 窗口

10.3.2　操作资源文件

在 Android 应用中，有一类特殊的文件称为资源文件。Android 的资源文件是在编译之前存放在 res 的特定目录下，系统提供了专用的 API。注意，资源文件只能读取，不能修改。

1. 读取 res/raw 中的文件

```
public void readRaw(View view){
        String res = "";
        try {
            //在 res/raw/hello.txt,
            InputStream in = getResources().openRawResource(R.raw.hello);
            int length = in.available();
            byte[] buffer = new byte[length];
            in.read(buffer);
            //选择合适的编码，如果不调整会乱码
            res = EncodingUtils.getString(buffer, "UTF-8");
            in.close();
        } catch (Exception e) {
            e.printStackTrace();
        }
        //把得到的内容显示在 TextView 上
        myTextView.setText(res);
}
```

2. 读取 res/asset 中的文件

```
public void readAsset(View view) {
        String fileName = "test.txt";
        String res = "";
        try {
            //res/assets/test.txt 有这样的文件存在
            InputStream in = getResources().getAssets().open(fileName);
            int length = in.available();
            byte[] buffer = new byte[length];
            in.read(buffer);
            res = EncodingUtils.getString(buffer, "UTF-8");
        } catch (Exception e) {
            e.printStackTrace();
        }
        //把得到的内容显示在 TextView 上
        myTextView.setText(res);
    }
```

10.3.3 操作 SD 卡上的文件

使用 Activity 的 openFileOutput()方法保存文件，文件是存放在手机空间上，一般手机的存储空间不是很大，如果要存放像视频这样的大文件，是不可行的。对于像视频这样的大文件，可以把它存放在 SDCard。SDCard 是干什么的？你可以把它看作是移动硬盘或 U 盘。

在模拟器中使用 SDCard，需要先创建一张 SDCard 卡（当然不是真的 SDCard，只是镜像文件）。创建 SDCard 可以在 Eclipse 创建模拟器时随同创建,操作界面如图 10.14 所示。

也可以使用 DOS 命令进行创建，过程如下。

在 Dos 窗口中进入 android SDK 安装路径的 tools 目录,输入以下命令创建一张容量为 2G 的 SDCard，文件后缀可以随便取，建议使用.img：

mksdcard 2048M D:\AndroidTool\sdcard.img

然后执行下面的命令将 SD 卡加载到模拟器中：

Emulator -sdcard D:\AndroidTool\sdcard.img -avd myavd

启动模拟器后，在 Android 主界面的选项菜单中选择"设置"，然后选择"存储"，在打开的主界面中查看"USB 存储器"的容量即可查看 SD 卡的加载情况。

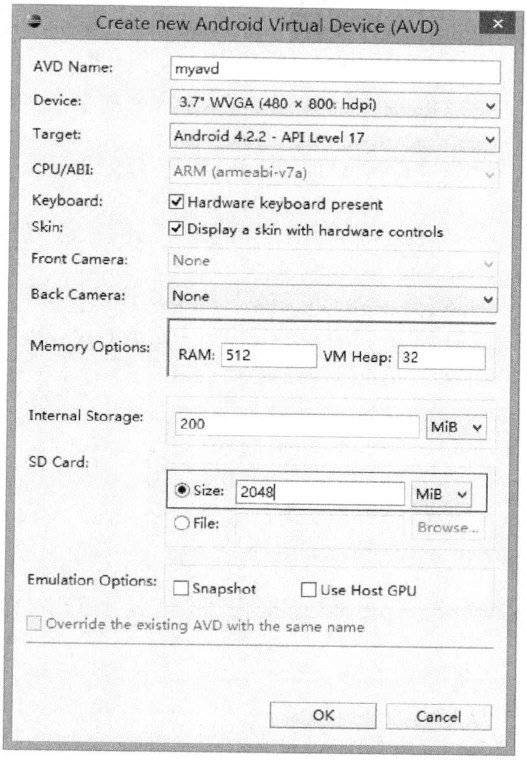

图 10.14　创建模拟器指定 SD 卡容量

在程序中访问 SDCard，需要具有访问 SDCard 的权限。在 AndroidManifest.xml 中加入访问 SDCard 的权限如下。

```
<!-- 在SDCard中创建与删除文件权限 -->
<uses-permission android:name="android.permission.MOUNT_UNMOUNT_FILESYSTEMS"/>
```

```xml
<!-- 往 SDCard 写入数据权限 -->
<uses-permission    android:name="android.permission.WRITE_EXTERNAL_STORAGE"/>
```

使用 SDCard 目录前,需要判断是否有 sdcard:Environment.getExternalStorageState() 方法用于获取 SDCard 的状态,如果手机装有 SDCard,并且可以进行读写,那么,方法返回的状态等于 Environment.MEDIA_MOUNTED。

Environment 类中还提供了以下方法用来获取系统文件夹。

`Environment.getExternalStorageDirectory()`方法用于获取 SDCard 的目录,当然要获取 SDCard 的目录,你也可以这样写:`File sdCardDir = new File("/mnt/sdcard");`

`Environment.getDataDirectory()`方法用于获取"/data"目录;

`Environment.getDownloadCacheDirectory()`方法用于获取"/cache"目录;

`Environment.getRootDirectory()`方法用于获取"/system"目录;

`Environment.getExternalStoragePublicDirectory(Environment.DIRECTORY_MUSIC)`方法用于获取"/mnt/sccard/Music"目录;

`Environment.getExternalStoragePublicDirectory(Environment.DIRECTORY_ALARMS)`方法用于获取"/mnt/sccard/Alarms"目录;

`Environment.getExternalStoragePublicDirectory(Environment.DIRECTORY_DCIM)`方法用于获取"/mnt/sccard/DCIM"目录;

`Environment.getExternalStoragePublicDirectory(Environment.DIRECTORY_DOWNLOADS)`方法用于获取"/mnt/sccard/Download"目录;

`Environment.getExternalStoragePublicDirectory(Environment.DIRECTORY_MOVIES)`方法用于获取"/mnt/sccard/Movies"目录;

`Environment.getExternalStoragePublicDirectory(Environment.DIRECTORY_NOTIFICATIONS)`方法用于获取"/mnt/sccard/Notifications"目录;

`Environment.getExternalStoragePublicDirectory(Environment.DIRECTORY_PICTURES)`方法用于获取"/mnt/sccard/Pictures"目录;

`Environment.getExternalStoragePublicDirectory(Environment.DIRECTORY_PODCASTS)`方法用于获取"/mnt/sccard/Podcasts"目录;

`Environment.getExternalStoragePublicDirectory(Environment.DIRECTORY_RINGTONES)`方法用于获取"/mnt/sccard/Ringtones"目录;

下面,通过一个例子演示在 Android 中如何操作 SD 上的文件。Activity 的布局文件文件如下。

```xml
<LinearLayout xmlns:android="http://schemas.android.com/apk/res/android"
    xmlns:tools="http://schemas.android.com/tools"
    android:layout_width="match_parent"
    android:layout_height="match_parent"
    tools:context=".FileTestActivity"
    android:orientation="vertical" >
    <LinearLayout
        android:orientation="horizontal"
        android:layout_width="fill_parent"
        android:layout_height="wrap_content">
```

```xml
        <TextView
            android:layout_width="wrap_content"
            android:layout_height="wrap_content"
            android:text="文件名:"/>
        <EditText
            android:id="@+id/title"
            android:layout_height="wrap_content"
            android:layout_width="fill_parent"
            android:inputType="text"/>
    </LinearLayout>
    <TextView
        android:layout_width="wrap_content"
        android:layout_height="wrap_content"
        android:text="内容: " />
    <EditText
        android:id="@+id/content"
        android:layout_height="0dp"
        android:layout_width="fill_parent"
        android:layout_weight="1"
        android:gravity="top"
        android:inputType="textMultiLine"
        />
    <LinearLayout
        android:orientation="horizontal"
        android:layout_width="wrap_content"
        android:layout_height="wrap_content"
        android:layout_gravity="center_horizontal">
        <Button
            android:layout_width="wrap_content"
            android:layout_height="wrap_content"
            android:text="保存"
            android:onClick="saveFile" />
        <Button
            android:layout_width="wrap_content"
            android:layout_height="wrap_content"
            android:text="读取"
            android:onClick="readFile" />
    </LinearLayout>
</LinearLayout>
```

MainActivity 的代码如下。

```java
package com.example.sdfiletest;

import java.io.BufferedReader;
import java.io.BufferedWriter;
import java.io.File;
```

```java
import java.io.FileInputStream;
import java.io.FileOutputStream;
import java.io.IOException;
import java.io.InputStream;
import java.io.InputStreamReader;
import java.io.OutputStream;
import java.io.OutputStreamWriter;
import android.os.Bundle;
import android.os.Environment;
import android.app.Activity;
import android.view.View;
import android.widget.EditText;
import android.widget.Toast;

public class MainActivity extends Activity {
    private EditText etTitle, etContent;

    @Override
    protected void onCreate(Bundle savedInstanceState) {
        super.onCreate(savedInstanceState);
        setContentView(R.layout.activity_main);

        etTitle = (EditText) findViewById(R.id.title);
        etContent = (EditText) findViewById(R.id.content);
    }
    //保存文件
    public void saveFile(View view) {
        String title = etTitle.getText().toString();
        String content = etContent.getText().toString();

        OutputStream out = null;
        BufferedWriter bw = null;
        //判断是否插入SD卡
        if (Environment.getExternalStorageState().equals(
                Environment.MEDIA_MOUNTED)) {
            String folderName = Environment.getExternalStorageDirectory()
                    .getPath() + "/myTest";
            File folder = new File(folderName);
            if (folder == null || !folder.exists()) {
                //如果文件夹不存在，则创建
                folder.mkdir();
            }
            File saveFile = new File(folderName, title);

            try {
```

```java
            if (!saveFile.exists()) {
                saveFile.createNewFile();
            }
            out = new FileOutputStream(saveFile);
            bw = new BufferedWriter(new OutputStreamWriter(out, "UTF-8"));
            bw.write(content);
        } catch (IOException e) {
            e.printStackTrace();
        } finally {
            if (bw != null) {
                try {
                    bw.close();
                } catch (IOException e) {
                    e.printStackTrace();
                }
            }
            if(out != null){
                try {
                    out.close();
                } catch (IOException e) {
                    e.printStackTrace();
                }
            }
        }
    }
}
//根据文件名读取文件
public void readFile(View view) {
    String title = etTitle.getText().toString();

    StringBuilder sb = new StringBuilder();
    InputStream in = null;
    BufferedReader br = null;
    //判断是否插入SD卡
    if (Environment.getExternalStorageState().equals(
            Environment.MEDIA_MOUNTED)) {
        String folderName = Environment.getExternalStorageDirectory()
                .getPath() + "/myTest";
        File folder = new File(folderName);
        if (folder == null || !folder.exists()) {
            //如果文件夹不存在，则退出
            Toast.makeText(this, folderName + "文件夹不存在", Toast.LENGTH_SHORT)
```

```java
                    .show();
            return;
        }
        File saveFile = new File(folderName, title);
        try {
            if (!saveFile.exists()) {
                //如果文件不存在,则退出
                Toast.makeText(this, folderName + "/" + title + "文件夹不存在",
                        Toast.LENGTH_SHORT).show();
                return;
            }
            in = new FileInputStream(saveFile);
            br = new BufferedReader(new InputStreamReader(in, "UTF-8"));
            String tmp;
            while ((tmp = br.readLine()) != null) {
                sb.append(tmp);
            }
            etContent.setText(sb.toString());
        } catch (IOException e) {
            e.printStackTrace();
        } finally {
            if (br != null) {
                try {
                    br.close();
                } catch (IOException e) {
                    e.printStackTrace();
                }
            }
            if(in != null){
                try {
                    in.close();
                } catch (IOException e) {
                    e.printStackTrace();
                }
            }
        }
    }
}
```

最后,一定要在 AndroidManifest.xml 文件中添加 SD 卡的访问权限语句,如图 10.15 所示。

```xml
<?xml version="1.0" encoding="utf-8"?>
<manifest xmlns:android="http://schemas.android.com/apk/res/android"
    package="com.example.sdfiletest"
    android:versionCode="1"
    android:versionName="1.0" >

    <uses-sdk
        android:minSdkVersion="8"
        android:targetSdkVersion="18" />
    <!-- 添加访问SD卡的权限 -->
    <uses-permission android:name="android.permission.MOUNT_UNMOUNT_FILESYSTEMS"/>
    <uses-permission android:name="android.permission.WRITE_EXTERNAL_STORAGE"/>
    <application
        android:allowBackup="true"
        android:icon="@drawable/ic_launcher"
        android:label="@string/app_name"
        android:theme="@style/AppTheme" >
```

图 10.15　添加 SD 卡的访问权限

在模拟器上运行 SDFileTest，并输入文件名和内容（如图 10.16 所示），单击"保存"按钮，即可在 SD 卡上创建一个名为 myTest 的文件夹，并将文件 aaa 添加到该文件夹中。

图 10.16　示例运行界面

10.4　数　据　库

在 Android 系统中可以用 SQLite 数据库来存储应用的数据，在 Android 系统中的很多应用，如联系人、图库、音乐等都用 SQLite 数据库来存储数据，以后我们在开发应用时也经常会用到，这个知识点必须掌握。

之前我们已经讲过如何使用文件和 SharedPreferences 来存储数据，它们都比较适合存储数据量比较小且被访问频率不是很高的数据，如果要存储的数据比较多，并且以后很可能还会去检索那些数据，就要选择使用 SQLite 数据库存储数据。

10.4.1 SQLite 简介

SQLite 是一个开源的嵌入式关系数据库，它在 2000 年由 D. Richard Hipp 发布，它可以减少应用程序管理数据的开销，SQLite 可移植性好、很容易使用、很小、高效而且可靠。

目前，在 Android 系统中集成的是 SQLite3 版本，SQLite 不支持静态数据类型，而是使用列关系，这意味着它的数据类型不具有表列属性，而具有数据本身的属性。当某个值插入数据库时，SQLite 将检查它的类型。如果该类型与关联的列不匹配，则 SQLite 会尝试将该值转换成列类型。如果不能转换，则该值将作为其本身具有的类型存储。

SQLite 支持 NULL、INTEGER 、REAL、TEXT 和 BLOB 数据类型。

例如，可以在 Integer 字段中存放字符串，或者在布尔型字段中存放浮点数，或者在字符型字段中存放日期型值。但是有一种例外，如果主键是 INTEGER ，只能存储 64 位整数，当向这种字段中保存除整数以外的数据时，将会产生错误。另外，SQLite 在解析 CREATE TABLE 语句时，会忽略 CREATE TABLE 语句中跟在字段名后面的数据类型信息，如

```
CREATE TABLE person (_id integer primary key autoincrement, name varchar(20))
```

这个 SQLite 数据库在解析这个语句时，会忽略掉跟在 name 字段后面的 varchar（20），也就是说。可以往 nam e 字段填 SQLit e 支持的那五种数据类型的任意一种，而且 name 字段可以存任意长度的字符，长度 20 不起作用。也就是说，可以把 SQLite 数据库近似看作是一种无数据类型的数据库，你可以把任何类型的资料存放在非 Integer 类型的主键之外的其他字段上去，另外，字段的长度也是没有限度的。

SQLite 的特点如下。

（1）零配置

SQlite3 不用安装、不用配置、不用启动、关闭或者配置数据库实例。当系统崩溃后不用做任何恢复操作，在下次使用数据库时自动恢复。

（2）可移植

它是运行在 Windows、 Linux、 BSD、 Mac OS X 和一些商用 Unix 系统，如 Sun 的 Solaris、IBM 的 AIX ，同样，它也可以工作在许多嵌入式操作系统下，如 Android 、QNX、VxWorks、Palm OS、Symbin 和 Windows CE。

（3）紧凑

SQLite 是被设计成轻量级、自包含的。一个头文件、一个 lib 库，用户就可以使用关系数据库了，不用启动任何系统进程。

（4）简单

SQLite 有着简单易用的 API 接口。

（5）可靠

SQLite 的源码达到 100%分支测试覆盖率。

在 Android 的 SDK 的安装目录下，有一个 tools 文件夹，里面包含了 sqlite3.exe，下面简单了解一下，在 Windows 下如何使用 SQLite。首先打开命令提示符窗口，切换到 tools 目录下。

```
//创建数据库
D:\Java\android-sdk-windows\tools>sqlite3 mydb.db
SQLite version 3.7.11 2012-03-20 11:35:50
Enter ".help" for instructions
Enter SQL statements terminated with a ";"
//查看常用命令
sqlite> .help
.backup ?DB? FILE      Backup DB (default "main") to FILE
.bail ON|OFF           Stop after hitting an error.  Default OFF
.databases             List names and files of attached databases
.dump ?TABLE? ...      Dump the database in an SQL text format
                       If TABLE specified, only dump tables matching
                       LIKE pattern TABLE.
.echo ON|OFF           Turn command echo on or off
.exit                  Exit this program
.explain ?ON|OFF?      Turn output mode suitable for EXPLAIN on or off.
                       With no args, it turns EXPLAIN on.
.header(s) ON|OFF      Turn display of headers on or off
.help                  Show this message
.import FILE TABLE     Import data from FILE into TABLE
.indices ?TABLE?       Show names of all indices
                       If TABLE specified, only show indices for tables
                       matching LIKE pattern TABLE.
.log FILE|off          Turn logging on or off.  FILE can be stderr/stdout
.mode MODE ?TABLE?     Set output mode where MODE is one of:
                         csv      Comma-separated values
                         column   Left-aligned columns.  (See .width)
                         html     HTML <table> code
                         insert   SQL insert statements for TABLE
                         line     One value per line
                         list     Values delimited by .separator string
                         tabs     Tab-separated values
                         tcl      TCL list elements
.nullvalue STRING      Print STRING in place of NULL values
.output FILENAME       Send output to FILENAME
.output stdout         Send output to the screen
.prompt MAIN CONTINUE  Replace the standard prompts
.quit                  Exit this program
.read FILENAME         Execute SQL in FILENAME
```

```
.restore ?DB? FILE     Restore content of DB (default "main") from FILE
.schema ?TABLE?        Show the CREATE statements
                       If TABLE specified, only show tables matching
                       LIKE pattern TABLE.
.separator STRING      Change separator used by output mode and .import
.show                  Show the current values for various settings
.stats ON|OFF          Turn stats on or off
.tables ?TABLE?        List names of tables
                       If TABLE specified, only list tables matching
                       LIKE pattern TABLE.
.timeout MS            Try opening locked tables for MS milliseconds
.vfsname ?AUX?         Print the name of the VFS stack
.width NUM1 NUM2 ...   Set column widths for "column" mode
.timer ON|OFF          Turn the CPU timer measurement on or off
//创建表
sqlite> create table person(_id integer primary key autoincrement,name,age);
//查看数据库
sqlite> .databases
seq  name              file
---  ---------------   ---------------------------------
0    main              D:\Java\android-sdk-windows\tools\mydb.db
//查看表
sqlite> .tables
Person
//插入数据
sqlite> insert into person values(null,'zhang',15);
sqlite> insert into person values(null,'li',18);
//查询
sqlite> select * from person;
1|zhang|15
2|li|18
//修改记录
sqlite> update person set age=19 where name='zhang';
//删除记录
sqlite> delete from person where name='li';
//退出
sqlite> .quit
```

使用命令查看操作数据库比较麻烦,还可以使用图形界面的管理工具。现在网络上的 SQLite 管理工具很多,向大家推荐一款好用的工具:Navicat Premium。Navicat Premium 是一个可多重连接的数据库管理工具,它可让你以单一程序同时连接到 MySQL、Oracle、PostgreSQL、SQLite 及 SQL Server 数据库,方便管理不同类型的数据库。Navicat Premium 结合了其他 Navicat 成员的功能。有了不同数据库类型的连接能力,Navicat Premium 支持在 MySQL、Oracle、PostgreSQL、SQLite 及 SQL Server 之间传输数据。它支持大部分

MySQL、Oracle、PostgreSQL、SQLite 及 SQL Server 的功能。

打开 Navicat Premium，其主界面如图 10.17 所示。

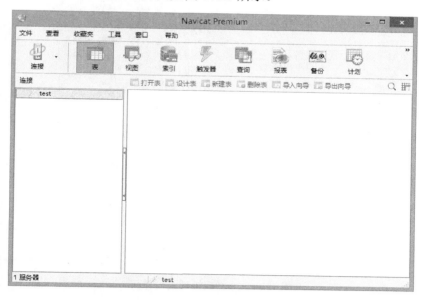

图 10.17　Navicat Premium 主界面

单击"连接"图标，选择"SQLite"，打开如图 10.18 所示窗口。

图 10.18　新建连接窗口

单击"确定"按钮，在图 10.17 所示的窗口左侧会新增一个"mydb"的连接，展开该连接（如图 10.19 所示），就可以在图形管理工具下进行数据库的操作了。

图 10.19 增加连接后的主界面

10.4.2 使用 SQLite 数据库

1. 创建数据库

在 Android 应用程序中使用 SQLite，必须自己创建数据库，然后创建表、索引，填充数据。Android 提供了 SQLiteOpenHelper 帮助用户创建一个数据库，只要继承 SQLiteOpenHelper 类，就可以轻松地创建数据库。SQLiteOpenHelper 类根据开发应用程序的需要，封装了创建和更新数据库使用的逻辑。SQLiteOpenHelper 的子类，至少需要实现三个方法：

构造函数，调用父类 SQLiteOpenHelper 的构造函数，这个方法需要四个参数：上下文环境（例如，一个 Activity），数据库名字，一个可选的游标工厂（通常是 Null），一个代表正在使用的数据库模型版本的整数。

onCreate()方法，它需要一个 SQLiteDatabase 对象作为参数，根据需要对这个对象填充表和初始化数据。

onUpgrade()方法，它需要三个参数，一个 SQLiteDatabase 对象，一个旧的版本号和一个新的版本号，这样就可以清楚如何把一个数据库从旧的模型转变到新的模型。

下面示例代码展示了如何继承 SQLiteOpenHelper 创建数据库。

```
package cn.wang.dbtest;

import android.content.Context;
import android.database.sqlite.SQLiteDatabase;
import android.database.sqlite.SQLiteOpenHelper;
```

```java
import android.util.Log;

public class DataBaseHelper extends SQLiteOpenHelper{

    public static final String DB_NAME = "mydb.db";
    public static final String TABLENAME = "person";
    public static final int DB_VERSION = 1;

    public static final String CREATETABLE = "create table "+TABLENAME+
        "(_id integer primary key,name text,age integer);";

    public DataBaseHelper(Context context) {
        super(context, DB_NAME, null, DB_VERSION);
    }

    @Override
    public void onCreate(SQLiteDatabase db) {
        db.execSQL(CREATETABLE);
    }

    @Override
    public void onUpgrade(SQLiteDatabase db,int oldVersion,int newVersion){
        //提示版本升级
        Log.i("Database update......", "Update database from "+oldVersion+" to "
            + newVersion);
        //删除旧的表
        db.execSQL("drop table if it exists "+TABLENAME);
        //创建新表
        onCreate(db);
    }
}
```

SQLiteOpenHelper 是一个抽象类，继承它需要实现其包含的两个函数，各函数的具体说明如下。

void onCreate(SQLiteDatabase db)：当数据库第一次生产时会调用该方法，一般在该方法中生成数据表。

void onUpgrade(SQLiteDatabase db, int oldVersion, int newVersion)：当数据库需要升级时，Android 系统会主动调用该方法，一般在该方法中删除数据表，并建立新表，或者对现有表进行修改。

使用 SQLiteOpenHelper 访问数据库，需要调用 getWritableDatabase() 或 getReadableDatabase()来获取一个可写或只读的数据库实例。如果数据库不存在，辅助类就会执行它的 onCreate 方法，如果数据库已经创建，则返回建好的数据库。也就是说，onCreate 方法，会在第一次创建数据库时自动运行。

如果不使用SQLiteOpenHelper，可以使用Context对象的openOrCreateDatabase方法来创建数据库：

```
SQLiteDatabase db = openOrCreateDatabase("test.db", MODE_PRIVATE, null);
```

2．操作数据库

对于添加、更新和删除来说，我们都可以使用

```
db.executeSQL(String sql);
db.executeSQL(String sql, Object[] bindArgs);//sql语句中使用占位符，然后第二个参数是实际的参数集。
```

除了统一的形式之外，它们还有各自的操作方法：

```
db.insert(String table, String nullColumnHack, ContentValues values);
db.update(String table, ContentValues values, String whereClause, String whereArgs);
db.delete(String table, String whereClause, String whereArgs);
```

以上三个方法的第一个参数都表示要操作的表名；insert中的第二个参数表示如果插入的数据每一列都为空，需要指定此行中某一列的名称，系统将此列设置为NULL，不至于出现错误；insert中的第三个参数是ContentValues类型的变量，是键值对组成的Map，key代表列名，value代表该列要插入的值；update的第二个参数也很类似，只不过它是更新该字段key为最新的value值，第三个参数whereClause表示WHERE表达式，比如"age > ? and age < ?"等，最后的whereArgs参数是占位符的实际参数值；delete方法的参数也是一样。

下面介绍查询操作。查询操作相对于上面的几种操作要复杂，因为我们经常要面对着各种各样的查询条件，所以系统也考虑到这种复杂性，为我们提供了较为丰富的查询形式。

```
db.rawQuery(String sql, String[] selectionArgs);
db.query(String table, String[] columns, String selection, String[] selectionArgs, String groupBy, String having, String orderBy);
db.query(String table, String[] columns, String selection, String[] selectionArgs, String groupBy, String having, String orderBy, String limit);
db.query(String distinct, String table, String[] columns, String selection, String[] selectionArgs, String groupBy, String having, String orderBy, String limit);
```

上面几种都是常用的查询方法，第一种最为简单，将所有的SQL语句都组织到一个字符串中，使用占位符代替实际参数，selectionArgs就是占位符实际参数集；下面的几种参数都很类似，columns表示要查询的列所有名称集，selection表示WHERE之后的条件语句，可以使用占位符，groupBy指定分组的列名，having指定分组条件，配合groupBy使用，orderBy指定排序的列名，limit指定分页参数，distinct可以指定"true"或"false"表示要不要过滤重复值。注意，selection、groupBy、having、orderBy、limit这几个参数中不包括WHERE、GROUP BY、HAVING、ORDER BY、LIMIT等SQL关键字。

最后，同时返回一个 Cursor 对象，代表数据集的游标，类似于 JavaSE 中的 ResultSet。下面是 Cursor 对象的常用方法。

```
cursor.move(int offset);          //以当前位置为参考,移动到指定行
cursor.moveToFirst();             //移动到第一行
cursor.moveToLast();              //移动到最后一行
cursor.moveToPosition(int position);   //移动到指定行
cursor.moveToPrevious();          //移动到前一行
cursor.moveToNext();              //移动到下一行
cursor.isFirst();                 //是否指向第一条
cursor.isLast();                  //是否指向最后一条
cursor.isBeforeFirst();           //是否指向第一条之前
cursor.isAfterLast();             //是否指向最后一条之后
cursor.isNull(int columnIndex);   //指定列是否为空(列基数为0)
cursor.isClosed();                //游标是否已关闭
cursor.getCount();                //总数据项数
cursor.getPosition();             //返回当前游标所指向的行数
cursor.getColumnIndex(String columnName);   //返回某列名对应的列索引值
cursor.getString(int columnIndex);          //返回当前行指定列的值
```

下面，通过一个例子演示在 Android 中如何操作数据库。Activity 的布局文件如下。

```xml
<LinearLayout xmlns:android="http://schemas.android.com/apk/res/android"
    android:layout_width="match_parent"
    android:layout_height="match_parent"
    android:orientation="vertical" >

    <LinearLayout
        android:layout_width="match_parent"
        android:layout_height="wrap_content"
        android:orientation="horizontal">
        <Button
            android:id="@+id/button1"
            android:layout_width="wrap_content"
            android:layout_height="wrap_content"
            android:text="Insert"
            android:onClick="insertData" />
        <Button
            android:id="@+id/button2"
            android:layout_width="wrap_content"
            android:layout_height="wrap_content"
            android:text="Insert2"
            android:onClick="insertData2" />
        <Button
            android:id="@+id/button3"
```

```xml
            android:layout_width="wrap_content"
            android:layout_height="wrap_content"
            android:text="Insert3"
            android:onClick="insertData3" />
    </LinearLayout>
    <LinearLayout
        android:layout_width="match_parent"
        android:layout_height="wrap_content"
        android:orientation="horizontal">
        <Button
            android:id="@+id/button4"
            android:layout_width="wrap_content"
            android:layout_height="wrap_content"
            android:text="Update"
            android:onClick="updateData" />
        <Button
            android:id="@+id/button5"
            android:layout_width="wrap_content"
            android:layout_height="wrap_content"
            android:text="Update2"
            android:onClick="updateData2" />
        <Button
            android:id="@+id/button6"
            android:layout_width="wrap_content"
            android:layout_height="wrap_content"
            android:text="Update3"
            android:onClick="updateData3" />
    </LinearLayout>
    <LinearLayout
        android:layout_width="match_parent"
        android:layout_height="wrap_content" >
        <Button
            android:id="@+id/button7"
            android:layout_width="wrap_content"
            android:layout_height="wrap_content"
            android:text="Delete"
            android:onClick="deleteData" />
        <Button
            android:id="@+id/button8"
            android:layout_width="wrap_content"
            android:layout_height="wrap_content"
            android:text="Delete2"
            android:onClick="deleteData2" />
```

```xml
        <Button
            android:id="@+id/button9"
            android:layout_width="wrap_content"
            android:layout_height="wrap_content"
            android:text="Delete3"
            android:onClick="deleteData3" />
    </LinearLayout>
    <LinearLayout
        android:layout_width="match_parent"
        android:layout_height="wrap_content" >
        <Button
            android:id="@+id/button10"
            android:layout_width="wrap_content"
            android:layout_height="wrap_content"
            android:text="Select"
            android:onClick="selectData" />
        <Button
            android:id="@+id/button11"
            android:layout_width="wrap_content"
            android:layout_height="wrap_content"
            android:text="Select2"
            android:onClick="selectData2" />
    </LinearLayout>
    <ListView
        android:id="@+id/listView1"
        android:layout_width="match_parent"
        android:layout_height="wrap_content" >
    </ListView>
</LinearLayout>
```

为了方便数据库的访问，定义 DataBaseHelper 类，该类继承了 SQLiteOpenHelper。因该类在前文中已经给出，在此不再重复。

DBTestActivity 的代码如下。

```java
package cn.wang.dbtest;

import android.os.Bundle;
import android.view.View;
import android.widget.ListView;
import android.widget.SimpleCursorAdapter;
import android.app.Activity;
import android.content.ContentValues;
import android.database.Cursor;
import android.database.sqlite.SQLiteDatabase;
```

```java
public class DBTestActivity extends Activity {
    private DataBaseHelper dbh;
    private SQLiteDatabase db;
    private ListView lv;
    @Override
    protected void onCreate(Bundle savedInstanceState) {
        super.onCreate(savedInstanceState);
        setContentView(R.layout.activity_dbtest);

        dbh = new DataBaseHelper(this);
        db = dbh.getWritableDatabase();

        lv = (ListView)findViewById(R.id.listView1);
    }
    //直接使用 SQL 语句插入
    public void insertData(View view){
        String str = "insert into person values(null,'zhangsan',20);";
        db.execSQL(str);
    }
    //使用带参数的 SQL 语句插入。? 是占位符。
    public void insertData2(View view){
        String str = "insert into person values(null,?,?);";
        Object[] obj = new Object[]{"lisi",21};
        db.execSQL(str,obj);
    }
    //使用 insert() 插入数据
    public void insertData3(View view){
        //ContentValues:是列名-值的映射
        ContentValues cv = new ContentValues();
        cv.put("name", "Jack");
        cv.put("age", 22);
        db.insert("person", null, cv);
    }
    //直接使用 SQL 语句修改
    public void updateData(View view){
        String str = "update person set age = 25 where name='zhangsan';";
        db.execSQL(str);
    }
    //使用带参数的 SQL 语句修改
    public void updateData2(View view){
        String str = "update person set age = ? where name=?;";
        Object[] obj = new Object[]{26,"lisi"};
        db.execSQL(str,obj);
    }
    //使用 update() 修改数据
```

```java
    public void updateData3(View view){
        ContentValues cv = new ContentValues();
        cv.put("age", 27);
        db.update("person", cv, "name=?", new String[]{"Jack"});
    }
    //直接使用SQL语句删除
    public void deleteData(View view){
        String str = "delete from person where _id =1";
        db.execSQL(str);
    }
    //使用带参数的SQL语句删除
    public void deleteData2(View view){
        String str = "delete from person where _id =?";
        Object[] obj = new Object[]{2};
        db.execSQL(str,obj);
    }
    //使用delete()删除数据
    public void deleteData3(View view){
        db.delete("person", "_id=?", new String[]{"3"});
        //db.delete("person", "_id=3", null);
    }
    //查询数据
    public void selectData(View view){
        String str = "select * from person";
        Cursor cursor = db.rawQuery(str, null);
        if(cursor!=null){
            //遍历游标
            int rowCount = cursor.getColumnCount();
            while(cursor.moveToNext()){
                for(int i=0;i<rowCount;i++){
System.out.print(cursor.getColumnName(i)+":"+cursor.getString(i)+"");
                }
                System.out.println("");
            }
            cursor.close();
        }
    }
    //查询数据
    public void selectData2(View view){
        Cursor cursor = db.query("person", new String[]{"_id", "name", "age"}, null,
                null, null, null, null);
        if(cursor!=null){
            //将数据显示到ListView中
            SimpleCursorAdapter sca = new SimpleCursorAdapter(this,
```

```
                    R.layout.my_list1, cursor,
                    new String[]{"_id","name","age"},
                    new           int[]{R.id.textView1,R.id.textView2,R.id.textView3},
                    SimpleCursorAdapter.NO_SELECTION);
            lv.setAdapter(sca);
        }
    }
}
```

my_list1.xml 的布局文件如下。

```
<?xml version="1.0" encoding="utf-8"?>
<LinearLayout xmlns:android="http://schemas.android.com/apk/res/android"
    android:layout_width="match_parent"
    android:layout_height="match_parent"
    android:orientation="horizontal" >
    <TextView
        android:id="@+id/textView1"
        android:layout_width="wrap_content"
        android:layout_height="wrap_content"
        android:text="TextView"
        android:textSize="20sp"
        android:layout_weight="1" />
    <TextView
        android:id="@+id/textView2"
        android:layout_width="wrap_content"
        android:layout_height="wrap_content"
        android:text="TextView"
        android:textSize="20sp"
        android:layout_weight="2" />
    <TextView
        android:id="@+id/textView3"
        android:layout_width="wrap_content"
        android:layout_height="wrap_content"
        android:text="TextView"
        android:textSize="20sp"
        android:layout_weight="1" />
</LinearLayout>
```

在模拟器中运行该程序，主界面如图 10.20 所示。

依次单击 Insert、Insert2 和 Insert3 三个按钮，往数据库中插入三条记录，然后单击 Select 按钮，通过 DDMS 的 LogCat 视图，可以查看到如图 10.21 所示的结果。单击 Select2 按钮，查询的结果显示到 ListView 中，如图 10.22 所示。

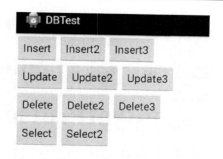

图 10.20　DBTestActivity 运行界面

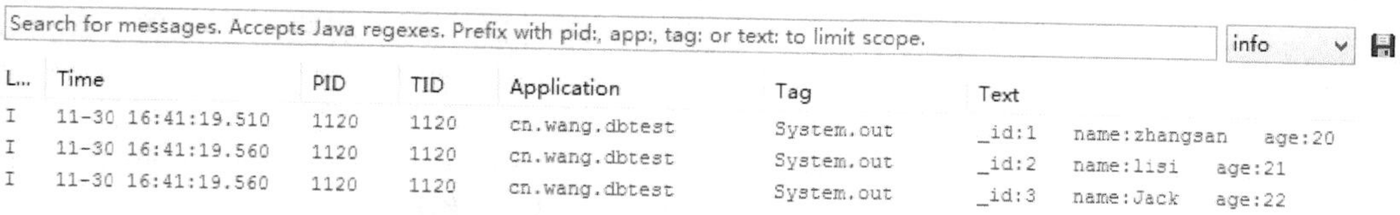

图 10.21　使用 LogCat 查看查询结果

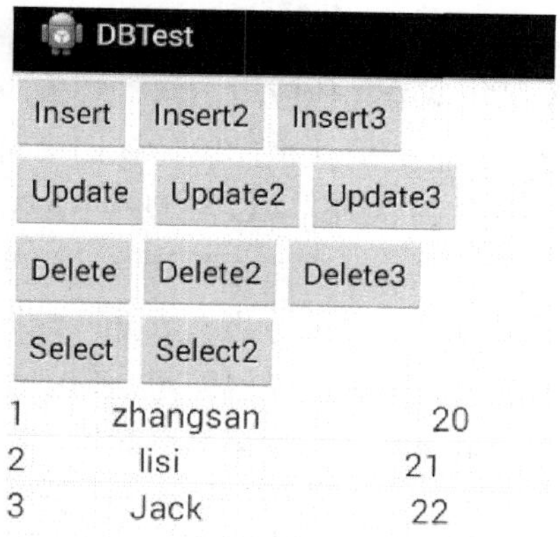

图 10.22　使用 ListView 显示查询结果

其他几个按钮的功能请读者自行验证，在此不再一一说明。

10.5　本 章 小 结

本章主要介绍了 Android 的 Preferences、文件和数据库 SQLite 这三种数据存储方式，要求掌握使用 SharedPreferences 工具类对简单的配置信息的读、写方法，以及 PreferenceActivity 的使用方法；由于文件操作是从 Java 中移植过来的，需要重点掌握 Android 系统为文件操作提供的新方法，以及对 SD 卡进行操作的方法；本章重点是 SQLite 数据库，Android 系统内置了 SQLite 数据库，并为之提供了大量方便的工具类，需要掌握并能熟练使用它们。

第 11 章 Android 的图书馆——ContentProvider

系统的多个应用程序之间，有时候需要进行数据的共享与交换，而 Android 系统采用的 Linux 内核，因此，继承了 Linux 的严格权限管理机制。前面讲过的 Intent 只适合用于传递数据量小的场合，对于大的数据文件交互明显不合适。为了解决这个难题，Android 提供了 ContentProvider 机制。

11.1 ContentProvider 概述

ContentProvider 在 Android 中的作用是对外共享数据，也就是说，可以通过 ContentProvider 把应用中的数据共享给其他应用访问，其他应用可以通过 ContentProvider 对应用中的数据进行增删改查。关于数据共享，以前我们学习过文件操作模式，知道通过指定文件的操作模式为 Context.MODE_WORLD_READABLE 或 Context.MODE_WORLD_WRITEABLE 同样也可以对外共享数据。那么，这里为何要使用 ContentProvider 对外共享数据呢？如果采用文件操作模式对外共享数据，数据的访问方式会因数据存储的方式而不同，导致数据的访问方式无法统一，如采用 XML 文件对外共享数据，需要进行 XML 解析才能读取数据；采用 SharedPreferences 共享数据，需要使用 SharedPreferences API 读取数据。使用 ContentProvider 对外共享数据的好处是统一了数据的访问方式。

ContentProvider 类实现了一组标准的方法接口，从而能够让其他的应用程序保存或读取此 ContentProvider 的各种数据类型。在程序内可以通过实现 ContentProvider 的抽象接口将自己的数据显示出来，而外界根本不用看到这个显示的数据在应用程序中是如何存储的，以及究竟是用文件存储还是用数据库存储的。外界正是通过这个统一的接口来实现数据的增删改查。

当应用需要通过 ContentProvider 对外共享数据时，第一步需要继承 ContentProvider 并重写下面方法：

```
public class PersonProvider extends ContentProvider{
    public boolean onCreate()
    public Uri insert(Uri uri, ContentValues values)
    public int delete(Uri uri, String selection, String[] selectionArgs)
    public int update(Uri uri, ContentValues values, String selection, String[] selectionArgs)
```

```
public Cursor query(Uri uri, String[] projection, String selection,
String[] selectionArgs, String sortOrder)
public String getType(Uri uri)
}
```

第二步需要在 AndroidManifest.xml 使用<provider>对该 ContentProvider 进行配置，为了能让其他应用找到该 ContentProvider，ContentProvider 采用了 authorities（主机名/域名）对它进行唯一标识，你可以把 ContentProvider 看作是一个网站，authorities 就是它的域名。

```
<manifest.... >
    <application
        android:icon="@drawable/icon"
        android:label="@string/app_name">
        <provider android:name=".PersonProvider"
            android:authorities="cn.wang.providers.personprovider"/>
    </application>
</manifest>
```

当外部应用需要对 ContentProvider 中的数据进行添加、删除、修改和查询操作时，可以使用 ContentResolver 类来完成，要获取 ContentResolver 对象，可以使用 Activity 提供的 getContentResolver()方法。

ContentResolver 类提供了与 ContentProvider 类相同签名的四个方法。

- public Uri insert(Uri uri, ContentValues values)：该方法用于往 ContentProvider 添加数据。
- public int delete(Uri uri, String selection, String[] selectionArgs)：该方法用于从 ContentProvider 删除数据。
- pubic int update(Uri uri, ContentValues values, String selection, String[] selectionArgs)：该方法用于更新 ContentProvider 中的数据。
- public Cursor query(Uri uri, String[] projection, String selection, String[] selectionArgs, String sortOrder)：该方法用于从 ContentProvider 中获取数据。

很显然，使用 ContentResolver 类访问 ContentProvider，Uri 起到了关键作用，因为它决定了去访问那个 ContentProvider。那么，什么是 Uri 呢？

Uri 代表了要操作的数据，Uri 主要包含了两部分信息：需要操作的 ContentProvider；对 ContentProvider 中的什么数据进行操作，一个 Uri 由以下几部分组成。

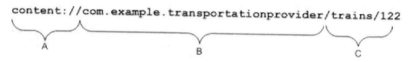

A：ContentProvider（内容提供者）的 scheme 已经由 Android 所规定，scheme 为 content://。

B：主机名（或叫 Authority）用于唯一标识这个 ContentProvider，外部调用者可以根据这个标识来找到它。

C：路径（path）可以用来表示要操作的数据，路径的构建应根据业务而定，如下所示。

- 要操作 person 表中 ID 为 10 的记录，可以构建这样的路径：/person/10；
- 要操作 person 表中 ID 为 10 的记录的 name 字段，person/10/name；
- 要操作 person 表中的所有记录，可以构建这样的路径：/person；
- 要操作 xxx 表中的记录，可以构建这样的路径：/xxx。

当然要操作的数据不一定来自数据库，也可以是文件、xml 或网络等其他存储方式，如下所示。

要操作 xml 文件中 person 节点下的 name 节点，可以构建这样的路径：/person/name。

如果要把一个字符串转换成 Uri，可以使用 Uri 类中的 parse()方法，如下。

```
Uri uri = Uri.parse("content://com.ljq.provider.personprovider/person");
```

在几乎所有的 Content Provider 的操作中都会用到 Uri，因此如果是自己开发的 Content Provider，最好将 Uri 定义为常量，这样在简化开发的同时也提高了代码的可维护性。

因为 Uri 代表了要操作的数据，所以经常需要解析 Uri，并从 Uri 中获取数据。Android 系统提供了两个用于操作 Uri 的工具类，分别为 UriMatcher 和 ContentUris。掌握它们的使用，会便于我们的开发工作。

UriMatcher：用于匹配 Uri，其用法如下。

（1）首先把需要匹配 Uri 路径全部给注册上，如下。

```
//常量UriMatcher.NO_MATCH 表示不匹配任何路径的返回码
UriMatcher  sMatcher = new UriMatcher(UriMatcher.NO_MATCH);
//如果match()方法匹配content://cn.wang.provider.personprovider/person路径，
返回匹配码为1
sMatcher.addURI("cn.wang.provider.personprovider", "person", 1);//添加需要
匹配uri，如果匹配就会返回匹配码
//如果 match()方法匹配 content://cn.wang.provider.personprovider/person/11
路径，返回匹配码为2
sMatcher.addURI("cn.wang.provider.personprovider", "person/#", 2);//#号为
通配符
```

（2）注册完需要匹配的 Uri 后，就可以使用 uriMatcher.match(uri)方法对输入的 Uri 进行匹配，如果匹配就返回匹配码，匹配码是调用 addURI()方法传入的第三个参数，如 sMatcher.match(Uri.parse("content://cn.wang.provider.personprovider/person/10"))，返回的匹配码为 2。

ContentUris：用于获取 Uri 路径后面的 ID 部分，它有两个比较实用的方法：

```
withAppendedId(uri, id)用于为路径加上 ID 部分
 Uri uri =Uri.parse("content://com.ljq.provider.personprovider/person");
 Uri resultUri =ContentUris.withAppendedId(uri, 10);
 //生成后的 Uri 为: content://com.ljq.provider.personprovider/person/10
·parseId(uri)方法用于从路径中获取 ID 部分
 Uri uri =Uri.parse("content://com.ljq.provider.personprovider/person/10");
 long personid = ContentUris.parseId(uri);//获取的结果为:10
```

11.2 自定义 ContentProvider

下面通过示例展示如何自定义 Content Provider。

1. 数据存储

数据存储常用的方式是文件和数据库，为了操作方便，采用 SQLite 数据库对数据进行存储。为了简单起见，将 10.4 创建的 DataBaseHelper 类复制到当前项目的包中，仍然使用 mydb.db 数据库和 person 表。

2. 继承 ContentProvider

新建一个类 PersonProvider，继承 ContentProvider，需要实现该类的六个方法，这些方法的作用如下。

public boolean onCreate()：该方法在 ContentProvider 创建后就会被调用，Android 开机后，ContentProvider 在其他应用第一次访问它时才会被创建。

public Uri insert(Uri uri, ContentValues values)：该方法用于供外部应用往 ContentProvider 添加数据。

public int delete(Uri uri, String selection, String[] selectionArgs)：该方法用于供外部应用从 ContentProvider 删除数据。

public int update(Uri uri, ContentValues values, String selection, String[] selectionArgs)：该方法用于供外部应用更新 ContentProvider 中的数据。

public Cursor query(Uri uri, String[] projection, String selection, String[] selectionArgs, String sortOrder)：该方法用于供外部应用从 ContentProvider 中获取数据。

public String getType(Uri uri)：该方法用于返回当前 Uri 所代表数据的 MIME 类型。

如果操作的数据属于集合类型，MIME 类型字符串应该以 vnd.android.cursor.dir/开头，例如，要得到所有 person 记录的 Uri 为 content://cn.wang.provider.personprovider/person，那么返回的 MIME 类型字符串应该为"vnd.android.cursor.dir/person"。

如果要操作的数据属于非集合类型数据，那么 MIME 类型字符串应该以 vnd.android.cursor.item/开头，例如，得到 id 为 10 的 person 记录，Uri 为 content://cn.wang.provider.personprovider/person/10，那么返回的 MIME 类型字符串为"vnd.android.cursor.item/person"。

PersonProvider 类文件如下。

```
package cn.wang.mypersonprovider;

import android.content.ContentProvider;
import android.content.ContentUris;
import android.content.ContentValues;
import android.content.UriMatcher;
```

```java
import android.database.Cursor;
import android.database.sqlite.SQLiteDatabase;
import android.net.Uri;

public class PersonProvider extends ContentProvider {
    private DataBaseHelper dbh = null;
    //1.发布 Content Provider 的 Uri 地址
    private static final String AUTHORITY = "cn.wang.personprovider";
    public static final Uri CONTENT_URI = Uri.parse(
            "content://cn.wang.personprovider/persons");

    //2.注册需要匹配的 Uri
    private static UriMatcher uriMatcher = new UriMatcher(UriMatcher.NO_MATCH);
    static{
        uriMatcher.addURI(AUTHORITY, "persons", 1);
        uriMatcher.addURI(AUTHORITY, "persons/#", 2);
    }
    @Override
    public boolean onCreate() {
        //3.实例化 dbh
        dbh = new DataBaseHelper(getContext());
        return false;
    }

    @Override
    public Cursor query(Uri uri, String[] projection, String selection,
            String[] selectionArgs, String sortOrder) {
        //4.实现查询
        SQLiteDatabase db = dbh.getReadableDatabase();
        Cursor cursor = null;

        switch (uriMatcher.match(uri)) {
        case 1://查询所有行
            cursor = db.query(
                    "person",        //表名
                    null,            //列的数组, null 代表所有列
                    selection,       //where 条件
                    selectionArgs,   //where 条件的参数值的数组
                    null,            //分组
                    null,            //having
                    sortOrder);      //排序规则
            break;
        case 2://查询指定 ID 的行
            //获取 ID
```

```java
            long id = ContentUris.parseId(uri);
            //在selection上增加条件_id=id
            if(selection == null){
                selection = "_id="+id;
            }else{
                selection = "_id="+id+" and ("+selection+")";
            }
            cursor = db.query(
                    "person",         //表名
                    null,             //列的数组，null代表所有列
                    selection,        //where条件
                    selectionArgs,    //where条件的参数值的数组
                    null,             //分组
                    null,             //having
                    sortOrder);       //排序规则
            break;
        default:
            break;
    }
    return cursor;
}

@Override
public String getType(Uri uri) {
    //5.返回当前Uri所代表数据的MIME类型
    switch (uriMatcher.match(uri)) {
    case 1:
        return "vnd.android.cursor.dir/person";
    case 2:
        return "vnd.android.cursor.item/person";
    }
    return null;
}

@Override
public Uri insert(Uri uri, ContentValues values) {
    //6.实现插入方法
    SQLiteDatabase db = dbh.getWritableDatabase();
    long id = db.insert("person", null, values);
    if (id>-1) {//插入数据成功
        //构建新插入行的Uri
        Uri insertUri = ContentUris.withAppendedId(CONTENT_URI, id);
        //通知所有的观察者，数据集已经改变
        getContext().getContentResolver().notifyChange(insertUri, null);
```

```java
            return insertUri;
        }
        return null;
    }

    @Override
    public int delete(Uri uri, String selection, String[] selectionArgs) {
        //7.实现删除方法
        SQLiteDatabase db = dbh.getWritableDatabase();
        int num = 0;//已经删除的记录数量
        switch (uriMatcher.match(uri)) {
        case 1:
            num = db.delete("person", selection, selectionArgs);
            break;
        case 2:
            //获取ID
            long id = ContentUris.parseId(uri);
            //在selection上增加条件_id=id
            if(selection == null){
                selection = "_id="+id;
            }else{
                selection = "_id="+id+" and ("+selection+")";
            }
            num = db.delete("person", selection, selectionArgs);
            break;
        default:
            break;
        }
        //通知所有的观察者，数据集已经改变
        getContext().getContentResolver().notifyChange(uri, null);
        return num;
    }

    @Override
    public int update(Uri uri, ContentValues values, String selection,
            String[] selectionArgs) {
        //8.实现修改方法
        SQLiteDatabase db = dbh.getWritableDatabase();
        int num = 0;//已经修改的记录数量
        switch (uriMatcher.match(uri)) {
        case 1:
            num = db.update("person", values, selection, selectionArgs);
            break;
        case 2:
            //获取ID
            long id = ContentUris.parseId(uri);
```

```
            //在selection上增加条件_id=id
            if(selection == null){
                selection = "_id="+id;
            }else{
                selection = "_id="+id+" and ("+selection+")";
            }
            num = db.update("person", values, selection, selectionArgs);
            break;
        default:
            break;
        }
        //通知所有的观察者，数据集已经改变
        getContext().getContentResolver().notifyChange(uri, null);
        return num;
    }
}
```

3. 注册 Provider

在 AndroidManifest.xml 使用<provider>对该 ContentProvider 进行配置，代码如下。

```
<provider
        android:name="cn.wang.mypersonprovider.PersonProvider"
        android:authorities="cn.wang.personprovider"
        android:exported="true"></provider>
```

4. 访问 Provider

在 MainActivity 的 onCreate 方法中增加如下代码：

```
ContentResolver cr = getContentResolver();
//增加记录
ContentValues values = new ContentValues();
values.put("name", "Mike");
values.put("age", 20);
cr.insert(PersonProvider.CONTENT_URI, values);
values.clear();
values.put("name", "Mary");
values.put("age", 18);
cr.insert(PersonProvider.CONTENT_URI, values);
//查询所有记录
Cursor cursor = cr.query(PersonProvider.CONTENT_URI,
        null, null, null, null);
Log.i("after inserted", "-----------------------------------------");
while(cursor.moveToNext()){
    Log.i("after inserted", "id:"+cursor.getString(0)+" name:" +
```

```
        cursor.getString(1)+" age:"+cursor.getString(2));
}
cursor.close();
//修改记录
values.clear();
values.put("age", 19);
//构建的 Uri 为: "content://cn.wang.personprovider/persons/2"
Uri uri = ContentUris.withAppendedId(PersonProvider.CONTENT_URI, 2);
//修改 id 为 2 的记录
cr.update(uri, values, null, null);
//查询 id 为 2 的记录
cursor = cr.query(uri, null, null, null, null);
Log.i("after updated", "--------------------------------------------");
while(cursor.moveToNext()){
    Log.i("after updated", "id:"+cursor.getString(0)+" name:"+
        cursor.getString(1)+" age:"+cursor.getString(2));
}
cursor.close();
//删除 id 为 2 的记录
cr.delete(uri, null, null);
//查询记录
cursor = cr.query(PersonProvider.CONTENT_URI,
        null, null, null, null);
Log.i("after deleted", "--------------------------------------------");
while(cursor.moveToNext()){
    Log.i("after deleted", "id:"+cursor.getString(0)+" name:"+
        cursor.getString(1)+" age:"+cursor.getString(2));
}
cursor.close();
```

运行应用程序，在 LogCat 窗口中看到的输出如图 11.1 所示。

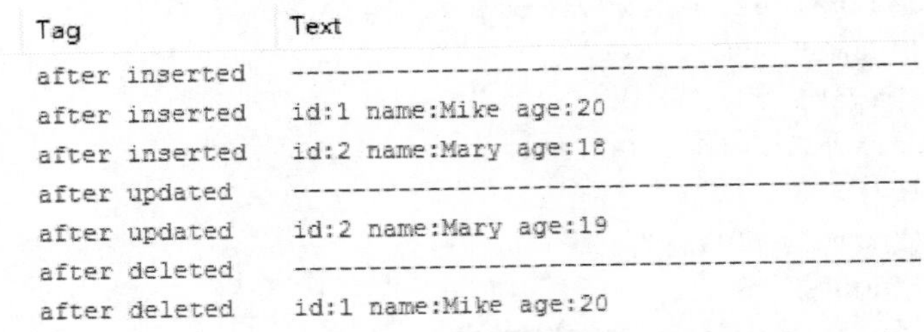

Tag	Text
after inserted	--------------------------------------
after inserted	id:1 name:Mike age:20
after inserted	id:2 name:Mary age:18
after updated	--------------------------------------
after updated	id:2 name:Mary age:19
after deleted	--------------------------------------
after deleted	id:1 name:Mike age:20

图 11.1 访问 PersonProvider 的结果

如果在其他应用程序中调用此 Provider，因为 PersonProvider.CONTENT_URI 无法访问，需要定义一个新的 Uri 变量，如

```
private static final Uri CONTENT_URI = Uri.parse("content://cn.wang.person
provider/persons");
```

然后将 PersonProvider.CONTENT_URI 用代替即可，监听 ContentProvider 中数据的变化。

在 PersonProvider 的 insert、update 和 delete 方法的最后，都调用了 getContext().getContentResolver().notifyChange(Uri, null)方法，该方法的作用可以在 ContentProvider 发生数据变化时通知注册在此 URI 上的访问者。其中，Uri 表示监听的 Uri，null 表示发送消息给任何人。

如果 ContentProvider 的访问者需要得到数据变化通知，必须使用 ContentObserver 对数据进行监听，当监听到数据变化通知时，系统就会调用 ContentObserver 的 onChange()方法。

为了监听指定的 ContentProvider 的数据变化，需要通过 ContentResolver 向指定 Uri 注册 ContentObserver 监听器。用如下方法来注册监听器。

registerContentObserver(Uri uri,boolean notifyForDescendents,ContentObserver observer);

notifyForDescendents：如果该参数设为 true，假如 Uri 为 content://abc，那么 Uri 为 content://abc/xyz, content://abc/xyz/foo 的数据改变时也会触发该监听器，如果参数为 false，那么只有 content://abc 的数据改变时会触发该监听器。

下面通过实例来演示如何监听 ContentProvider 中数据的变化。需要创建两个应用程序，第一个应用程序的 Activity 上放置一个按钮，该按钮的单击事件如下。

```java
public void cpInsert(View view){
    ContentResolver cr = getContentResolver();
    Uri uri = Uri.parse("content://cn.wang.personprovider/persons");
    cr.registerContentObserver(uri, true, new PersonObserver(new Handler()));
    //增加记录
    ContentValues values = new ContentValues();
    values.put("name", "Mike");
    values.put("age", 20);
    cr.insert(uri, values);
}
private class PersonObserver extends ContentObserver{//监听
    public PersonObserver(Handler handler) {
        super(handler);
    }
    //当ContentProvier数据发生改变,则触发该函数
    @Override
    public void onChange(boolean selfChange) {
        super.onChange(selfChange);
        Log.i("Test1", "数据改变");
    }
}
```

第二个应用程序的 MainActivity 的代码如下。

```java
public class MainActivity extends Activity {
```

```java
    private static final Uri CONTENT_URI = Uri.parse("content://cn.wang.
personprovider/persons");
    @Override
    protected void onCreate(Bundle savedInstanceState) {
        super.onCreate(savedInstanceState);
        setContentView(R.layout.activity_main);

        ContentResolver cr = getContentResolver();
        cr.registerContentObserver(CONTENT_URI, true, new PersonObserver(new
        Handler()));

    }

    private class PersonObserver extends ContentObserver{//监听
        public PersonObserver(Handler handler) {
            super(handler);
        }
        //当ContentProvier数据发生改变,则触发该函数
        @Override
        public void onChange(boolean selfChange) {
            super.onChange(selfChange);
            Log.i("Test2", "数据改变");
        }
    }
}
```

运行这两个应用程序,当单击第一个应用程序的按钮时,会增加一条记录,ContentProvider 就会通知所有的监听者,数据已经发生改变,应用程序的 onChange 方法就会触发,输出结果如图 11.2 所示。

```
I  12-08 23:03:42.414    1340    1340    com.example.cpdemo    Test1    数据改变
I  12-08 23:03:42.434    1297    1297    cn.wang.cpdemo2       Test2    数据改变
```

图 11.2　ContentProvider 监听

11.3　系统 ContentProvider

Android 提供了一些主要数据类型的 Contentprovider,如音频、视频、图片和私人通讯录等。可在 android.provider 包下面找到一些 android 提供的 Contentprovider。可以获得这些 Contentprovider,查询它们包含的数据,当然前提是已获得适当的读取权限。

11.3.1 使用 Contacts Contract Content Provider

Contacts Contract Content Provider 提供了一个可扩展的联系人信息数据库。它允许开发人员任意扩展为每个联系人的存储数据，甚至可以为联系人管理提供可替代提供程序。该数据库保存的路径为/data/data/com.android.providers.contacts/databases/contacts2.db。可以使用上一章介绍的方法来查看该数据库，常用的几张表如下。

（1）Data 表：储存所有与 Raw_Contacts 相关具体的信息。表的每一条记录对应一个特定信息，如名字、电话、E-mail 地址、头像和组信息等。每一记录通过一个 mimetype_id 字段表明该行所记录的数据类型。如果 row 的数据类型是 Phone.CONTENT_ITEM_TYPE，那么第一个字段应该保存电话号码，如果数据类型为 Email.CONTENT_ITEM_TYPE，那么这一记录的字段应该保存邮件地址。

Data 表的字段如表 11.1 所示。

表 11.1 Data 表的字段

字段	说明
mimetype_id	表示该行存储的信息的类型
raw_contact_id	表示该行所属的 Raw_Contact
is_primary	多个 data 数据组成一个 raw_contact，该字段表示此 data 是否是其所属的 raw contact 的主 data，即其 display name 会作为 raw contact 的 display name
is_super_primary	该 data 是否是其所属的 contact 的主 data，如果 is_super_primary 为 1 则 is_primary 一定为 1
data1~data15	15 个数据字段，对于不同类型的信息，表示不同的含义，ContactsContract.CommomDataKinds 类中定义了与常用的数据类型相对应的一些类，这些类中分别定义了相应数据类型中这些字段表示的含义。一般 data1 表是主信息（如电话、E-mail 地址等），data2 表示副信息，data15 表示 Blob 数据
data_sync1~data_sync4	sync_adapter 要用的字段（sync_adapter 用于数据的同步，如你手机中的 Gmail 账户与 Google 服务器的同步）
data_version	数据的版本，用于数据的同步

（2）Raw_Contacts 表中的一行存储 Data 表中一些数据行的集合及一些其他的信息，表示一个联系人某一特定账户的信息，如 Facebook 或 Exchange 的一个联系人。

当插入一个 Raw Contact 或当一个 Raw Contact 所属的一个 Data 改变时，系统会检查这个 Raw Contact 跟其他的 Raw Contact 是否可以匹配（比如，如果两个 Raw Contact 的 Data 包含相同的电话号码或名字），如果匹配，它们就会被综合到一起，也就是说，会属于同一个 Contact，表现为在 Raw_Contacts 表中它们引用的 cantact_id 是相同的。

（3）Contacts 表中的一行表示一个联系人，它是 Raw_Contacts 表中的一行或多行的数据的组合，这些 Raw_Contacts 表中的行表示同一个人的不同的账户信息。Contacts 中的数据由系统组合 Raw_Contacts 表中的数据自动生成，不可以直接向这个表中插入数据，当一个 raw contact 被插入时，系统会首先查找 Contacts 表看是否有记录跟插入的 raw contact 表示同一个人，如果找到了，则把找到的这个 contact 的_ID 插入 raw contact 记录的

CONTACT_ID 字段，如果没有找到，则系统自动插入一个 Contact 记录并把它的_ID 插入新插入的 raw contact 的 CONTACT_ID 列。

从 Android 2.0 SDK 开始有关联系人 provider 的类由 android.provider.Contacts 变成了 android.provider.ContactsContract，虽然老的 android.provider.Contacts 能用，但是在 SDK 中标记为 deprecated 将被放弃或不推荐的方法。

使用 ContentResolver 对通信录中的数据进行添加、删除、修改和查询操作，需要加入读写联系人信息的权限。

```
<uses-permissionandroid:name="android.permission.READ_CONTACTS" />
<uses-permissionandroid:name="android.permission.WRITE_CONTACTS" />
```

ContactsContract.Contacts 以静态形式提供了访问 Contacts 表中数据的常量，如列名、Uri 等。下面的代码实现了查询通讯录中联系人的 ID 和姓名，并将结果输出到 LogCat 窗口中。

```
//创建一个数组，将结果 Cursor 限制为所需的列
String[] projection = {
ContactsContract.Contacts._ID,
    ContactsContract.Contacts.DISPLAY_NAME
};
Cursor cursor = getContentResolver().query(
    ContactsContract.Contacts.CONTENT_URI, projection,
    null, null, null);
while(cursor.moveToNext()){
    String id = cursor.getString(0);
    String name = cursor.getString(1);
    Log.i("Read Contact", "id:"+id+",name:"+name);
}
cursor.close();
```

ContactsContract.Data 以静态形式提供了访问 Data 表中数据的常量，存储在表中数据由 MIMETYPE 确定，例如，如果行的数据类型是 Phone.CONTENT_ITEM_TYPE，那么，DATA1 应该保存电话号码，如果数据类型为 Email.CONTENT_ITEM_TYPE，那么 DATA1 应该保存邮件地址。

为了简化操作，ContactsContract 定义了一些数据种类，如 ContactsContract.CommonDataKinds.Phone, ContactsContract.CommonDataKinds.Email。为了操作方便，这些类为 DATA1 定义了新的别名，如 Phone.NUMBER，等同于 Data.DATA1。下面通过一个完整的实例演示如何来读取联系人信息，并把结果通过一个 ExpandableListView 进行显示。

MainActivity 的布局文件如下。

```
<RelativeLayout xmlns:android="http://schemas.android.com/apk/res/android"
    xmlns:tools="http://schemas.android.com/tools"
    android:layout_width="match_parent"
    android:layout_height="match_parent"
```

```xml
        android:paddingBottom="@dimen/activity_vertical_margin"
        android:paddingLeft="@dimen/activity_horizontal_margin"
        android:paddingRight="@dimen/activity_horizontal_margin"
        android:paddingTop="@dimen/activity_vertical_margin"
        tools:context=".MainActivity" >

    <ExpandableListView
        android:id="@+id/expandableListView1"
        android:layout_width="match_parent"
        android:layout_height="wrap_content"
        android:layout_alignParentLeft="true"
        android:layout_alignParentTop="true" >
    </ExpandableListView>

</RelativeLayout>
```

MainActivity 的代码如下。

```java
package cn.wang.readcontactsdemo;
import java.util.ArrayList;
import android.os.Bundle;
import android.provider.ContactsContract;
import android.view.Gravity;
import android.view.View;
import android.view.ViewGroup;
import android.widget.AbsListView;
import android.widget.BaseExpandableListAdapter;
import android.widget.ExpandableListAdapter;
import android.widget.ExpandableListView;
import android.widget.TextView;
import android.app.Activity;
import android.database.Cursor;

public class MainActivity extends Activity {
    private ExpandableListView elv = null;
    @Override
    protected void onCreate(Bundle savedInstanceState) {
        super.onCreate(savedInstanceState);
        setContentView(R.layout.activity_main);

        elv = (ExpandableListView)findViewById(R.id.expandableListView1);
        //定义两个 List 来封装系统的联系人信息、指定联系人的电话号码、E-mail 等详情
        final ArrayList<String> names = new ArrayList<String>();
        final ArrayList<ArrayList<String>> details
            = new ArrayList<ArrayList<String>>();
        //使用 ContentResolver 查找联系人数据
```

```java
Cursor cursor = getContentResolver().query(
    ContactsContract.Contacts.CONTENT_URI, null, null,
    null, null);
//遍历查询结果,获取系统中所有联系人
while (cursor.moveToNext())
{
    //获取联系人 ID
    String contactId = cursor.getString(cursor
        .getColumnIndex(ContactsContract.Contacts._ID));
    //获取联系人的名字
    String name = cursor.getString(cursor.getColumnIndex(
        ContactsContract.Contacts.DISPLAY_NAME));
    names.add(name);
    //使用 ContentResolver 查找联系人的电话号码
    Cursor phones = getContentResolver().query(
        ContactsContract.CommonDataKinds.Phone.CONTENT_URI,
        null,
        ContactsContract.CommonDataKinds.Phone.CONTACT_ID
            + " = " + contactId, null, null);
    ArrayList<String> detail = new ArrayList<String>();
    //遍历查询结果,获取该联系人的多个电话号码
    while (phones.moveToNext())
    {
        //获取查询结果中电话号码列中数据
        String phoneNumber = phones.getString(phones
            .getColumnIndex(ContactsContract
            .CommonDataKinds.Phone.NUMBER));
        detail.add("电话号码:" + phoneNumber);
    }
    phones.close();
    //使用 ContentResolver 查找联系人的 Email 地址
    Cursor emails = getContentResolver().query(
        ContactsContract.CommonDataKinds.Email.CONTENT_URI,
        null,
        ContactsContract.CommonDataKinds.Email.CONTACT_ID
            + " = " + contactId, null, null);
    //遍历查询结果,获取该联系人的多个 Email 地址
    while (emails.moveToNext())
    {
        //获取查询结果中 Email 地址列中数据
        String emailAddress = emails.getString(emails
            .getColumnIndex(ContactsContract
            .CommonDataKinds.Email.DATA));
        detail.add("邮件地址:" + emailAddress);
    }
```

```java
        emails.close();
        details.add(detail);
    }
    cursor.close();

    //创建一个ExpandableListAdapter对象
    ExpandableListAdapter adapter =
        new BaseExpandableListAdapter()
    {
        //获取指定组位置、指定子列表项处的子列表项数据
        @Override
        public Object getChild(int groupPosition,
            int childPosition)
        {
            return details.get(groupPosition).get(
                childPosition);
        }

        @Override
        public long getChildId(int groupPosition,
            int childPosition)
        {
            return childPosition;
        }

        @Override
        public int getChildrenCount(int groupPosition)
        {
            return details.get(groupPosition).size();
        }

        private TextView getTextView()
        {
            AbsListView.LayoutParams lp = new AbsListView
                .LayoutParams(ViewGroup.LayoutParams.MATCH_PARENT
                , 64);
            TextView textView = new TextView(
                MainActivity.this);
            textView.setLayoutParams(lp);
            textView.setGravity(Gravity.CENTER_VERTICAL
                | Gravity.LEFT);
            textView.setPadding(36, 0, 0, 0);
            textView.setTextSize(20);
            return textView;
        }
```

```java
//该方法决定每个子选项的外观
@Override
public View getChildView(int groupPosition,
    int childPosition, boolean isLastChild,
    View convertView, ViewGroup parent)
{
    TextView textView = getTextView();
    textView.setText(getChild(groupPosition,
        childPosition).toString());
    return textView;
}

//获取指定组位置处的组数据
@Override
public Object getGroup(int groupPosition)
{
    return names.get(groupPosition);
}

@Override
public int getGroupCount()
{
    return names.size();
}

@Override
public long getGroupId(int groupPosition)
{
    return groupPosition;
}

//该方法决定每个组选项的外观
@Override
public View getGroupView(int groupPosition,
    boolean isExpanded, View convertView,
    ViewGroup parent)
{
    TextView textView = getTextView();
    textView.setText(getGroup(groupPosition)
        .toString());
    return textView;
}

@Override
public boolean isChildSelectable(int groupPosition,
```

```
            int childPosition)
        {
            return true;
        }

        @Override
        public boolean hasStableIds()
        {
            return true;
        }
    };
    //为ExpandableListView设置Adapter对象
    elv.setAdapter(adapter);
}
```

运行应用程序,获取的联系人如图 11.3 所示,单击联系人可以看到联系人的详细信息,如图 11.4 所示。

图 11.3　获取联系人

图 11.4　联系人详细信息

11.3.2　读取短信

Telephony Provider 提供了电话、短信和彩信相关数据的共享,可以通过该数据提供者访问手机中的短信休息。该数据库保存的路径为/data/data/com.android.providers.telephony/databases/mmssms.db,如图 11.5 所示。读者可以自行查看该数据库的结构,在此不再赘述。

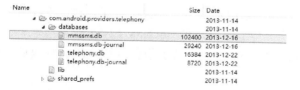

图 11.5　Telephony Provider 的数据库

在 mmssms.db 数据库中，短信存储在 sms 表中，该表的结果如图 11.6 所示。

图 11.6　sms 表结构

其中，_id 表示该短信的 ID。
thread_id 表示该短信所属的会话的 ID，每个会话代表和一个联系人之间短信的群组。
address 表示该短信的发件人地址，手机号码如+8613666666666。
person 表示该短信的发件人，返回一个数字就是联系人列表里的序号，陌生人为 null。
date 表示该短信的接收日期。
date_sent 表示该短信的发送日期。
protocol 协议，0 表示 SMS_RPOTO，1 表示 MMS_PROTO。
read 表示该短信是否已读。
type 表示该短信的类型，例如，1 表示接收类型，2 表示发送类型，3 表示草稿类型。
body 表示短信的内容。
通过查看 API 和源文件（\sources\android-19\android\provider\Telephony.java），主要的 URI 如下。

```
content://sms/                   所有短信
    content://sms/inbox          收件箱
    content://sms/sent           已发送
    content://sms/draft          草稿
    content://sms/outbox         发件箱
    content://sms/failed         发送失败
    content://sms/queued         待发送列表
```

下面通过实例演示如何读取短信信息，并把结果通过一个 TextView 进行显示。
（1）创建一个 Android 项目：SmsReadDemo。
（2）修改 MainActivity.java 文件，增加短信读取代码，修改后文件如下。

```
package cn.wang.smsreaddemo;
```

```java
import java.text.SimpleDateFormat;
import java.util.Date;
import java.util.Locale;

import android.net.Uri;
import android.os.Bundle;
import android.app.Activity;
import android.database.Cursor;
import android.database.sqlite.SQLiteException;
import android.util.Log;
import android.view.Menu;
import android.widget.ScrollView;
import android.widget.TextView;

public class MainActivity extends Activity {

    @Override
    protected void onCreate(Bundle savedInstanceState) {
        super.onCreate(savedInstanceState);
        //setContentView(R.layout.activity_main);
        TextView tv = new TextView(this);
        tv.setText(getSmsInPhone());

        ScrollView sv = new ScrollView(this);
        sv.addView(tv);

        setContentView(sv);
    }

    @Override
    public boolean onCreateOptionsMenu(Menu menu) {
        //Inflate the menu; this adds items to the action bar if it is present.
        getMenuInflater().inflate(R.menu.main, menu);
        return true;
    }

    public String getSmsInPhone() {
       final String SMS_URI_ALL = "content://sms/";

       StringBuilder smsBuilder = new StringBuilder();

       try {
           Uri uri = Uri.parse(SMS_URI_ALL);
           String[] projection = new String[] { "_id", "address", "person",
           "body", "date", "type" };
```

```java
Cursor cur = getContentResolver().query(uri, projection, null,
null, "date desc");        //获取手机内部短信

if (cur.moveToFirst()) {
    int index_Address = cur.getColumnIndex("address");
    int index_Person = cur.getColumnIndex("person");
    int index_Body = cur.getColumnIndex("body");
    int index_Date = cur.getColumnIndex("date");
    int index_Type = cur.getColumnIndex("type");

    do {
        String strAddress = cur.getString(index_Address);
        int intPerson = cur.getInt(index_Person);
        String strbody = cur.getString(index_Body);
        long longDate = cur.getLong(index_Date);
        int intType = cur.getInt(index_Type);

        SimpleDateFormat dateFormat = new SimpleDateFormat("yyyy-
MM-dd hh:mm:ss",Locale.US);
        Date d = new Date(longDate);
        String strDate = dateFormat.format(d);

        String strType = "";
        if (intType == 1) {
            strType = "接收";
        } else if (intType == 2) {
            strType = "发送";
        } else {
            strType = "null";
        }

        smsBuilder.append(strAddress + ", ");
        smsBuilder.append(intPerson + ", ");
        smsBuilder.append(strbody + ", ");
        smsBuilder.append(strDate + ", ");
        smsBuilder.append(strType);
        smsBuilder.append("\n");
    } while (cur.moveToNext());

    if (!cur.isClosed()) {
        cur.close();
        cur = null;
    }
} else {
    smsBuilder.append("没有短信!");
```

```
        } //end if

        smsBuilder.append("-------End!-------");

    } catch (SQLiteException ex) {
        Log.d("SQLiteException in getSmsInPhone", ex.getMessage());
    }

    return smsBuilder.toString();
    }
}
```

（3）修改 AndroidManifest.xml 文件，增加 android.permission.READ_SMS 权限，否则将报错，代码如下。

```
<uses-permission android:name="android.permission.READ_SMS"/>
```

运行应用程序，获取的短信如图 11.7 所示。

图 11.7 读取短信

11.4 本 章 小 结

本章主要介绍了 Android 系统中 ContentProvider 组件的功能和用法，ContentProvider 的本质就像是一个"网站"，它可以把应用程序的数据按照"固定规范"暴露出来，其他应用程序就可以通过 ContentProvider 暴露出来的接口操作内部的数据了。本章需要重点掌握三个 API 的用法：ContentResolver、ContentProvider 和 ContentObserver，其中 ContentProvider 是所有 ContentProvider 组件的基类，ContentResolver 用于操作 ContentProvider 提供的数据，而 ContentObserver 用于监听 ContentProvider 的数据改变。

第 12 章　Android 多媒体应用开发

本章主要介绍 Android 平台的多媒体应用开发方面的基础知识。多媒体主要包括音频和视频，为方便起见，本章分开介绍音频和视频的录制与播放相关的功能，此外，本章还将探索在线音频和视频的播放等高级功能。

12.1　音　频　录　制

本节主要介绍录制音频的 3 种主要方法，每种方法都有它们各自的使用范围，第一种方法，使用 Intent 调用系统功能进行录制，该方法简单方便，但灵活性不足；第二种方法，使用 MediaRecorder 录制音频，该方法使用难度适中，并且灵活性较强；第三种方法，使用 AudioRecord 类录制音频，该方法难于掌握，灵活性很强，但功能十分强大。

12.1.1　使用 Intent 录制音频

录制音频最简单的方法是使用 Intent 调用系统已有的录制应用程序提供的录制功能。Android 平台默认提供一个录音机应用程序，当然也可以使用第三方提供的其他录音应用程序。使用 Intent 录制音频的基础代码如下所示。

```
Intent audio_recording_intent =
        new Intent(MediaStore.Audio.Media.RECORD_SOUND_ACTION);
startActivity(audio_recording_intent);
```

其中，MediaStore.Audio.Media.RECORD_SOUND_ACTION 常量表示启动录音机应用程序的动作意图。

调用 startActivity(audio_recording_intent)可以启动录音机应用程序进行音频录制，但是控制权已经交给了录音机应用程序，在录制完音频后，音频数据无法返回给调用程序，下面给出一个较完整的示例，提供使用 Intent 录制音频的功能。

```
import android.app.Activity;
import android.content.Intent;
import android.media.MediaPlayer;
import android.media.MediaPlayer.OnCompletionListener;
import android.net.Uri;
import android.os.Bundle;
import android.provider.MediaStore;
import android.view.View;
```

```java
import android.view.View.OnClickListener;
import android.widget.Button;

public class IntentAudioRecorder extends Activity implements OnClickListener,
        OnCompletionListener {

    public static int RECORD_REQUEST = 0;

    Button createRecording, playRecording;

    Uri audioFileUri;

    @Override
    public void onCreate(Bundle savedInstanceState) {
        super.onCreate(savedInstanceState);
        setContentView(R.layout.main);

        createRecording = (Button) this.findViewById(R.id.RecordButton);
        createRecording.setOnClickListener(this);

        playRecording = (Button) this.findViewById(R.id.PlayButton);
        playRecording.setOnClickListener(this);
        playRecording.setEnabled(false);
    }

    public void onClick(View v) {
        if (v == createRecording) {
            Intent intent = new Intent(
                    MediaStore.Audio.Media.RECORD_SOUND_ACTION);
            startActivityForResult(intent, RECORD_REQUEST);
        } else if (v == playRecording) {

            MediaPlayer mediaPlayer = MediaPlayer.create(this, audioFileUri);
            mediaPlayer.setOnCompletionListener(this);
            mediaPlayer.start();
            playRecording.setEnabled(false);
        }
    }

    protected void onActivityResult(int requestCode, int resultCode, Intent data) {
        if (resultCode == RESULT_OK && requestCode == RECORD_REQUEST) {
            audioFileUri = data.getData();
            playRecording.setEnabled(true);
        }
```

    }

    public void onCompletion(MediaPlayer mp) {
        playRecording.setEnabled(true);
    }
}

12.1.2　使用 MediaRecorder 录制音频

如何使用 MediaRecorder 类来录制音频。MediaRecorder 类是 Android SDK 提供的录制音频和视频的功能类，使用它可以建立功能更加完善的音频录制应用，比如，可以控制录制的时长等。MediaRecorder 类内部维护一个状态机来管理录制音频和视频中的各种状态，该状态机如图 12.1 所示，它描述了音频和视频录制过程中的各种状态和每个状态下可以调用的方法。

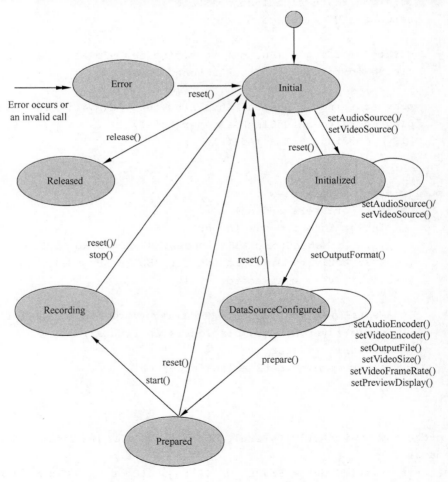

图 12.1　MediaRecorder 的状态机图

MediaRecorder 类提供音频和视频录制的功能，这里只介绍它的音频录制相关的 API，视频录制相关的 API 将在 12.3 中进行介绍。MediaRecorder 类中录制音频的主要 API 如表 12.1 所示，使用这几个主要的 API 就可以构建出功能较完善的音频录制应用。

表 12.1　音频录制的主要 API

方法名	描述
MediaRecorder.setAudioSource()	设置录制的音频源
MediaRecorder.setOutputFormat()	设置输出文件的格式
MediaRecorder.setOutputFile()	设置音频数据的存储文件
MediaRecorder.setAudioEncoder()	设置音频数据的编码器
MediaRecorder.prepare()	开始录制前准备 MediaRecorder 对象
MediaRecorder.start()	开始录制并存储音频数据到指定的文件
MediaRecorder.stop()	停止录制
MediaRecorder.release()	释放 MediaRecorder 对象占用的资源

录制音频的过程大致遵循以下 9 个步骤。

（1）创建 android.media.MediaRecorder 对象实例，以后所有的工作都是围绕该对象展开。

```
MediaRecorder mRecorder = new MediaRecorder();
```

（2）设置音频录制时采用的音频源（audio source）。这里，通过调用前面介绍的 android.media.MediaRecorder 对象的 MediaRecorder.setAudioSource() 方法完成。MediaRecorder.AudioSource 是 MediaRecorder 的内部类，主要定义智能手机常用的音频源资源，MediaRecorder.AudioSource.MIC 是最常用的音频源，代表手机的麦克风外设。除了 MediaRecorder.AudioSource.MIC 外，还提供 VOICE_CALL，VOICE_DOWNLINK，VOICE_UPLINK 音频源，可以通过这些音频源进行语音通话的录制。

```
mRecorder.setAudioSource(MediaRecorder.AudioSource.MIC);
```

（3）设置输出音频数据的存储文件格式。这里，通过调用前面介绍的 MediaRecorder.setOutputFormat() 方法完成。Android 平台支持的音频文件格式由 MediaRecorder.OutputFormat 内部类定义。以下为 Android 支持的主要文件格式。

- MediaRecorder.OutputFormat.AMR_NB：该常量表示输出文件是 AMR-NB 格式的语音文件，主要对人声进行编码。
- MediaRecorder.OutputFormat.MPEG_4：该常量表示输出文件是 MPEG-4 格式的多媒体文件，其中，可能同时包含音频和视频信息。
- MediaRecorder.OutputFormat.THREE_GPP：该常量表示输出文件是 3GPP 格式的文件，其中，可能同时包含音频和视频信息。

```
mRecorder.setOutputFormat(MediaRecorder.OutputFormat.AMR_NB);
```

（4）指定输出音频数据存放的文件。这里，通过调用前面介绍的 MediaRecorder.setOutputFile()方法完成，该方法可以接受两种参数：文件描述符（FileDescriptor）和文件路径字符串。

```
FileDescriptor fd = ...;
mRecorder.setOutputFile(fd);
```

或者

```
String mFileName = ...;
mRecorder.setOutputFile(mFileName);
```

(5) 设置音频编码器。这里,通过调用前面介绍的 MediaRecorder.setAudioEncoder() 方法完成。Android 平台支持的音频编码器由 MediaRecorder.AudioEncoder 内部类定义。最常用的音频编码器是 MediaRecorder.AudioEncoder.AMR_NB。AMR_NB 是自适应多速率窄带音频编解码器,该编解码器主要针对语音数据进行优化,使其不适应语音之外的其他音频信息。

```
mRecorder.setAudioEncoder(MediaRecorder.AudioEncoder.AMR_NB);
```

(6) 调用 MediaRecorder.prepare()方法,准备工作就绪,可以开始录制音频。

```
mRecorder.prepare();
```

(7) 调用 MediaRecorder.start()方法,开始录制。

```
mRecorder.start();
```

(8) 录制完成后,要停止录制,需要调用 MediaRecorder.stop()方法。

```
mRecorder.stop();
```

(9) 需要调用 MediaRecorder.release()方法,释放所占用的资源。到此整个录制过程就完成了。

下面给出一个完整的录制音频的程序示例,详细介绍设置、启动、停止录制音频的过程。

```java
import android.app.Activity;
import android.widget.LinearLayout;
import android.os.Bundle;
import android.os.Environment;
import android.view.ViewGroup;
import android.widget.Button;
import android.view.View;
import android.view.View.OnClickListener;
import android.content.Context;
import android.util.Log;
import android.media.MediaRecorder;
import android.media.MediaPlayer;
import java.io.IOException;
public class AudioRecordTest extends Activity
{
    private static final String LOG_TAG = "AudioRecordTest";
    private static String mFileName = null;
    private RecordButton mRecordButton = null;
    private MediaRecorder mRecorder = null;
```

```java
    private PlayButton   mPlayButton = null;
    private MediaPlayer  mPlayer = null;

    private void onRecord(boolean start) {
        if (start) {
            startRecording();
        } else {
            stopRecording();
        }
    }
    private void onPlay(boolean start) {
        if (start) {
            startPlaying();
        } else {
            stopPlaying();
        }
    }

    private void startPlaying() {
        mPlayer = new MediaPlayer();
        try {
            mPlayer.setDataSource(mFileName);
            mPlayer.prepare();
            mPlayer.start();
        } catch (IOException e) {
            Log.e(LOG_TAG, "prepare() failed");
        }
    }

    private void stopPlaying() {
        mPlayer.release();
        mPlayer = null;
    }
    private void startRecording() {
        mRecorder = new MediaRecorder();
        mRecorder.setAudioSource(MediaRecorder.AudioSource.MIC);
        mRecorder.setOutputFormat(MediaRecorder.OutputFormat.THREE_GPP);
        mRecorder.setOutputFile(mFileName);
        mRecorder.setAudioEncoder(MediaRecorder.AudioEncoder.AMR_NB);
        try {
            mRecorder.prepare();
        } catch (IOException e) {
            Log.e(LOG_TAG, "prepare() failed");
        }
        mRecorder.start();
```

```java
    }
    private void stopRecording() {
        mRecorder.stop();
        mRecorder.release();
        mRecorder = null;
    }

    class RecordButton extends Button {
        boolean mStartRecording = true;
        OnClickListener clicker = new OnClickListener() {
            public void onClick(View v) {
                onRecord(mStartRecording);
                if (mStartRecording) {
                    setText("Stop recording");
                } else {
                    setText("Start recording");
                }
                mStartRecording = !mStartRecording;
            }
        };

        public RecordButton(Context ctx) {
            super(ctx);
            setText("Start recording");
            setOnClickListener(clicker);
        }
    }
    class PlayButton extends Button {
        boolean mStartPlaying = true;
        OnClickListener clicker = new OnClickListener() {
            public void onClick(View v) {
                onPlay(mStartPlaying);
                if (mStartPlaying) {
                    setText("Stop playing");
                } else {
                    setText("Start playing");
                }
                mStartPlaying = !mStartPlaying;
            }
        };
        public PlayButton(Context ctx) {
            super(ctx);
            setText("Start playing");
            setOnClickListener(clicker);
        }
```

```java
    }

    public AudioRecordTest() {
        mFileName = Environment.getExternalStorageDirectory().
            getAbsolutePath();
            mFileName += "/audiorecordtest.3gp";
    }
    @Override
    public void onCreate(Bundle icicle) {
      super.onCreate(icicle);
      LinearLayout ll = new LinearLayout(this);
      mRecordButton = new RecordButton(this);
       ll.addView(mRecordButton,
          new LinearLayout.LayoutParams(
             ViewGroup.LayoutParams.WRAP_CONTENT,
                ViewGroup.LayoutParams.WRAP_CONTENT, 0));
      mPlayButton = new PlayButton(this);
      ll.addView(mPlayButton, new LinearLayout.LayoutParams(
             ViewGroup.LayoutParams.WRAP_CONTENT,
             ViewGroup.LayoutParams.WRAP_CONTENT, 0));
      setContentView(ll);
      }
      @Override
      public void onPause() {
         super.onPause();
            if (mRecorder != null) {
                mRecorder.release();
                mRecorder = null;
         }
            if (mPlayer != null) {
                mPlayer.release();
                mPlayer = null;
         }
      }
}
```

12.2 音 频 播 放

Android 平台提供强大的媒体播放功能，它支持相当广泛的音频和视频格式。本节将首先介绍 Android 平台支持的常见音频格式，然后介绍播放音频的两种主要方法。同音频录制类似，第一种方法，使用 Intent 调用系统功能进行录制，该方法简单方便但灵活性不足；第二种方法，使用 MediaPlayer 播放音频，该方法使用难度适中并且灵活性较强。

12.2.1 常见的音频格式

Android 支持多种音频格式和编解码器，下面介绍几种常见的音频格式。

（1）AMR：自适应多速率编解码器，包括 AMR 窄带 AMR-NB 和 AMR 宽带 AMR-WB，文件扩展名是.3gp(audio/3gpp)或.amr(audio/amr)。AMR 是 3GPP 使用的基本音频编解码标准。AMR 主要应用于手机上的语音通话应用，并得到手机厂商的广泛支持，该编码标准适应于简单的语音编码，不适用处理更复杂的音频数据，如音乐等。

（2）AAC：全称是 Advanced Audio Coding。一种专为声音数据设计的文件压缩格式，与 MP3 不同，它采用了全新的算法进行编码，更加高效，具有更高的"性价比"。利用 AAC 格式，可使人感觉声音质量没有明显降低的前提下，更加小巧。Android 除了支持 AAC 外，还支持新添加到 AAC 规范中的高效 AAC（High Efficiency AAC）格式。

（3）MP3：是一种音频压缩技术，其全称是动态影像专家压缩标准音频层面 3（Moving Picture Experts Group Audio Layer III），简称为 MP3。它被设计用来大幅度地降低音频数据量。利用 MPEG Audio Layer 3 的技术，将音乐以 1:10 甚至 1:12 的压缩率，压缩成容量较小的文件，而对于大多数用户来说，重放的音质与最初的不压缩音频相比没有明显下降。它是在 1991 年由位于德国埃尔朗根的研究组织 Fraunhofer-Gesellschaft 的一组工程师发明和标准化的。用 MP3 形式存储的音乐叫作 MP3 音乐，能播放 MP3 音乐的机器叫作 MP3 播放器。MP3 是目前互联网上使用最广泛的音频编解码器之一。

（4）Ogg：全称是 OGG Vorbis，是一种新的音频压缩格式，类似于 MP3 的音乐格式。但有一点不同的是，它是完全免费、开放和没有专利限制的。OGG Vorbis 有一个特点是支持多声道。Vorbis 是这种音频压缩机制的名字，而 Ogg 则是一个计划的名字，该计划意图设计一个完全开放性的多媒体系统。Ogg Vorbis 文件的扩展名是.og，这种文件的设计格式是非常先进的。创建的 Ogg 文件可以在未来的任何播放器上播放，因此，这种文件格式可以不断地进行大小和音质的改良，而不影响旧有的编码器或播放器。

12.2.2 使用 Intent 播放音频

播放音频最简单的方法是使用 Intent 调用系统已有的音乐播放应用程序提供的播放功能。Android 平台默认提供一个音乐播放应用程序，当然也可以使用第三方提供的其他音乐播放应用程序。使用 Intent 播放音频的基础代码如下所示。

```
Intent audio_playing_intent =
new Intent(android.content.Intent.ACTION_VIEW);
    audio_playing_intent.setDataAndType(audioFileURI, "audio/mpeg");
startActivity(audio_playing_intent);
```

其中，android.content.Intent.ACTION_VIEW 常量表示启动音乐播放应用程序的动作意图，然后设置 intent 对象的参数：播放音乐的 URI 和 MIME 类型。

调用 startActivity(audio_playing_intent)可以启动音乐播放应用程序进行音频播放，但是控制权已经交给了音乐播放应用程序。

MIME 代表多用途 Internet 邮件扩展（Multipurpose Internet Mail Extension）。最初被用于帮助电子邮件客户端发送和接受附件。现在它的使用范围已经超出电子邮件，扩展到了许多其他通信协议，包括 HTTP 或 Web 服务。当解析一个 Intent 时，Android 使用 MIME 类型来帮助确定应该处理该 Intent 的应用程序。每种文件类型都有一个特定的 MIME 类型。使用至少两部分（由斜杠分隔开）来指定类型。第一部分是更一般的类型，如 "audio"，第二部分是更具体的类型，如 "mpeg"。一般类型 "audio" 和具体类型 "mpeg" 构成一个 MIME 类型 "audio/mpeg"，这通常用于 MP3 文件的 MIME 类型。

下面给出一个较完整的示例，提供音频播放的功能。

该程序由一个 Activity 组成，并实现 OnClickListener 接口，以监听用户是否按下播放按钮。

```
public class AudioPlayerbyIntent extends Activity implements OnClickListener{
    Button playButton;
    @Override
    public void onCreate(Bundle savedInstanceState){
        super.onCreate(savedInstanceState);
        setContentView(R.layout.main);
```

设置内容视图后，获得按钮对象的引用，并设置按钮的 OnClickListener 为该 Activity 对象。

```
playButton = (Button) this.findViewById(R.id.Button01);
playButton.setOnClickListener(this);
```

当用户单击播放按钮时，系统调用 onClick 方法，该方法提供使用 Intent 播放音频的功能。首先使用上面介绍的 android.content.Intent.ACTION_VIEW 创建 Intent 对象，然后创建 File 对象，引用 SD 卡上的音频文件。

```
public void onClick(View v){
    Intent intent = new Intent(android.content.Intent.ACTION_VIEW);

    File sdcard = Environment.getExternalStorageDirectory();
    File audioFile = new File(sdcard.getPath()
                            + "/Music/goodmorningandroid.mp3");
                            }
```

设置 Intent 的两个参数：音频文件对象的 URI 和它的 MIME 类型（audio/mpeg）。最后，通过调用 startActivity(intent)来播放音频。

```
intent.setDataAndType(Uri.fromFile(audioFile), "audio/mp3");
startActivity(intent);
```

12.2.3　使用 MediaPlayer 播放音频

在介绍了如何使用 Intent 播放音频后，接下来介绍如何使用 MediaPlayer 类来播放音频。MediaPlayer 类是 Android SDK 提供的播放音频和视频的功能类，这里仅使用其音频播放功能，使用它可以建立功能更加完善的音频播放应用。

MediaPlayer 播放音频的最简单情况是播放与应用程序本身一起打包的音频文件。音频文件放置在应用程序的原始资源中。具体操作是在项目的 res 文件夹中创建一个新文件夹，命名为 raw，把音频文件放置入该 raw 文件夹中，ADT 将自动更新 R.java 文件（位于 gen 文件夹中），为该音频文件生成资源 ID，使用 R.raw.file_name_without_extension 语法访问该音频文件。

播放与应用程序一起打包的音频文件非常简单。使用 MediaPlayer 类的静态方法 create 实例化一个 MediaPlayer 对象，传入上下文 this 及音频文件的资源 ID。

```
MediaPlayer mediaPlayer =
    MediaPlayer.create(this, R.raw.audio_file_name_without_extension);
```

由于调用 MediaPlayer 的静态方法 create 创建 MediaPlayer 对象成功后，系统自动调用 prepare()方法，不再需要手动调用，MediaPlayer 对象已经处于 Prepared 状态，因此，只需调用 MediaPlayer 对象的 start()方法即可播放该音频文件。

```
mediaPlayer.start();
```

MediaPlayer 类内部维护一个状态机来管理播放音频和视频中的各种状态，该状态机如图 12.2 所示（摘自 Android API 参考手册），它描述了音频和视频播放过程中的各种状态和每个状态下可以调用的方法。

这里需要着重强调的是 MediaPlayer 是基于状态的。当写代码时，必须始终注意 MediaPlayer 所处的状态，因为 MediaPlayer 的方法都有其可以正常执行的有效状态。如果在一个错误的状态执行一个方法时，系统会抛出异常或产生不可预料的行为。

上面所述的 MediaPlayer 的状态机图明确指出哪些方法可以把 MediaPlayer 从一个状态迁移到另一个状态。例如，当新建一个 MediaPlayer 对象时（使用 new 操作），它处于 Idle 状态。在 Idle 状态时，通过调用 setDataSource()方法初始化 MediaPlayer 对象，迁移到 Initialized 状态。之后，使用 prepare()或 prepareAsync()方法准备 MediaPlayer 对象。当 MediaPlayer 对象准备就绪时，它进入 Prepared 状态，这时可以调用 start()方法来播放媒体文件。此时，MediaPlayer 对象处于 Started 状态，正如上图所示，可以通过调用 start()、pause() 和 seekTo()方法，在 Started、Paused 和 PlaybackCompleted 三个状态之间进行迁移。当调用 stop()方法停止播放后，就不能再调用 start()方法播放媒体文件了，除非再次调用 prepare() 方法准备 MediaPlayer 对象。

再强调一次，编写媒体播放代码时，一定注意 MediaPlayer 对象的状态，因为在错误的状态调用方法会造成许多 bugs。

使用 MediaPlayer 播放时，需要注意不要在应用程序的 UI 主线程中准备 MediaPlayer。调用 prepare()方法通常会花费一定时间，因为它需要获取并解码媒体数据。因此，只要是任何需要花费一定时间执行的方法，都不要在应用程序的主 UI 线程中调用。否则将会挂起 UI 主线程直到该方法执行完毕，这是一个非常差的用户体验，会造成应用程序不响应（Application Not Responding）错误。即使是很短的媒体数据，加载可能很快，但是记住任何花费超过 100 毫秒的操作都会造成 UI 界面发生明显停顿或应用程序响应很慢。

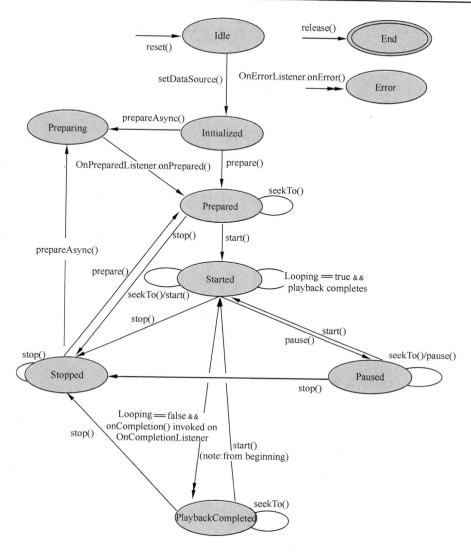

图 12.2　MediaPlayer 的状态机图

为了避免挂起 UI 主线程,创建一个新的线程准备 MediaPlayer,当准备就绪时,通知主 UI 线程。其实,我们不用自己编写新的线程逻辑,MediaPlayer 本身提供了一个非常方便的方法可以完成该任务,即 prepareAsync()方法。该方法在后台准备媒体数据,准备就绪后立刻返回。当媒体数据准备好后,MediaPlayer.OnPreparedListener 接口的 onPrepared()回调方法会自动被调用,以便继续后续处理。我们可以通过 MediaPlayer 的 setOnPreparedListener()方法注册监听器。

MediaPlayer 可能会耗尽系统资源。因此,我们应该采取措施避免 MediaPlayer 一直占用系统资源。当使用 MediaPlayer 播放完毕时,应该总是调用 release()方法,确保任何分配给它的系统资源被合适地释放。例如,如果应用的 Activity 收到 onStop()回调方法的调用,我们必须在其中释放 MediaPlayer,因为当该 Activity 不和用户进行交互时,MediaPlayer 还占用系统资源已经没有意义,除非正在后台播放。当 Activity 被继续或重启动时,需要创建一个新的 MediaPlayer，并再次准备它，之后才能继续播放。

```
mediaPlayer.release();
mediaPlayer = null;
```

下面代码实现自定义 MediaPlayer 进行音频资源的播放。

```
package com.apress.proandroidmedia.ch5.customaudio;

import android.app.Activity;
import android.media.MediaPlayer;
import android.media.MediaPlayer.OnCompletionListener;
import android.os.Bundle;

public class CustomAudioPlayer extends Activity implements OnCompletionListener {
    MediaPlayer mediaPlayer;

    @Override
    public void onCreate(Bundle savedInstanceState) {
        super.onCreate(savedInstanceState);
        setContentView(R.layout.main);
    }

    public void onCompletion(MediaPlayer mp) {
        mediaPlayer.start();
    }

    public void onStart() {
        super.onStart();
        mediaPlayer = MediaPlayer.create(this, R.raw.goodmorningandroid);
        mediaPlayer.setOnCompletionListener(this);
        mediaPlayer.start();
    }

    public void onStop() {
        super.onStop();
        //Log.v("CustomAudioPlayer","onStop Called");
        mediaPlayer.stop();
        mediaPlayer.release();
    }
}
```

12.3 视频录制

本节主要介绍录制视频的两种主要方法。每种方法都有它们各自的使用范围，第一种方法，使用 Intent 调用系统摄像头进行录制，该方法简单方便但灵活性不足；第二种方法，使用 MediaRecorder 录制视频，该方法使用难度适中并且灵活性较强。

12.3.1 使用 Intent 录制视频

正如音频录制一样，Android 平台上录制视频的最简单方式是通过 Intent 调用系统已有的摄像应用程序。Android 平台默认提供一个 Camera 应用程序，当然也可以使用第三方提供的其他视频录制应用程序。使用 Intent 录制视频的基础代码如下所示。

```
Intent video_recording_intent =
        new Intent(MediaStore.ACTION_VIDEO_CAPTURE);
startActivity(video_recording_intent);
```

其中，MediaStore.ACTION_VIDEO_CAPTURE 常量表示启动视频录制应用程序的动作意图。调用 startActivity(video_recording_intent)可以启动视频录制应用程序进行视频录制，但是控制权已经交给了视频录制应用程序，在录制完视频后，视频数据无法返回给调用程序。

调用 startActivityForResult(video_recording_intent)可以把录制的视频数据返回给调用程序。用户 Android 手机上可能有多个视频应用程序将该字符串常量注册为一个 Intent 筛选器，这时系统会弹出提示，以供用户选择使用哪个应用程序来执行该项操作。

```
Intent video_recording_intent =
        new Intent(MediaStore.ACTION_VIDEO_CAPTURE);
startActivityForResult (video_recording_intent, VIDEO_CAPTURED);
```

其中，VIDEO_CAPTURED 是定义的常量，当 Camera 应用程序调用 onActivityResult 方法将结果返回给 Activity 时，可以使用该常量进行确认。

onActivityResult 方法返回给 Activity 的 Intent 中包含录制的视频文件的 URI，该文件由 Camera 应用程序创建。

```
protected void onActivityResult(int requestCode, int resultCode, Intent video_data){
    if (requestCode == VIDEO_CAPTURED && resultCode == RESULT_OK) {
        Uri videoFileUri = video_data.getData();
    }
}
```

下面给出一个较完整的示例，提供使用 Intent 录制视频的功能。

```
import android.app.Activity;
import android.content.Intent;
import android.net.Uri;
import android.os.Bundle;
import android.view.View;
import android.view.View.OnClickListener;
import android.widget.Button;
import android.widget.VideoView;
```

```java
public class VideoCaptureIntent extends Activity implements OnClickListener {
    //创建 Activity 的 VIDEO_CAPTURED 常量,将在调用 onActivityResult 时返回
    public static int VIDEO_CAPTURED = 1;

    /*本 Activity 的两个按钮,一个用户触发 Intent,即 recordVideoBtn,另一个用于
    *播放录制的视频,即 playVideoBtn
    */
    Button recordVideoBtn;
    Button playVideoBtn;
    //使用 VideoView 播放视频
    VideoView videoView;
    Uri videoFileUri;

    @Override
    public void onCreate(Bundle savedInstanceState) {
        super.onCreate(savedInstanceState);
        setContentView(R.layout.main);

        recordVideoBtn = (Button) this
                .findViewById(R.id.RecordVideoBtn);
        playVideoBtn = (Button) this.findViewById(R.id.PlayVideoBtn);

        recordVideoBtn.setOnClickListener(this);
        playVideoBtn.setOnClickListener(this);

        //初始化时设置 playVideoBtn 为不可用状态,因为这时还没有录制好视频
        playVideoBtn.setEnabled(false);

        videoView = (VideoView) this.findViewById(R.id.VideoView);
    }

    public void onClick(View v) {
        if (v == recordVideoBtn) {
            //按下录制按钮,开始录制视频
            Intent captureVideoIntent = new Intent(
                    android.provider.MediaStore.ACTION_VIDEO_CAPTURE);
            startActivityForResult(captureVideoIntent, VIDEO_CAPTURED);
        } else if (v == playVideoBtn) {
            //按下播放按钮,播放录制的视频
            videoView.setVideoURI(videoFileUri);
            videoView.start();
        }
    }
}
```

```
/*当 Camera 应用程序返回时，将调用 onActivityResult()回调方法。首先检查
*requestCode 是否为传入 startActivityForResult 的值 VIDEO_CAPTURED 和
*resultCode 是否为常量 RESULT_OK，然后获取录制的视频文件的 URI，接着启*用
playVideoBtn 按钮，从而用户可以单击它来播放视频。
*/
protected void onActivityResult(int requestCode, int resultCode, Intent data) {
    if (resultCode == RESULT_OK && requestCode == VIDEO_CAPTURED) {

        videoFileUri = data.getData();
        playVideoButton.setEnabled(true);
    }
}
```

12.3.2 使用 MediaRecorder 录制视频

在介绍了使用 Intent 录制视频后，接下来介绍如何使用 MediaRecorder 类来录制视频。MediaRecorder 类是 Android SDK 提供的录制音频和视频的功能类，使用它可以建立功能更加完善的视频录制应用。MediaRecorder 类内部维护一个状态机来管理录制音频和视频中的各种状态，该状态机如图 12.1 所示，它描述了音频和视频录制过程中的各种状态和每个状态下可以调用的方法。

为了将 MediaRecorder 用于视频录制，必须采用与音频录制相似的步骤，同时加上与视频相关的特殊步骤。

首先需要创建 MediaRecorder 对象，然后依次进行后续的操作。

```
MediaRecorder video_recorder = new MediaRecorder();
```

(1) 设置音频和视频源

创建 MediaRecorder 对象后，需要设置音频和视频源。可以使用 setAudioSource()方法设置音频源，传入一个想要使用的音频源常量，设置方法已在 12.1 节音频录制中介绍过，此处不再重复叙述。为了设置视频源，可以使用 setVideoSource()方法。可能的视频源的值定义在 MediaRecorder.VideoSource 类的常量，其中只包含两个常量：CAMERA 和 DEFAULT。其实这两个常量表示含义相同，都是指设备上的主摄像头。

```
video_recorder.setVideoSource(MediaRecorder.VideoSource.DEFAULT);
```

(2) 输出格式

设置音频和视频源之后，可以使用 MediaRecorder 的 setOutputFormat()方法设置输出格式，传入要使用的格式。

```
video_recorder.setOutputFormat(MediaRecorder.OutputFormat.DEFAULT);
```

可能的格式定义在 MediaRecorder.OutputFormat 中的常量。
- DEFAULT：使用默认的输出格式。默认的输出格式根据设备的不同而不同。

- **MPEG_4**：指定音频和视频被录制在一个 MPEG-4 格式的文件中，扩展名是.mp4。MPEG-4 文件通常包含 H.264、H.263 或 MPEG-4 Part 2 编码的视频，以及 AAC 或 MP3 编码的音频。MPEG-4 文件被广泛用于许多其他在线视频技术或消费电子设备上。
- **THREE_GPP**：指定音频和视频将被录制到一个 3GP 格式的文件中，扩展名是.3gp。3GPP 文件通常包含使用 H.264、H.263 或 MPEG-4 Part 2 编码的视频和使用 AMR 或 AAC 编码的音频。

(3) 设置音频和视频编解码器

设置输出格式后，需要指定想要使用的音频和视频编解码器。可以使用 MediaRecorder 的 setVideoEncoder()方法设置视频编解码器。

```
video.setVideoEncoder(MediaRecorder.VideoEncoder.DEFAULT);
```

可以使用的编解码器定义在 MediaRecorder.VideoEncoder 中的常量。

- **DEFAULT**：使用默认的视频编解码器。多数情况下是 H.263，Android 设备上必须唯一支持的编解码器。
- **H263**：指定 H.263 为视频编解码器。H.263 是在 1995 年发布的编解码器，专门为低比特率视频传输而开发。它是许多早期 Internet 视频技术的基础，如 Flash 和 RealPlayer 早期使用的技术。在 Android 平台是必须支持的编码格式，因此，可以可靠地使用。
- **H264**：指定 H.264 为视频编解码器。H.264 是当前最先进的编解码器，广泛应用于各种技术，从 BlueRay 到 Flash。
- **MPEG_4_SP**：指定视频编解码器为 MPEG-4 SP。MPEG-4 SP 是 MPEG-4 Part 2 Simple Profile。它发布于 1999 年，为需要低比特率视频且不需要大处理器能力的技术而开发。

对于音频部分，可以使用 setAudioEncoder()方法设置音频编解码器，设置方法已在 12.1 节音频录制中介绍过，此处不再重复叙述。

(4) 设置音频和视频比特率

使用 MediaRecorder 的 setVideoEncodingBitRate()方法设置视频编码比特率。视频的低比特率设置在 256000 位/秒（256kbps）范围之内，而高比特率在 3000000 位/秒（3mbps）范围之内。

```
video_recorder.setVideoEncodingBitRate(150000);
```

使用 MediaRecorder 的 setAudioEncodingBitRate()方法设置音频编码比特率。8000 位/秒是一个非常低的比特率，适合于在慢速网络上实时传输的音频，而 196000 位/秒在 MP3 文件中很常见。

```
video_recorder.setAudioEncodingBitRate (8000);
```

(5) 设置音频采样率

和编码比特率一样，音频采样率对于音频的质量也非常重要。可以使用 MediaRecorder 的 setAudioSamplingRate()方法设置音频采样率。采样率以 Hz 为单位，表示每秒采样的数量。采样率越高，则在录制音频文件中可以表示的音频频率的范围越大。一个低端的采样率 8000Hz 适合于录制低质量的音频,而高端的采样率 48000Hz 可用于 DVD 和许多其他高质量的视频格式。

```
video_recorder. setAudioSamplingRate (8000);
```

（6）设置音频通道

可以使用 setAudioChannels() 方法指定将要录制的音频通道的数量。目前,音频大都限制为大多数 Android 设备上的单一通道麦克风,因此,使用一个以上的通道不会有益处。对于通道数量,一般是单声道为一个通道,而立体声为两个通道。

```
video_recorder. setAudioChannels (1);
```

（7）设置视频帧速率

可以使用 setVideoFrameRate() 方法来控制每秒录制的视频帧数。每秒 12～15 帧之间的值通常足以表示运动。具体使用的实际帧率取决于设备的能力。

```
video_recorder. setVideoFrameRate(15);
```

（8）设置视频大小

可以通过 setVideoSize() 方法并传入宽高值来控制录制的视频的宽度和高度。标准大小的范围是 176×144～640×480,许多设备甚至支持更高的分辨率。

```
video_recorder. setVideoSize (640, 480);
```

（9）设置最大文件大小

使用 setMaxFileSize() 方法设置最大文件大小,单位为字节。

```
video_recorder. setMaxFileSize (10000000); //10MB
```

为了确定是否已达到最大文件大小,需要在 Activity 中实现 MediaRecorder.OnInfoListener,同时在 MediaRecorder 中注册它。然后系统会调用 onInfo 方法,检查 what 参数是否等于 MediaRecorder.MEDIA_RECORDER_INFO_FILESIZE_REACHED。

（10）设置持续时间

使用 setMaxDuration() 方法设置最长持续时间,单位为毫秒。

```
video_recorder. setMaxDuration(10000); //10 秒
```

为了确定是否已达到最长持续时间。需要在 Activity 中实现 MediaRecorder.OnInfoListener,同时在 MediaRecorder 中注册它。当已达到最长持续时间时就会触发 onInfo() 方法,检查 what 参数是否等于 MediaRecorder.MEDIA_RECORDER_INFO_MAX_DURATION_REACHED。

（11）概要

MediaRecorder 有一个 setProfile() 方法,接受 CamcorderProfile 实例作为参数。使用该方法允许根据预设值设置整个配置变量集。其中,CamcorderProfile.QUALITY_LOW 指低质量视频捕获设置,CamcorderProfile.QUALITY_HIGH 指高质量视频捕获设置。

（12）输出文件

使用 setOutputFile() 方法设置输出文件的位置。

```
video_recorder. setOutputFile ("/sdcard/video_recorded.mp4"); //10 秒
```

(13) 预览表面

由于是视频录制，录制过程中需要看到画面。因此，需要为 MediaRecorder 指定一个取景器以预览要绘制的图像。需要使用 SurfaceView 和 SurfaceHolder.Callback，将在例子中进行介绍。

(14) 准备录制

设置好 MediaRecorder 实例后，就可以使用 prepare() 方法准备 MediaRecorder 了。

```
video_recorder.prepare();
```

(15) 开始录制

MediaRecorder 实例准备好后，就可以开始录制了。

```
video_recorder.start();
```

(16) 停止录制

录制过程中，可以通过 stop() 方法停止录制。

```
video_recorder.stop();
```

(17) 释放资源

最后，不要忘记需要释放占用的资源。

```
video_recorder.release();
```

下面代码实现 MediaRecorder 进行视频的录制。

```
import java.io.IOException;
import android.app.Activity;
import android.content.pm.ActivityInfo;
import android.media.CamcorderProfile;
import android.media.MediaRecorder;
import android.os.Bundle;
import android.util.Log;
import android.view.SurfaceHolder;
import android.view.SurfaceView;
import android.view.View;
import android.view.Window;
import android.view.WindowManager;
import android.view.View.OnClickListener;

//实现 OnClickListener 以便响应单击启动和停机录制按钮，实现
//SurfaceHolder.Callback 以处理和图像预览的操作
public class VideoRecorder extends Activity implements OnClickListener,
        SurfaceHolder.Callback {
    MediaRecorder recorder;
    SurfaceHolder holder;
//该布尔量 recording 用于表示当前是否正在录制
    boolean recording = false;
```

```java
        public static final String TAG = "VIDEOCAPTURE";

        @Override
        public void onCreate(Bundle savedInstanceState) {
            super.onCreate(savedInstanceState);

//将以全屏和横屏模式运行
            requestWindowFeature(Window.FEATURE_NO_TITLE);
            getWindow().setFlags(WindowManager.LayoutParams.FLAG_FULLSCREEN,
                    WindowManager.LayoutParams.FLAG_FULLSCREEN);

setRequestedOrientation(ActivityInfo.SCREEN_ORIENTATION_LANDSCAPE);

//实例化 MediaRecorder 对象
            recorder = new MediaRecorder();
//初始化 MediaRecorder 对象
            initRecorder();
            setContentView(R.layout.main);

//获取 SurfaceView 和 SurfaceHolder 的引用,同时注册该 Activity 为
//SurfaceHolder.Callback
            SurfaceView cameraView = (SurfaceView) findViewById(R.id.CameraView);
            holder = cameraView.getHolder();
            holder.addCallback(this);
            holder.setType(SurfaceHolder.SURFACE_TYPE_PUSH_BUFFERS);
//设置 SurfaceView 为可单击,注册监听器
            cameraView.setClickable(true);
            cameraView.setOnClickListener(this);
        }
//处理所有和 MediaRecorder 设置有关的操作
        private void initRecorder() {
            recorder.setAudioSource(MediaRecorder.AudioSource.DEFAULT);
            recorder.setVideoSource(MediaRecorder.VideoSource.DEFAULT);

            CamcorderProfile cpHigh = CamcorderProfile
                    .get(CamcorderProfile.QUALITY_HIGH);
            recorder.setProfile(cpHigh);
            recorder.setOutputFile("/sdcard/videocapture_example.mp4");
            recorder.setMaxDuration(50000); //50 seconds
            recorder.setMaxFileSize(5000000); //Approximately 5 megabytes
        }

        private void prepareRecorder() {
//设置 MediaRecorder 的预览表面
            recorder.setPreviewDisplay(holder.getSurface());
```

```java
        try {
            recorder.prepare();
        } catch (IllegalStateException e) {
            e.printStackTrace();
            finish();
        } catch (IOException e) {
            e.printStackTrace();
            finish();
        }
    }

    public void onClick(View v) {
        if (recording) {
            recorder.stop();
            recording = false;
            Log.v(TAG, "Recording Stopped");
            initRecorder();
            prepareRecorder();
        } else {
            recording = true;
            recorder.start();
            Log.v(TAG, "Recording Started");
        }
    }

//一旦创建表面,就会调用此方法
    public void surfaceCreated(SurfaceHolder holder) {
        Log.v(TAG, "surfaceCreated");
        prepareRecorder();
    }

    public void surfaceChanged(SurfaceHolder holder, int format, int width,
            int height) {
    }

//当销毁表面时,如果正在录制,那么停止录制。这可能会在Activity不可见时发生
    public void surfaceDestroyed(SurfaceHolder holder) {
        Log.v(TAG, "surfaceDestroyed");
        if (recording) {
            recorder.stop();
            recording = false;
        }
        recorder.release();
        finish();
    }
}
```

12.4 视频播放

近些年,随着 iPhone 和 Android 智能手机的流行,视频播放功能已经成为智能手机的标配。接下来将探讨 Android 的视频播放功能,本节重点介绍在 Android 上播放视频的各种方法,以及所支持的视频格式。

12.4.1 常见的视频格式

在深入介绍如何播放视频的具体机制之前,先来了解一下可以播放的视频类型。Android 支持多种视频格式和编解码器,并且支持的类型还在不断增加,下面介绍几种常见的视频格式。

(1)H.263,自适应多速率编解码器,包括 AMR 窄带 AMR-NB 和 AMR 宽带 AMR-WB,文件扩展名是.3gp(audio/3gpp)或.amr(audio/amr)。AMR 是 3GPP 使用的基本音频编解码标准。AMR 主要应用于手机上的语音通话应用,并得到手机厂商的广泛支持,该编码标准适应于简单的语音编码,不适用处理更复杂的音频数据,如音乐等。

(2)H.264 AVC,全称是 Advanced Audio Coding。一种专为声音数据设计的文件压缩格式,与 MP3 不同,它采用了全新的算法进行编码,更加高效,具有更高的"性价比"。利用 AAC 格式,可使人感觉声音质量没有明显降低的前提下,更加小巧。Android 除了支持 AAC 之外,还支持新添加到 AAC 规范中的高效 AAC(High Efficiency AAC)格式。

(3)MPEG-4 SP,是一种音频压缩技术,其全称是动态影像专家压缩标准音频层面 3(Moving Picture Experts Group Audio Layer III),简称为 MP3。它被设计用来大幅度地降低音频数据量。利用 MPEG Audio Layer 3 的技术,将音乐以 1:10 甚至 1:12 的压缩率,压缩成容量较小的文件,而对于大多数用户来说,重放的音质与最初的不压缩音频相比没有明显下降。它是在 1991 年由位于德国埃尔朗根的研究组织 Fraunhofer-Gesellschaft 的一组工程师发明和标准化的。用 MP3 形式存储的音乐就叫作 MP3 音乐,能播放 MP3 音乐的机器就叫作 MP3 播放器。MP3 是目前互联网上使用最广泛的音频编解码器之一。

(4)VP8,全称是 OGG Vorbis,是一种新的音频压缩格式,类似于 MP3 的音乐格式。但有一点不同的是,它是完全免费、开放和没有专利限制的。OGG Vorbis 有一个特点是支持多声道。Vorbis 是这种音频压缩机制的名字,而 Ogg 则是一个计划的名字,该计划意图设计一个完全开放性的多媒体系统。Ogg Vorbis 文件的扩展名是.ogg。这种文件的设计格式是非常先进的。创建的 Ogg 文件可以在未来的任何播放器上播放,因此,这种文件格式可以不断地进行大小和音质的改良,而不影响旧有的编码器或播放器。

12.4.2 使用 Intent 播放视频

正如本章已经探讨的播放音频的功能,Android 可以很容易的通过使用 Intent 调用内置的媒体播放器来实现简单的视频播放功能。

为了通过创建 Intent 来调用内置的媒体播放器应用程序的播放功能，可以使用 Intent.ACTION_VIEW 来构建 Intent，并通过 setDataAndType()方法传入视频文件的 URI 和 MIME 类型。

下面给出一个较完整的示例，提供使用 Intent 来视频播放的功能。

```
import android.app.Activity;
import android.content.Intent;
import android.net.Uri;
import android.os.Bundle;
import android.os.Environment;
import android.view.View;
import android.view.View.OnClickListener;
import android.widget.Button;

public class VideoPlayerIntent extends Activity implements OnClickListener {
    Button playButton;

    @Override
    public void onCreate(Bundle savedInstanceState) {
        super.onCreate(savedInstanceState);
        setContentView(R.layout.main);
        playButton = (Button) this.findViewById(R.id.PlayButton);
        playButton.setOnClickListener(this);
    }

    public void onClick(View v) {
        Intent intent = new Intent(android.content.Intent.ACTION_VIEW);
        //Download "Test_Movie_iPhone.m4v from
        //http://www.mobvcasting.com/android/video/Test_Movie_iPhone.m4v
        //and save to the root of your device's SD card.
        Uri data = Uri.parse(Environment.getExternalStorageDirectory()
                .getPath()
                + "/Test_Movie_iPhone.m4v");
        intent.setDataAndType(data, "video/mp4");
        startActivity(intent);
    }
}
```

12.4.3 使用 VideoView 播放视频

VideoView 是一个带有视频播放功能的视图，可以直接在布局中使用，使用起来非常简单。下面给出一个具体的示例程序，介绍它的功能。

```
//首先导入所需要的包
import android.app.Activity;
```

```java
import android.net.Uri;
import android.os.Bundle;
import android.os.Environment;
import android.widget.VideoView;

/*由于该程序由一个Activity组成,在线性布局(LinearLayout)中放置一个VideoView
 *视图
 */
public class ViewTheVideo extends Activity {
    VideoView vv;

    @Override
    public void onCreate(Bundle savedInstanceState) {
        super.onCreate(savedInstanceState);
        setContentView(R.layout.main);
/*调用 setContentView(R.layout.main)方法设置 Activity 的界面布局后,可以通过
 *VideoView 的资源 ID,调用 findViewById(R.id.VideoView)方法获取该视图的引用
 */
        vv = (VideoView) this.findViewById(R.id.VideoView);
/*通过Uri 类的parse方法获取视频文件的URI,该视频文件放置于设备SD卡的根目录下
 */
        Uri videoUri = Uri.parse(Environment.getExternalStorageDirectory()
            .getPath()
            + "/Test_Movie_iPhone.m4v");

/*最后,通过VideoView 的 setVideoView方法设置视频URI,并调用 start()方法播放
 *该视频文件
 */
        vv.setVideoURI(videoUri);
        vv.start();
    }
}
```

然而,VideoView 控制视频播放的功能相对较少,它只有 start()和 pause()方法。为了提供更多的控制,可以实例化一个 MediaController,并通过 setMediaController 方法把它设置为 VideoView 的控制器。

默认的 MediaController 有后退(rewind)、暂停(pause)、播放(play)和快进(fast forward)按钮,还有一个清除和控制条组合空间,可以用来定位到视频中的任何一个位置。

下面是对上面 VideoView 示例的更新,在通过 setContentView 方法设置内容视图之后,通过调用 setMediaPlayer 方法设置 MediaController 为 VideoView 的控制器。

```
@Override
public void onCreate(Bundle savedInstanceState) {
    super.onCreate(savedInstanceState);
    setContentView(R.layout.main);

    vv = (VideoView) this.findViewById(R.id.VideoView);

    vv.setMediaController(new MediaController(this));

    Uri videoUri = Uri.parse(Environment.getExternalStorageDirectory()
            .getPath() + "/Test_Movie_iPhone.m4v");

    vv.setVideoURI(videoUri);
    vv.start();
}
```

12.4.4 使用 MediaPlayer 播放视频

12.2.3 节音频播放是介绍了 MediaPlayer 类，同样，MediaPlayer 类也可以通过类似的方式用于视频播放。与使用 Intent 和 VideoView 播放视频相比，将 MediaPlayer 用于视频播放能够为播放视频文件提供更大的灵活性。事实上，在通过 Intent 和 VideoView 播放视频时，处理视频播放的内部机制也有由 MediaPlayer 完成的。

```
import java.io.IOException;

import android.app.Activity;
import android.os.Bundle;
import android.os.Environment;
import android.util.Log;
import android.view.Display;
import android.widget.LinearLayout;

import android.media.MediaPlayer;
import android.media.MediaPlayer.OnCompletionListener;
import android.media.MediaPlayer.OnErrorListener;
import android.media.MediaPlayer.OnInfoListener;
import android.media.MediaPlayer.OnPreparedListener;
import android.media.MediaPlayer.OnSeekCompleteListener;
import android.media.MediaPlayer.OnVideoSizeChangedListener;

import android.view.SurfaceHolder;
import android.view.SurfaceView;
```

```java
public class CustomVideoPlayer extends Activity implements
        OnCompletionListener, OnErrorListener, OnInfoListener,
        OnPreparedListener, OnSeekCompleteListener, OnVideoSizeChangedListener,
        SurfaceHolder.Callback {
    Display currentDisplay;
    SurfaceView surfaceView;
    SurfaceHolder surfaceHolder;
    MediaPlayer mediaPlayer;
    int videoWidth = 0;
    int videoHeight = 0;
    boolean readyToPlay = false;
    public final static String LOGTAG = "CUSTOM_VIDEO_PLAYER";

    @Override
    public void onCreate(Bundle savedInstanceState) {
        super.onCreate(savedInstanceState);
        setContentView(R.layout.main);

        surfaceView = (SurfaceView) this.findViewById(R.id.SurfaceView);
        surfaceHolder = surfaceView.getHolder();
        surfaceHolder.addCallback(this);
        surfaceHolder.setType(SurfaceHolder.SURFACE_TYPE_PUSH_BUFFERS);

        mediaPlayer = new MediaPlayer();
        mediaPlayer.setOnCompletionListener(this);
        mediaPlayer.setOnErrorListener(this);
        mediaPlayer.setOnInfoListener(this);
        mediaPlayer.setOnPreparedListener(this);
        mediaPlayer.setOnSeekCompleteListener(this);
        mediaPlayer.setOnVideoSizeChangedListener(this);

        //Download "Test_Movie_iPhone.m4v from
        //http://www.mobvcasting.com/android/video/Test_Movie_iPhone.m4v
        //and save to the root of your device's SD card.
        String filePath = Environment.getExternalStorageDirectory().getPath()
                + "/Test_Movie_iPhone.m4v";

        try {
            mediaPlayer.setDataSource(filePath);
        } catch (IllegalArgumentException e) {
            Log.v(LOGTAG, e.getMessage());
            finish();
        } catch (IllegalStateException e) {
            Log.v(LOGTAG, e.getMessage());
            finish();
```

```java
        } catch (IOException e) {
            Log.v(LOGTAG, e.getMessage());
            finish();
        }

        currentDisplay = getWindowManager().getDefaultDisplay();
}

public void surfaceCreated(SurfaceHolder holder) {
    Log.v(LOGTAG, "surfaceCreated Called");

    mediaPlayer.setDisplay(holder);

    try {
        mediaPlayer.prepare();
    } catch (IllegalStateException e) {
        Log.v(LOGTAG, e.getMessage());
        finish();
    } catch (IOException e) {
        Log.v(LOGTAG, e.getMessage());
        finish();
    }
}

public void surfaceChanged(SurfaceHolder holder, int format, int width,
        int height) {
    Log.v(LOGTAG, "surfaceChanged Called");
}

public void surfaceDestroyed(SurfaceHolder holder) {
    Log.v(LOGTAG, "surfaceDestroyed Called");
}

public void onCompletion(MediaPlayer mp) {
    Log.v(LOGTAG, "onCompletion Called");
    finish();
}

public boolean onError(MediaPlayer mp, int whatError, int extra) {
    Log.v(LOGTAG, "onError Called");

    if (whatError == MediaPlayer.MEDIA_ERROR_SERVER_DIED) {
        Log.v(LOGTAG, "Media Error, Server Died " + extra);
    } else if (whatError == MediaPlayer.MEDIA_ERROR_UNKNOWN) {
        Log.v(LOGTAG, "Media Error, Error Unknown " + extra);
```

```java
        }

        return false;
    }

    public boolean onInfo(MediaPlayer mp, int whatInfo, int extra) {
        if (whatInfo == MediaPlayer.MEDIA_INFO_BAD_INTERLEAVING) {
            Log.v(LOGTAG, "Media Info, Media Info Bad Interleaving " + extra);
        } else if (whatInfo == MediaPlayer.MEDIA_INFO_NOT_SEEKABLE) {
            Log.v(LOGTAG, "Media Info, Media Info Not Seekable " + extra);
        } else if (whatInfo == MediaPlayer.MEDIA_INFO_UNKNOWN) {
            Log.v(LOGTAG, "Media Info, Media Info Unknown " + extra);
        } else if (whatInfo == MediaPlayer.MEDIA_INFO_VIDEO_TRACK_LAGGING) {
            Log.v(LOGTAG, "MediaInfo, Media Info Video Track Lagging " + extra);
            /*
             * Android Version 2.0 and Higher } else if (whatInfo ==
             * MediaPlayer.MEDIA_INFO_METADATA_UPDATE) {
             * Log.v(LOGTAG,"MediaInfo, Media Info Metadata Update " + extra);
             */
        }
        return false;
    }

    public void onPrepared(MediaPlayer mp) {
        Log.v(LOGTAG, "onPrepared Called");
        videoWidth = mp.getVideoWidth();
        videoHeight = mp.getVideoHeight();

        if (videoWidth > currentDisplay.getWidth()
                || videoHeight > currentDisplay.getHeight()) {
            float heightRatio = (float) videoHeight
                    / (float) currentDisplay.getHeight();
            float widthRatio = (float) videoWidth
                    / (float) currentDisplay.getWidth();

            if (heightRatio > 1 || widthRatio > 1) {
                if (heightRatio > widthRatio) {
                    videoHeight = (int) Math.ceil((float) videoHeight
                            / (float) heightRatio);
                    videoWidth = (int) Math.ceil((float) videoWidth
                            / (float) heightRatio);
                } else {
                    videoHeight = (int) Math.ceil((float) videoHeight
                            / (float) widthRatio);
                    videoWidth = (int) Math.ceil((float) videoWidth
```

```
                            / (float) widthRatio);
            }
        }
    }

    surfaceView.setLayoutParams(new LinearLayout.LayoutParams(videoWidth,
            videoHeight));
    mp.start();
}

public void onSeekComplete(MediaPlayer mp) {
    Log.v(LOGTAG, "onSeekComplete Called");
}

public void onVideoSizeChanged(MediaPlayer mp, int width, int height) {
    Log.v(LOGTAG, "onVideoSizeChanged Called");
}
}
```

12.5 本章小结

　　本章主要对基于 Android 平台的音视频录制和播放功能进行介绍，重点讲解了两种方法，使用 Intent 进行录制和播放与使用系统 API 进行录制和播放。此外，本章并没有探讨在线音视频的播放，但是通过本章介绍的知识，读者完全有能力自己开发包含在线播放功能的应用程序。

第 13 章 Android 的网络编程

本章主要介绍 Android 平台下进行网络编程的相关知识，内容包括 Android 网络编程的常用接口，基于 HTTP 协议的 GET 请求和 POST 请求的两种网络编程方式，套接字 Socket 的基本概念及服务器和客户端网络编程，WebView 的常用组件及方法和基于 WebView 的浏览器开发方法。

13.1 Android 网络编程基础

Android 是基于 Linux 内核为基础的操作系统，继承了 Linux 优秀的联网功能。在网络通信应用的开发中，Android 平台有三种网络编程接口可以使用。

（1）标准的 Java 网络接口

Java 的标准网络接口 java.net.*中提供了网络编程的相关类，主要包括流操作、数据包、套接字 Socket 以及 HTTP 协议处理，通过设置参数，能够完成连接服务器、和服务器进行数据交换等网络操作，有 Java 开发基础的人员可以使用这些包快速的创建 Android 平台下的网络应用程序。

（2）Apache 网络接口

HTTP 协议是目前互联网中应用最为广泛的网络协议，虽然标准的 Java 网络接口提供了对 HTTP 协议的支持，但灵活性不足。HttpClient 是 Apache 下的子项目，是一个功能强大并全面支持 HTTP 协议的客户端编程工具包，它实际是对 java.net.*进行了封装和扩展。Android 系统中已经集成了 HttpClient 的开源包 org.apache.http.*，可以在 Android 中直接使用 HttpClient 访问网络。

（3）Android 网络接口

Android 的网络接口 android.net.*是通过对 Apache 网络接口进行封装来实现的，它同样提供了对 HTTP 协议的支持,该包下的类常用来开发 WiFi 连接、手机邮件处理等 Android 系统特有的网络应用。

在 Android 平台下进行网络应用程序的开发必须遵守 Android 系统的规范，否则应用程序将无法访问服务器或出错，下面的这些规范适用于本章的实例，以后不再逐一描述。

- Android 的网络程序需要为其添加网络访问权限，否则将无法连接服务器，必须在项目的 AndroidManifest.xml 文件的 Manifast 标签中加入如下指令：

`<uses-permission android:name="android.permission.INTERNET"/>`

- 在 Android 系统中，主线程的任务是 UI 操作和交互，为了提升用户感受和响应度，不允许主线程进行网络和磁盘读写等响应周期长的操作，如果将访问网络的代码放在主线程中，将会出现如下的异常信息。

android.os.NetworkOnMainThreadException

解决该问题有两个方法：第一个方法是在主线程中引入系统提供的开发工具 StrictMode 类，该类用来捕获磁盘访问或网络访问中与主线程交互产生的潜在问题，提示开发者对其进行修复。在主线程的 onCreate()方法中加入下面的代码。

```
StrictMode.setThreadPolicy(new StrictMode.ThreadPolicy.Builder()
    .detectDiskReads()
    .detectDiskWrites()
    .detectNetwork()
    .penaltyLog()
    .build());
StrictMode.setVmPolicy(new StrictMode.VmPolicy.Builder()
    .detectLeakedSqlLiteObjects()
    .penaltyLog()
    .penaltyDeath()
    .build());
```

加入上述代码后，运行程序依然会有异常抛出，但程序能够正常执行。由于该方法并不符合 Android 平台的的规范和要求，不建议使用。

第二种方法是启动一个新的线程去执行访问网络的代码。

```
new Thread(new Runnable() {
    public void run() {
        ……;
        网络任务;
        ……;
    }
}).start();
```

本章的所有实例将采用第二种方法。

一般情况下，在网络任务完成以后，主线程需要进行 UI 交互和更新，在网络通信异常的情况下，主线程需要给出提示或警告，因此，执行网络任务的线程需要使用 Handler 消息通信机制向主线程发送消息，通知主线程网络任务的状态。

13.2 基于 HTTP 协议的网络编程

Android 平台提供了使用 HTTP 协议进行网络编程的大量接口，比较重要的是标准的 Java 接口和 HttpClient 接口，本节主要针对这些接口的网络编程方式进行介绍。

13.2.1 HTTP 介绍

HTTP 协议是超文本传送协议的简称，是 Internet 上使用最为广泛的协议，在 TCP/IP

体系中处于应用层，是一个标准的客户端服务器模型，由一个客户端程序和一个 Web 服务器端的服务程序构成，客户端程序通过统一资源定位器 URL 向服务器发送 HTTP 请求报文，服务程序在 80 号端口监听客户端的请求并向客户端发送 HTTP 响应报文。HTTP 协议永远都是客户端发起请求，服务器回送响应，是一个无状态的协议。

HTTP 协议中定义了八种方法和服务器经行交互，其中，GET 和 HEAD 方法是服务器必须实现的，也是最常用的。GET 方法请求的数据会附在 URL 之后，以?分割 URL 和传输数据，参数之间以&相连，如 login.action ? name = Administrator & password = 123456，GET 方法提交的数据量跟 URL 的长度有直接关系，不同的浏览器对 URL 的长度要求不一样。POST 把提交的数据则放置在 HTTP 协议的包体中，理论上对数据长度没有要求。POST 的安全性要比 GET 的安全性高，通过 GET 提交数据，如果对数据不做修改和加密，用户名和密码将以明文的形式出现在 URL 上。HTTP 协议默认采用的是 GET 方法。

目前广泛使用的 Web 服务器是 Microsoft 公司的 IIS 和 Apache 软件基金会的 Tomcat，由于本节的实例需要，首先架设一个 Tomcat 服务器。搭建 Tomcat 服务器的步骤如下所示。

（1）进入 Apache 的官方网站 http://tomcat.apache.org，在该网站上可以免费获取各版本的 Tomcat 软件，根据操作系统的版本选择对应的下载点下载。

（2）将下载的压缩包解压缩到本地磁盘任意目录，以 D:\Tomcat 为例。

（3）配置环境变量，打开环境变量窗口，找到 CLASSPATH 变量，在变量值的最后添加如下内容：

`;D:\tomcat\lib\jsp-api.jar; D:\tomcat\lib\servlet-api.jar`

（4）进入 D:\Tomcat\bin 目录，双击执行 startup.bat 启动 Tomcat。

（5）打开浏览器，输入网址：http://localhost:8080 若出现图 13.1 则代表安装成功。

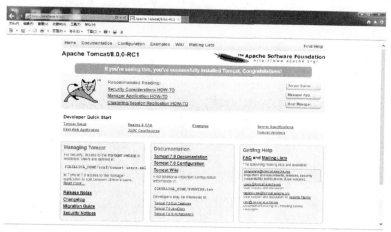

图 13.1　启动 Tomcat 成功

13.2.2　使用 HttpURLConnection 访问网络

HttpURLConnection 类位于 JAVA.NET 包中，它是 Java 中提供的访问网络资源的接口，使用这些接口可以方便地编写网络应用程序。由于该类是一个抽象类，需要使用 URL 对象的 openConnection()方法实例化，创建一个 HttpURLConnection 对象的代码如下。

```
URL myUrl = new URL(stringURL);
HttpURLConnection myConn =(HttpURLConnection)myUrl.openConnection();
```

创建了 HttpURLConnection 对象之后，就可以使用该对象向服务器发送 GET 请求和 POST 请求。

1. GET 请求

HttpURLConnection 对象默认使用 GET 请求，下面通过实例来说明。

【例 13.1】 在 Eclipse 中新建一个名称为 UrlGet 的 Android 的项目，该项目向服务器发送 Get 请求，从服务器获取一个文本文件的内容，并将该文件的内容显示在屏幕的 EditText 控件中。

（1）在 Tomcat\webapps 目录下新建一个文件夹 AndroidTest，将文本文件 Message.txt 拷贝到该文件夹下面作为服务器资源。

（2）修改项目的布局文件 activity_url_get.xml，使用相对布局管理器，添加一个向服务器发送 GET 请求的按钮控件和一个显示文本内容的 EditText 控件，代码如下。

```
<RelativeLayout xmlns:android="http://schemas.android.com/apk/res/android"
    android:layout_width="match_parent"
    android:layout_height="match_parent"
    android:paddingBottom="@dimen/activity_vertical_margin"
    android:paddingLeft="@dimen/activity_horizontal_margin"
    android:paddingRight="@dimen/activity_horizontal_margin"
    android:paddingTop="@dimen/activity_vertical_margin"
    >                                           <!-- 声明一个相对布局 -->
 <EditText
        android:id="@+id/et"
        android:layout_width="fill_parent"
        android:layout_height="wrap_content"
        android:inputType="none"
        android:cursorVisible="false"
        android:layout_alignParentTop="true"
  />                                            <!-- 声明一个 EditText 控件 -->
    <Button
        android:id="@+id/btn"
        android:text="@string/btn"
        android:layout_width="fill_parent"
        android:layout_height="wrap_content"
        android:layout_alignParentBottom="true"
  />                                            <!-- 声明一个 Button 控件 -->
</RelativeLayout>
```

（3）在 UrlGetActivity 类中，创建要使用的变量，同时本例中使用了消息通信机制 Handler，执行网络任务的线程向主线程发送消息，通知主线程更新 EditText 控件，需要重写 Handler 对象的 handleMessage()方法，代码如下。

```java
    private static final int SUCCESS = 0;
    private static final int FAILURE = 1;
    private EditText et;
    private Button btn;
    private String txt;
    //统一资源定位器 URL 字符串,192.168.1.120 为服务器地址
    String stringURL = "http://192.168.1.120:8080/AndroidTest/Message.txt";
    private Handler mHandler = new Handler(){
        //重写 handleMessage 方法
        public void handleMessage (Message msg){
            switch(msg.what) {
                case SUCCESS:
                    et.setText(txt);
                    break;
                case FAILURE:
                    et.setText("Download the file Failure!");
                    break;
            }
        }
    };
```

(4) 重写 onCreate()方法,获取布局管理器中的对象,并为按钮对象添加事件监听器,调用自定义的 GetURLResources(),代码如下。

```java
protected void onCreate(Bundle savedInstanceState) {
    super.onCreate(savedInstanceState);
    setContentView(R.layout.activity_url_get);
    et = (EditText)findViewById(R.id.et);
    btn = (Button)findViewById(R.id.btn);
    //为按钮添加事件监听器
    btn.setOnClickListener(new View.OnClickListener() {
        @Override
        public void onClick(View v) {
            GetURLResources();
        }
    });
}
```

(5) 编写无返回值的方法 GetURLResources(),该方法启动一个新线程创建一个 HTTP 连接,并用 Get 请求从服务器获取文本文件内容,如果获取成功,则向主线程发送 what 值为 0 的消息,否则发送 what 值为 1 的消息,代码如下。

```java
public void GetURLResources(){
    //创建一个新线程,读取服务器资源
    new Thread(new Runnable(){
        public void run(){
            try{
                URL myUrl = new URL(stringURL);
                //创建一个 HttpURLConnection 对象,打开链接
```

```
            HttpURLConnection myConn =(HttpURLConnection)myUrl.openConnection();
            //设置连接超时
            myConn.setConnectTimeout(3000);
            //获取输入流,得到读取的内容
            InputStreamReader in = new InputStreamReader(myConn.getInputStream());
            BufferedReader buffer = new BufferedReader(in);
            String inputLine = null;
            StringBuffer pageBuffer = new StringBuffer();
            while((inputLine = buffer.readLine())!= null){
                pageBuffer.append(inputLine +"\n");
            }
            //设置字符的编码格式
            txt =  new String(pageBuffer.toString().getBytes("UTF-8"));
            //去掉最后一行的换行符
            txt = txt.substring(0,txt.length()-1);
            mHandler.sendEmptyMessage(0);
            in.close();
            buffer.close();
            //关闭连接
            myConn.disconnect();
        }
        catch(Exception e){
            mHandler.sendEmptyMessage(1);
            e.printStackTrace();
        }
    }
}).start();
}
```

启动 Tomcat 服务器,运行本实例,结果如图 13.2 所示。

图 13.2　利用 GET 方法获取文件内容实例图

2. POST 请求

由于 HttpURLConnection 对象默认使用 GET 方法，在发送 POST 请求时，需要调用 setRequestMethod()方法来进行指定，同时 POST 方法还必须进行一些必要的设置，常用的设置如下表所示。

表 13.1 POST 常用设置

方法	描述
SetRequestMethod(String)	设置连接方式
SetDoInput(Boolean)	设置输入是否允许
setDoOutput(boolean)	设置输出是否允许
setUseCaches(boolean)	是否使用 Cache
setInstanceFollowRedirects(boolean)	是否使用 HTTP 重定向
setRequestProperty(String, String)	配置请求属性

【例 13.2】 在 Eclipse 中新建一个名称为 UrlPost 的 Android 的项目，该项目向服务器发送 Post 请求，Post 请求提交的数据是参数 username 及对应的值 Android User，服务器获取参数 username 的值，并向客户端返回欢迎信息在屏幕的 EditText 控件中显示。

（1）编写一个 UrlPost.jsp 文件,代码如下所示,将该文件保存在文件夹 Tomcat\webapps\AndroidTest\下面。

```jsp
<%@ page contentType="text/html;charset=GBK"%>
<%
    request.setCharacterEncoding("GBK");
    String name = (String)request.getParameter("username");
    if (name != null) {
        out.print("Welcome: "+name);
    }
%>
```

（2）修改项目的布局文件 activity_url_post.xml，使用相对布局管理器，添加一个向服务器发送 POST 请求的按钮控件和一个显示欢迎信息的 EditText 控件，代码同 activity_url_get_.xml。

（3）在 UrlPostActivity 类中，创建要使用的变量，本例仍然使用了消息通信机制 Handle 完成执行网络任务的线程和主线程之间的通信，重写 Handler 对象的 handleMessage()方法，代码如下。

```java
private static final int Success = 0;
private static final int ServerError = 1;
private static final int LinkFailure = 2;
private Button btn;
private EditText et;
private String txt;
//统一资源定位器 URL 字符串
String stringURL = "http://192.168.1.120:8080/AndroidTest/UrlPost.jsp";
```

```java
private Handler mHandler = new Handler(){
    //重写 handleMessage 方法
    public void handleMessage (Message msg) {
        switch(msg.what) {
            case Success:
                et.setText(txt);
                break;
            case ServerError:
                et.setText("The response of Server is error!");
                break;
            case LinkFailure:
                et.setText("Link Server failed!");
                break;
        }
    }
};
```

（4）重写 onCreate()方法，获取布局管理器中的对象，并为按钮对象添加事件监听器，调用自定义的 GetURLResources()，代码如下。

```java
protected void onCreate(Bundle savedInstanceState) {
    super.onCreate(savedInstanceState);
    setContentView(R.layout.activity_url_post);
    btn = (Button)findViewById(R.id.btn);
    et = (EditText)findViewById(R.id.et);
    //为按钮添加事件监听器
    btn.setOnClickListener(new View.OnClickListener() {
        @Override
        public void onClick(View v) {
            GetURLResources();
        }
    });
}
```

（5）编写无返回值的方法 GetURLResources()，该方法启动一个新线程创建一个 HTTP 连接，并用 Post 请求向服务器发送数据同时从服务器接收数据，如果接收数据成功，则向主线程发送 what 值为 0 的消息；如果服务器响应异常，发送 what 值为 1 的消息；如果服务器连接异常，发送 what 值为 2 的消息，代码如下。

```java
public void GetURLResources(){
    new Thread(new Runnable(){
        public void run(){
            try{
                String Param = "username=" + "Android User";    //发送的数据
                byte[] PostData = Param.getBytes();
                URL myUrl = new URL(stringURL);
```

```java
            //创建一个 HttpURLConnection 对象,打开链接
            HttpURLConnection myConn = (HttpURLConnection)myUrl.openConnection();
            myConn.setConnectTimeout(3000);          //设置连接超时
            myConn.setDoInput(true);                 //设置输入允许
            myConn.setDoOutput(true);                //设置输出允许
            myConn.setRequestMethod("POST");         //设置 POST 方式请求
            myConn.setUseCaches(false);              //POST 方法不能使用 Cache
            myConn.setInstanceFollowRedirects(true); //允许 HHTP 重定向
            myConn.setRequestProperty("Content-Type", "application/x-www-form-
            urlencoded");                            //配置请求设置
            myConn.connect();
            //发送数据
            DataOutputStream out = new DataOutputStream(myConn.getOutputStream());
            out.write(PostData);
            out.flush();
            out.close();
            if(myConn.getResponseCode()==200){       //判断连接状态
                InputStreamReader in = new InputStreamReader(myConn.getInputStream());
                BufferedReader buffer = new BufferedReader(in);
                String inputLine = null;
                StringBuffer pageBuffer = new StringBuffer();
                while((inputLine = buffer.readLine())!= null){
                    pageBuffer.append(inputLine +"\n");
                }
                //设置字符编码格式
                txt = new String(pageBuffer.toString().getBytes("UTF-8"));
                mHandler.sendEmptyMessage(0);
                in.close();
                buffer.close();
                myConn.disconnect();
            }
            else{
                mHandler.sendEmptyMessage(1);
            }
        }
        catch(Exception e){
            mHandler.sendEmptyMessage(2);
            e.printStackTrace();
        }
    }
}).start();
}
```

启动 Tomcat 服务器,运行本实例,结果如图 13.3 所示。

图 13.3　POST 方法接收信息

13.2.3　使用 HttpClient 访问网络

一般情况下，对于比较复杂的网络访问操作，使用 Java 的标准接口程序繁琐，工作量比较大，在 Android 系统中，最常用的是 Apache 提供的 HttpClient，它对 java.net 进行了抽象和封装，GET 请求封装成了 HttpGet 类，Post 请求封装成了 HttpPost 类，服务器的响应封装成了 HttpResponse 类，发送和接收 HTTP 报文的实体封装成了 HttpEntity 类，下面分别介绍利用 HttpClient 完成的 GET 方法和 POST 方法。

1. GET 请求

使用 HttpClient 发送 GET 请求的步骤大致如下。

（1）利用 DefaultHttpClient 类创建 HttpClient 对象。

```
HttpClient httpclient = new DefaultHttpClient();
```

（2）利用 HttpGet 类生成 HttpGet 对象。

```
HttpGet httpGet = new HttpGet(stringURL);
```

（3）如果需要向服务器发送请求参数，生成 HTTP 请求报文的 HttpEntity 对象，使用 httpGet 的 setEntity()方法设置发送请求参数。

```
httpGet.setEntity(httpentity);
```

（4）调用 httpclient 的 execute()方法发送 GET 请求，该方法返回一个服务器的响应 HttpResponse 对象。

```
HttpResponse httpresponse = httpclient.execute(httpGet);
```

（5）调用 HttpResponse 对象的 getEntity()方法得到服务器 HTTP 响应报文的 HttpEntity

对象，该对象包含了服务器的应答信息。

【例 13.3】 在 Eclipse 中新建一个名称为 HttpClientGet 的 Android 项目，该项目向服务器发送 GET 请求，从服务器获取图片，并在客户端屏幕的 ImageView 控件中显示。

（1）将图片文件 Flower.PNG 保存在文件夹 Tomcat\webapps\AndroidTest\下面。

（2）修改项目的布局文件 activity_httpclient_get.xml，使用相对布局管理器，添加一个向服务器发送 GET 请求的按钮控件、一个用于显示图片信息的 ImageView 控件和一个显示状态信息的 EditText 控件。

（3）在 HttpClientGetActivity 类中，创建要使用的变量，重写 Handler 对象的 handleMessage()方法，如果接收到的消息的 what 值为 0，将 ImageView 控件的对象设置为消息中的图片对象，代码如下。

```java
private static final int SUCCESS = 0;
private static final int FAILURE = 1;
private Button btn;
private EditText et;
private ImageView iv;
//统一资源定位器 URL 字符串
String stringURL = "http://192.168.1.120:8080/AndroidTest/Flower.PNG";
private  Handler mHandler = new Handler(){
        //重写 handleMessage 方法
public void handleMessage (Message msg) {
        switch(msg.what) {
            case SUCCESS:
                et.setText("Download the Picture Success!");
                //设置图片控件中显示的图片对象
                iv.setImageBitmap((Bitmap) msg.obj);
                break;
            case FAILURE:
                et.setText("Download the Picture Failure!");
                break;
        }
    }
};
```

（4）重写 onCreate()方法，获取布局管理器中的对象，并为按钮对象添加事件监听器，调用自定义的 GetURLResources()，代码如下。

```java
protected void onCreate(Bundle savedInstanceState) {
    super.onCreate(savedInstanceState);
    setContentView(R.layout.activity_httpclient_get);
    btn = (Button)findViewById(R.id.btn);
    et = (EditText)findViewById(R.id.et);
    iv = (ImageView)findViewById(R.id.iv);
    //为按钮添加事件监听器
    btn.setOnClickListener(new View.OnClickListener() {
```

```
        @Override
        public void onClick(View v) {
            GetURLResources();
        }
    });
}
```

(5)编写无返回值的方法 GetURLResources(),该方法启动一个新线程使用 HttpClient 向服务器发送 GET 请求,从服务器获取图片,如果获取成功,则向主线程发送 what 值为 0,并且包含一个图片对象的消息,否则发送 what 值为 1 的消息,代码如下。

```
public void GetURLResources(){
  new Thread(new Runnable(){
    public void run(){
      try{
          HttpClient httpclient = new DefaultHttpClient();  //创建httpclient 对象
          HttpGet httpGet = new HttpGet(stringURL);    //创建连接
          httpclient.getParams().setParameter(CoreConnectionPNames.CONNECTION_
              TIMEOUT, 3000);                         //设置超时
          HttpResponse httpresponse = httpclient.execute(httpGet); //执行GET请求
          //判断连接状态
          if(httpresponse.getStatusLine().getStatusCode() == HttpStatus.SC_OK){
              HttpEntity httpentity = httpresponse.getEntity();  //获取响应实体
              InputStream in = httpentity.getContent();         //获取输入流
              Bitmap bmp = BitmapFactory.decodeStream(in);
              mHandler.obtainMessage(0, bmp).sendToTarget(); //发送包含对象的消息
              in.close();
          }
      }
      catch(Exception e){
          mHandler.obtainMessage(1).sendToTarget();        //连接服务器失败消息
          e.printStackTrace();
      }
    }
  }).start();
}
```

启动 Tomcat 服务器,运行本实例,结果如图 13.4 所示。

2. POST 请求

POST 请求和 GET 请求的步骤基本一致,大致流程如下。
(1)利用 DefaultHttpClient 类创建 HttpClient 对象。
(2)利用 HttpPost 类生成 HttpGet 对象。
(3)如果需要向服务器发送请求参数,生成 HTTP 请求报文的 HttpEntity 对象,使用 HttpPost 的 setEntity()方法设置发送请求参数。

图 13.4　GET 获取图片信息

（4）调用 httpClient 的 execute()方法发送 POST 请求，该方法返回一个服务器的响应 HttpResponse 对象。

（5）调用 HttpResponse 对象的 getEntity()方法得到服务器 HTTP 响应报文的 HttpEntity 对象，该对象包含了服务器的应答信息。

下面通过实例说明 POST 请求的过程。

【例 13.4】　在 Eclipse 中新建一个名称为 HttpClientPost 的 Android 项目，该项目模拟登录的过程，客户端输入用户名和密码，向服务器发送 POST 请求，如果验证通过，则打开一个新的 activity。

（1）编写一个 HttpClientPost.jsp 文件，代码如下所示，将该文件保存在文件夹 Tomcat\webapps\AndroidTest\ 下面。

```
<%@ page contentType="text/html;charset=GBK"%>
<%
request.setCharacterEncoding("GBK");
String name = (String)request.getParameter("username");
String pass = (String)request.getParameter("password");
if (name.equals("Android User") && pass.equals("123456")) {
    out.print("ok");
}
else{
    out.print("error");
}
%>
```

（2）修改项目的布局文件 activity_httpclient_post.xml，使用相对布局管理器，添加一个向服务器发送 POST 请求的按钮控件、两个用于输入用户名和密码的 EditText 控件。新建一个布局文件 activity_response.xml，添加一个 TextView 控件，用于显示欢迎信息。

（3）在 HttpClientPostActivity 类中，创建要使用的变量，重写 Handler 对象的 handleMessage()方法，如果接收到的消息的 what 值为 0，打开一个新的 Activity，代码如下。

```java
private static final int SUCCESS = 0;
private static final int FAILURE = 1;
private static final int ServerError = 2;
private static final int LinkFailure = 3;
private Button btn;
private EditText editname;
private EditText editpass;
private String repeat;
//统一资源定位器 URL 字符串
String stringURL="http://192.168.1.120:8080/AndroidTest/HttpClientPost.jsp";
private Handler mHandler = new Handler(){
//重写 handleMessage 方法
public void handleMessage (Message msg) {
    switch(msg.what) {
      case SUCCESS:
          //打开一个新的 activity
          Intent _intent = new Intent(HttpClientPostActivity.this,
          ResponseActivity.class);
          startActivity(_intent);
          break;
      case FAILURE:
          Toast.makeText(getApplicationContext(),"Login failed!", Toast.
          LENGTH_SHORT).show();
          break;
      case ServerError:
          Toast.makeText(getApplicationContext(),"The response of Server
          is error! ", Toast.LENGTH_SHORT).show();
          break;
      case LinkFailure:
          Toast.makeText(getApplicationContext(),"Link  Server  failed!
             ", Toast.LENGTH_SHORT).show();
          break;
    }
  }
};
```

（4）重写 onCreate()方法，获取布局管理器中的对象，并为按钮对象添加事件监听器，调用自定义的 PostMessage()，代码如下。

```java
protected void onCreate(Bundle savedInstanceState) {
    super.onCreate(savedInstanceState);
    setContentView(R.layout.activity_httpclient_post);
```

```
btn = (Button)findViewById(R.id.btn);
editname = (EditText)findViewById(R.id.EditName);
editpass = (EditText)findViewById(R.id.EditPass);
//为按钮添加事件监听器
btn.setOnClickListener(new View.OnClickListener() {
    @Override
    public void onClick(View v) {
        PostMessage();
    }
});

}
```

（5）编写无返回值的方法 PostMessage()，该方法启动一个新线程使用 HttpClient 向服务器发送 POST 请求，POST 请求提交的数据是参数 username 和 password 及对应的值，服务器获取参数的值，判断用户名和密码是否合法，向客户端返回响应报文。客户端接收到响应报文后，根据报文内容向主线程发送消息。如果登录成功，则向主线程发送 what 值为 0 的消息，登录失败，则向主线程发送 what 值为 1 的消息；如果服务器响应异常，发送 what 值为 2 的消息；如果服务器连接异常，发送 what 值为 3 的消息，代码如下。

```
public void PostMessage(){
  new Thread(new Runnable(){
    public void run(){
        try{
            HttpClient httpclient = new DefaultHttpClient();//创建httpclient对象
            HttpPost httppost = new HttpPost(stringURL);         //创建连接
            String name = editname.getText().toString();         //获取用户名
            String pass = editpass.getText().toString();         //获取密码
            List<NameValuePair> params = new ArrayList<NameValuePair>();
            params.add(new BasicNameValuePair("username", name)); //添加参数
            params.add(new BasicNameValuePair("password", pass));//添加参数
            //创建 HttpEntity 对象，添加参数并设置编码格式
            HttpEntity httpentity = new UrlEncodedFormEntity(params, HTTP.UTF_8);
            //设置 Post 请求的参数实体
            httppost.setEntity(httpentity);
            httpclient.getParams().setParameter(CoreConnectionPNames.
            CONNECTION_TIMEOUT, 3000);            //设置网络延时
            HttpResponse httpresponse = httpclient.execute(httppost);
                                            //发送 Post 请求
            //判断连接状态
            if(httpresponse.getStatusLine().getStatusCode() == HttpStatus.SC_OK){
                httpentity = httpresponse.getEntity();  //获取响应实体
                repeat = EntityUtils.toString(httpentity);
                repeat = repeat.replace("\r\n", ""); //去掉响应字符串的回车和换行
                if(repeat.equals("ok")){
```

```
                mHandler.sendEmptyMessage(0);    //验证成功消息
            }
            else{
                mHandler.sendEmptyMessage(1);    //验证失败消息
            }
        }
        else{
            mHandler.sendEmptyMessage(2);        //服务器响应异常消息
        }
    }
    catch(Exception e){
        mHandler.sendEmptyMessage(3);            //连接服务器失败消息
        e.printStackTrace();
    }
}
}).start();
}
```

启动 Tomcat 服务器，运行本实例如图 13.5 所示，登录成功后打开一个新的 activity，登录失败如图 13.6 所示。

图 13.5 登录界面　　　　　　　　　图 13.6 登录失败

13.3 基于 Socket 的网络编程

在 HTTP 通信中，客户端需要首先发起连接请求，和服务器建立连接后接收服务器的响应信息，在请求结束之后连接主动释放，这种连接称为短连接，是 HTTP 协议中最常使用的方式，这种方式能够使服务器的资源充分利用，但是也带来了一定的问题，如果客户端需要服务器更新的数据，必须再次发送连接请求，服务器更新的数据不会主动发送给客户端，在一些比较复杂的网络通信应用中，需要客户端和服务器建立连接后能够相互发送数据传递消息，这就需要使用基于 Socket 的网络编程来完成。

13.3.1 套接字 Socket

TCP/IP 协议是互联网的基础，其核心部分是传输层的 TCP 协议和 UDP 协议及网际层的 IP 协议，通常在操作系统中实现，网络应用程序一般不直接使用 TCP/IP 协议，而是使用一组调用 TCP/IP 协议的应用程序接口函数，这就是 Socket，也称为套接字，通过 Socket 可以很容易地编写网络应用程序完成不同主机之间的相互通信，需要说明的是，TCP/IP 本身并没有对应用程序接口函数进行标准化，不同的操作系统提供的应用程序接口函数形式不一样。

Socket 有两种方式，基于 TCP 协议的面向连接的流 Socket 和基于 UDP 协议的面向无连接的数据包 Socket。流 Socket 在发送数据之前，客户端的 Socket 和服务器 Socket 建立连接，效率不高但可以保证数据安全有序地到达接收方，通信完成后需要关闭 Socket 来断开连接；数据包 Socket 在发送数据之前，不需要建立连接，效率高但不保证数据安全有序地到达接收方，推荐使用流 Socket。

Java.NET 中提供了两个类，ServerSocket 和 Socket。ServerSocket 类用于创建在服务器的指定端口进行监听的对象，监听客户端的连接请求，Socket 类用于创建在服务器端和客户端相互连接的对象，该对象已经封装了一个输入流和输出流，一旦客户端的 Socket 对象和服务器端的 Socket 对象建立了连接，只要把数据写出到相关联的 Socket 对象的输出流中就可以完成发送，读取相关联的 Socket 对象的输入流中就可以完成接收。

ServerSocket 类常用的构造函数如下。

ServerSocket(int port)：创建绑定到指定端口进行监听的对象，参数 port 是端口号。

ServerSocket(int port, int backlog)：创建绑定到指定端口进行监听的对象，参数 port 是端口号，参数 backlog 是请求等待队列的最大长度。

服务器端和客户端必须约定好使用的通信端口，端口不一致将不能通信。在选择端口号时需要注意，计算机中的端口号在 0~65535 之间，其中，0~1023 为系统保留，用户程序尽量不要使用，否则可能影响系统服务。

Socket 类常用的构造函数如下。

Socket(Inet address, int port)：创建一个连接指定 IP 地址和端口号的流 Socket，参数 address 是 IP 地址，参数 port 是端口号。

Socket(String host, int port)：创建一个连接指定主机名和端口号的流 Socket，参数 host 是主机名或域名，参数 port 是端口号。

13.3.2 Socket 编程

客户端和服务器要完成一次通信，首先，客户端必须知道远程主机的 IP 地址（或者主机名）和端口号，其次，客户端要发送一个连接请求，建立连接。下面分别介绍服务器端和客户端的网络通信的步骤。

1．服务器端

（1）指定端口实例化一个 ServerSocket。

```
ServerSocket serversocket = new ServerSocket(5678);
```

（2）调用 ServerSocket 的 accept()方法，该方法是一个阻塞方法，调用该方法会使服务器在指定的端口监听客户端的连接请求，如果没有连接请求，该方法被阻塞，如果有连接请求，该方法返回一个 Socket 对象，该对象用于和客户端的 Socket 建立连接。

```
Socket socket = serversocket.accept();
```

（3）调用 Socket 对象的 getInputStream()方法获取输入流，接收客户端的 Socket 发送的信息，调用 getOutputStream()方法获取输出流，向客户端 Socket 发送信息。

```
DataInputStream in = new DataInputStream(socket.getInputStream());
DataOutputStream out = new DataOutputStream(socket.getOutputStream());
```

（4）在通信完成后关闭流和 Socket。

```
socket.close();
```

2．客户端

（1）通过 IP 地址和端口号实例化一个 Socket 对象，该对象会向指定 IP 和端口的远程主机发送连接请求。

```
Socket socket = new Socket("192.168.1.120", 5678);
```

（2）调用 Socket 对象的 getOutputStream()方法获取输出流，向服务器的 Socket 对象发送信息，调用 getInputStream()方法获取输入流，接收服务器的 Socket 发送的信息。

```
DataInputStream in = new DataInputStream(socket.getInputStream());
DataOutputStream out = new DataOutputStream(socket.getOutputStream());
```

（3）在通信完成后关闭流和 Socket。

```
socket.close();
```

【例 13.5】 本例的服务器端是 PC，运行在 Java SE 平台下，客户端是 Android 系统，通过 Socket 完成信息传送。

（1）在 Eclipse 中，新建一个 Java 项目 SocketServer，main()方法中的代码如下。

```
try{
    ServerSocket serversocket = new ServerSocket(5678);//实例化 ServerSocket
    System.out.println("Listening...");
    while(true){
        Socket socket = serversocket.accept();//监听端口，得到服务器 Socket
        System.out.println("Client Connected...");
```

```
            //得到输入流
            DataInputStream in = new DataInputStream(socket.getInputStream());
            String InStr = in.readUTF();              //获取客户端发送的信息
            System.out.println(InStr);
            DataOutputStream out = new DataOutputStream(socket.getOutputStream());
            //得到输出流
            out.writeUTF("Link Server Success!");     //向客户端发送信息
            out.flush();
            in.close();
            out.close();
            socket.close();                            //关闭 Socket
        }
    }
    catch(Exception e){
        e.printStackTrace();
    }
}
```

（2）在 Eclipse 中新建一个名称为 SocketClient 的 Android 项目，修改项目的布局文件，使用线性布局管理器，添加一个 EditText 控件，用来显示服务器发送的消息。

（3）在 SocketClient 类中，创建要使用的变量，重写 onCreate()方法，获取布局管理器中的对象，同时创建 Socket 对象，和服务器交换信息，代码如下。

```
private EditText et;
@Override
protected void onCreate(Bundle savedInstanceState) {
    super.onCreate(savedInstanceState);
    setContentView(R.layout.activity_socket_client);
    et = (EditText)findViewById(R.id.et);           //获得 EditText 对象
    new Thread(new Runnable() {
      public void run() {
        try{
            Socket socket = new Socket("192.168.1.120", 5678); //创建 Socket 对象
            //获得输出流
            DataOutputStream out = new DataOutputStream(socket.getOutputStream());
            out.writeUTF("This is a Message Send by Client !");
                                                    //发送信息 out.flush();
            DataInputStream in = new DataInputStream(socket.getInputStream());
            String str = in.readUTF();              //读取服务端发来的消息
            et.setText(str);                        //设置 EditText 对象
            out.close();
            in.close();
            socket.close();                         //关闭 Socket
        }
        catch(Exception e){
            e.printStackTrace();
        }
      }
    }).start();
}
```

首先运行服务器端程序，然后启动客户端程序，连接成功后，服务器端的运行结果如图 13.7 所示，客户端的运行结果如图 13.8 所示。

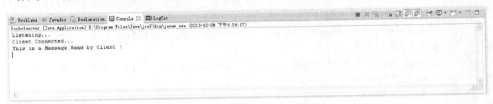

图 13.7 连接成功服务器端界面

图 13.8 连接成功后客户端界面

13.4 基于 WebView 的简单浏览器

Android 平台内置了开源的 WebKit 浏览器引擎，WebKit 有着很强大的网络功能，不仅能够浏览网页，收发电子邮件，还支持音视频节目的在线播放，通过 WebView 组件能够方便地使用 WebKit 来开发支持 JavaScript、AJAX 等功能的浏览器。WebView 组件常用的方法如表 13.2 所示。

表 13.2 WebView 常用的方法

方法	描述
loadUrl（String url）	加载 URL 指定的页面
reload()	刷新当前加载的页面
stopLoading()	停止正在加载的页面
goBack()	返回至当前页面上一次加载的页面
goForwoard()	前进至当前页面下一次加载的页面

在实际应用开发过程中,一般还需要使用 WebView 组件中的 WebSettings 对象、WebViewClient 对象与 WebChromeClient 对象设置浏览器的特性及重写事件的方法来丰富浏览器的功能。下面对其进行简单介绍。

WebSettings 对象使用 Webview 对象的 getSettings()方法可以得到,该对象主要是对 WebView 的属性和配置进行设置,常用的方法如表 13.3 所示。

表 13.3 WebSettings 常用方法

方法	描述
setJavaSciptEnabled(boolean flag)	是否支持 JavaScript 脚本
setBlockNetworkImage(boolean flag)	是否显示网络图片
setBuiltInZoomControls(boolean flag)	是否支持页面缩放
setDefaultTextEncodingName(String Encode)	设置解码时默认编码
setDefaultFontSize(int size)	设置默认的字体大小
setSupportMultipleWindows(boolean flag)	是否支持多屏幕显示
setPluginsEnabled(boolean flag)	是否支持插件

Webview 对象主要负责页面的解析和渲染工作,如果需要丰富浏览器的功能,就必须用到 WebViewClient 对象与 WebChromeClient 对象,WebViewClient 的方法会在影响页面渲染的动作发生时被调用,比如,页面开始加载、页面加载完毕、资源加载、页面中的 URL 请求等,WebChromeClient 的方法在一些影响页面的交互动作发生时被调用,如 WebView 的打开和关闭、页面加载进展、弹出 JavaScript 确认框和警告框等。

(1) WebViewClient 常用的方法如下。

·public void onPageStarted(WebView view,String url,Bitmap favicon)

该方法在页面加载时会调用。

·public void onPageFinished(WebView view,String url)

该方法会在页面加载完毕时会调用。

·public void onLoadResource(WebView view,String url)

该方法在加载页面的资源时会调用。

·public boolean shouldOverrideUrlLoading(WebView view,String url)

该方法在单击页面内的链接时调用,重写此方法并返回 true 表明单击页面内的链接还是在当前的 WebView 里打开,不跳转到系统默认的浏览器。

(2) WebChromeClient 常用的方法如下。

·public void onProgressChanged(WebView view, int progress)

该方法在加载进度改变时会调用。

·public Boolean onCreateWindow(WebView view, boolean dialog, boolean userGesture, Message resultMsg)

该方法在创建 WebView 时会调用。

·public void onCloseWindow(WebView window)

该方法在关闭 WebView 的时候会调用。

·public void onReceivedIcon(WebView view, Bitmap icon)

该方法在网页图标更改时会调用。

·public void onReceivedTitle(WebView view, String title)

该方法在网页标题更改时会调用。

下面通过实例利用 WebView 组件开发一个简单的浏览器。

【例 13.6】 在 Eclipse 中新建一个名称为 WebViewExplorer 的 Android 项目,该项目实现一个具有打开、前进、后退功能,支持 JavaScipt 的简单浏览器。

(1) 修改项目的布局文件 activity_webviewexplorer.xml,使用线性布局管理器,添加三个按钮控件,分别完成前进、后退和打开功能,添加一个 EditText 控件,用于输入 URL 地址,添加一个 WebView 控件显示网页的内容,代码如下。

```xml
<LinearLayout xmlns:android="http://schemas.android.com/apk/res/android"
    android:orientation="vertical"
    android:layout_width="fill_parent"
    android:layout_height="fill_parent"
    >                                               <!-- 声明一个线性布局 -->
    <LinearLayout
        android:orientation="horizontal"
        android:layout_width="fill_parent"
        android:layout_height="wrap_content"
        >                                           <!-- 声明一个线性布局 -->
        <Button
            android:id="@+id/btnForward"
            android:layout_width="wrap_content"
            android:layout_height="wrap_content"
            android:text="@string/buttonforward"
            />                                      <!-- 声明一个 Button 控件 -->
        <Button
            android:id="@+id/btnBack"
            android:layout_width="wrap_content"
            android:layout_height="wrap_content"
            android:text="@string/buttonback"
            />                                      <!-- 声明一个 Button 控件 -->
        <EditText
            android:id="@+id/editurl"
            android:layout_width="170dp"
            android:layout_height="wrap_content"
            android:selectAllOnFocus="true"
            android:singleLine="true"
            android:inputType="text"
            android:text="@string/editurl"
```

```xml
                    />                             <!-- 声明一个 EditText 控件 -->
        <Button
            android:id="@+id/btnGo"
            android:layout_width="50dp"
            android:layout_height="wrap_content"
            android:layout_gravity="right"
            android:text="@string/buttongo"
            />                                     <!-- 声明一个 Button 控件 -->
    </LinearLayout>
    <WebView
        android:id="@+id/webview"
        android:layout_width="fill_parent"
        android:layout_height="fill_parent"
        />                                         <!-- 声明一个 WebView 控件 -->
</LinearLayout>
```

（2）在 WebviewExplorerActivity 类中，创建要使用的变量如下。

```
private WebView webview;
private EditText editurl;
private Button btnForward;
private Button btnBack;
private Button btnGo;
```

（3）重写 onCreate()方法，首先获取布局管理器中的对象，然后创建 WebViewClient 对象并进行浏览器基本属性的配置，重写 WebChromeClient 的 onProgressChanged 方法，实现打开网页时的进度条功能，重写 WebViewClient 的 shouldOverrideUrlLoading 方法，实现在页面内单击链接时仍然使用当前的 WebView 打开，最后为打开、前进、后退添加监听器实现对应功能，代码如下。

```java
protected void onCreate(Bundle savedInstanceState) {
    super.onCreate(savedInstanceState);
    //使用 Android 系统自带的进度条
    this.getWindow().requestFeature(Window.FEATURE_PROGRESS);
    setContentView(R.layout.activity_webviewexplorer);
    editurl = (EditText)findViewById(R.id.editurl);
    btnForward = (Button)findViewById(R.id.btnForward);
    btnBack = (Button)findViewById(R.id.btnBack);
    btnGo = (Button)findViewById(R.id.btnGo);
    webview = (WebView)findViewById(R.id.webview);
    WebSettings browserSetting = webview.getSettings();  //创建 WebSettings 对象
    browserSetting.setSupportMultipleWindows(false);     //不支持对窗口
    browserSetting.setJavaScriptEnabled(true);           //支持 JavaScript 脚本
    webview.setWebChromeClient(new WebChromeClient(){
        @Override
        //重写 onProgressChanged 方法，实现打开页面时的进度条
        public void onProgressChanged(WebView view, int progress){
```

```java
            //设置打开页面时滚动条的文字
            WebviewExplorerActivity.this.setTitle("Loading...");
            //设置滚动条的进度
            WebviewExplorerActivity.this.setProgress(progress * 100);
            if(progress == 100)
                WebviewExplorerActivity.this.setTitle(R.string.app_name);
        }
    });
    webview.setWebViewClient(new WebViewClient() {
        @Override
        //重写shouldOverrideUrlLoading方法,保证在当前WebView中打开页面的链接
        public boolean shouldOverrideUrlLoading(WebView view, String url) {
            //加载页面内单击链接的请求页面
            view.loadUrl(url);
            return true;
        }
    });
    //前进按钮添加监听器
    btnForward.setOnClickListener(new View.OnClickListener() {
        @Override
        public void onClick(View v) {
            webview.goForward();            //返回当前页面下一次打开的页面
        }
    });
    //后退按钮添加监听器
    btnBack.setOnClickListener(new View.OnClickListener() {
        @Override
        public void onClick(View v) {
            webview.goBack();               //返回当前页面上一次打开的页面
        }
    });
    //打开按钮添加监听器
    btnGo.setOnClickListener(new View.OnClickListener() {
        @Override
        public void onClick(View v) {
            //获取EditText中的URL地址
            String url = editurl.getText().toString().trim();
            if(URLUtil.isNetworkUrl(url)){          //判断URL的是否正确
                webview.loadUrl(url);               //打开当前的链接
            }
            else{
                Toast.makeText(WebviewExplorerActivity.this, "The NetAddress
```

第 13 章 Android 的网络编程

```
                    is Error!", Toast.LENGTH_SHORT).show();    //URL 不正确,给出提示
                    editurl.requestFocus();
                }
            }
        });
        //EditText 添加监听器
        editurl.setOnKeyListener(new View.OnKeyListener() {
            @Override
            public boolean onKey(View v, int keyCode, KeyEvent event) {
                if(keyCode == KeyEvent.KEYCODE_ENTER) {//是否是 ENTER 键
                    String url = editurl.getText().toString().trim();
                    if(URLUtil.isNetworkUrl(url)){
                        webview.loadUrl(url);                //打开当前的链接
                        return true;
                    }
                    else{
                        Toast.makeText(WebviewExplorerActivity.this, "The
                        NetAddress is Error!", Toast.LENGTH_SHORT).show();
                        //URL 不正确,给出提示
                        editurl.requestFocus();
                    }
                }
                return false;
            }
        });
}
```

运行本实例,加载页面的的运行结果如图 13.9 所示。

图 13.9　浏览器浏览网页

13.5 本章小结

　　本章主要介绍了 Android 平台下进行网络编程的方法，首先介绍了基于 HTTP 协议访问网络的两种方法：一种是基于标准的 Java 接口；另一种是基于 Apache 的 HttpClient，后者在前者的基础上进行了进一步的封装和完善，在比较复杂的网络通信应用中更加方便高效，然后介绍基于 SOCKET 的网络通信，它是网络通信的基础，最后介绍了 WebView 组件及该组件中常用的对象及方法,利用该组件可以方便地实现一个具有基本功能的浏览器。

第 14 章 定位服务和地图服务

本章主要介绍在 Android 平台下开发基于 GPS 的定位服务应用程序和基于 Google 的地图服务应用程序，内容包括通过 GPS 获取用户位置、处理位置变化事件及利用 Google Map API 开发融合了地图服务的应用程序。

14.1 定位服务相关类

在 Android 系统中，可以通过 GPS 和 Network 两种方式主动获取用户的位置，通过 GPS 定位只能在户外，费电而且响应慢，但是精度高；通过 Network 定位对用户位置没有要求，省电而且响应快，但是精度低。它们被称为位置服务提供者（LocationProvider），开发人员可以指定两种位置服务提供者中的一种来开发具有定位服务的应用程序，也可以设定定位的条件，由系统选择两种位置服务提供者中的一种来完成定位服务。

定位服务的实现需要使用下面的类和接口。

（1）Criteria 类

该类用来对定位的条件进行设置，系统根据设定的定位条件作为选择位置服务提供者，在定位条件中，用户最关心的是耗电量和精度，该类中设置了常量来进行耗电量和精度的描述，具体如表 14.1 所示。

表 14.1 耗电量和精度常量

常量	描述	常量	描述
ACCERACY_HIGH	精度高	POWER_HIGH	耗电量高
ACCERACY_MEDIUM	精度中等	POWER__MEDIUM	耗电量等
ACCERACY_LOW	精度低	POWER_LOW	耗电量低

除了耗电量和精度外，还能对是否产生费用，海拔高度和速度等条件进行设置，该类常用的方法如表 14.2 所示。

表 14.2 Criteria 类常用的方法

方法	描述
setAccuracy(int accuracy)	设置精确度
setPowerRequirement (boolean powerRequirement)	设置用电量
setAltitudeRequired (boolean altitudeRequired)	设置是否需要高度信息
setBearingRequired (boolean bearingRequired)	设置是否需要方位信息
setSpeedAccuracy (int accuracy)	设置是否需要速度信息
setCostAllowed (boolean costAllowed)	设置是否产生费用

（2）Location 类

该类封装了位置服务提供者描述当前设备的一些物理数据，包括经纬度、高度、速度、海拔等数据，通过该类定义的一系列 Get 方法，可以返回这些数据供应用程序使用，该类定义的常用方法如表 14.3 所示。

表 14.3　Location 类常用方法

方法	描述
public float getAccuracy ()	返回定位的精确度
public double getAltitude ()	返回设备的高度数据
public float getBearing ()	返回设备方向
public double getLatitude ()	返回设备的经度
public double getLongitude ()	返回设备的纬度
public float getSpeed ()	返回设备的速度

（3）LocationProvider 类

该类用来描述位置服务提供者，设置位置提供者的一些属性，这些属性一般采用系统默认的参数。

（4）LocationManager 类

该类是实现设备的定位、追踪，是定位服务的重要的类，注意，该类不能被实例化，它是通过 getSystemService 方法来获得的，下面的代码能够获得一个 LocationManager 的实例：

```
LocationManager manager = (LocationManager)getSystemService(LOCATION_SERVICE);
```

该类中定义了两个用于描述位置服务提供者的字符串常量，GPS_PROVIDER 表示使用 GPS 方式，NETWORK_PROVIDER 表示使用网络方式，通过这些字符串常量可以直接指定位置服务提供者，也可以通过调用该类的 getBestProvider 方法，通过 Criteria 类设定的定位条件由系统确定最合适的位置服务提供者，该方法的返回值是字符串，表示位置服务提供者，该方法如下所示：

```
public string getBestProvider(Criteria criteria, boolean enabledOnly);
```

确定了位置服务提供者之后，就可以使用表示位置服务提供者的字符串作为参数调用该类的 getLastKnownLocation 方法来获取当前设备的定位信息，该方法返回一个 Location 对象，通过该对象能够得到设备的诸如经纬度、速度、高度等信息，该方法如下所示：

```
public Location getLastKnownLocation(String provider);
```

通过上面的步骤我们只能主动地获得设备的定位信息，如果需要在定位信息或状态发生变化时主动通知系统，就需要为 LocationManager 添加一个 LocationListener 监听器，调用该类的 requestLocationUpdates 方法就可以添加一个监听器，该方法如下所示。

```
public void requestLocationUpdates(String provider, long minTime, float minDistance, LocationListener listener);
```

该方法的第二个参数是位置更新的最短时间间隔，第三个参数是位移变化的最短距离。

（5）LocationListener 监听器

在 LocationListener 监听器中定义了 4 个方法，实现监听器需要重写这几个方法，这 4 个方法如表 14.4 所示。

表 14.4 LocationListener 监听器方法

方法	描述
onLocationChanged(Location location)	设备位置变化时调用该方法
onProviderDisabled(String Provider)	设备禁用时调用该方法
onProviderEnabled(String Provider)	设备启用时调用该方法
OnStatusChanged(String Provider,int status,Bundle extras)	当设备状态变化时调用该方法

14.2 定位实例

Android 平台下的定位服务开发的一般步骤如下。

（1）通过 getSystemService 方法实例化一个 LocationManager 类的对象：

```
LocationManager manager = (LocationManager)getSystemService(LOCATION_SERVICE);
```

（2）如果不需要设置定位条件，直接转入第 4 步，否则实例化一个 Criteria 类的对象，设置查询条件如下。

```
Criteria criteria = new Criteria();
criteria.setAccuracy(Criteria.ACCERACY_HIGH);      //设置精度
criteria.setPowerRequirement(Criteria.POWER_LOW);  //设置耗电量
criteria.setAltitudeRequired(false);               //设置是否需要海拔高度
criteria.setBearingRequired(false);                //设置是否需要方向
criteria.setSpeedRequired(false);                  //设置是否需要速度
criteria.setCostAllowed(false);                    //设置是否允许产生费用
```

（3）通过 LocationManager 类的 getBestProvider 方法得到位置服务提供者：

```
String provider = manager.getBestProvider(criteria, true);
```

（4）通过 LocationManager 类的 getLastKnownLocation 方法来获取当前设备的定位信息：

```
Location location = manager.getLastKnownLocation(provider);
```

如果没有设定定位条件，使用 LocationManager 的常量 GPS_PROVIDER 或者 NETWORK_PROVIDER 作为参数：

```
Location location = manager.getLastKnownLocation(LocationManager.GPS_PROVIDER);
```

注意：模拟器不提供网络定位服务，无法使用 NETWORK_PROVIDER。

（5）由于定位设备的定位需要一定时间，在使用得到的 Location 对象时，可能出现 Location 对象为 NULL 的情况，导致信息不正确，应该对 Location 对象进行判断，然后再获取定位信息进行应用。

（6）为 LocationManager 绑定 LocationListener 监听器并重写对应的方法：

```
manager.requestLocationUpdates(provider,10000,10, new LocationListener(){
//重写对应的方法
});
```

【例 14.1】在 Eclipse 中新建一个名称为 GpsTest 的 Android 的项目，获取更新后的经度和纬度信息，在 EditText 中进行显示。

（1）在项目的 AndroidManifest.xml 文件的 Manifast 标签中加入如下指令。

```
<uses-permission android:name="android.permission.ACCESS_FINE_LOCATION" />
<uses-permission android:name="android.permission.ACCESS_COARSE_LOCATION"/>
```

（2）修改项目的布局文件 activity_gps_test.xml，使用线性布局管理器，添加一个显示经度和纬度的 EditText 控件。

（3）在 GpsTest 类中，创建要使用的变量，重写 onCreate()方法，代码如下。

```
private LocationManager locationmanager;
private Location location;
private EditText edittext;
@Override
protected void onCreate(Bundle savedInstanceState) {
  super.onCreate(savedInstanceState);
  setContentView(R.layout.activity_gps_test);         //获取布局管理器
  edittext = (EditText)findViewById(R.id.et);
  //获取 locationmanager 的实例
  locationmanager = (LocationManager)getSystemService(LOCATION_SERVICE);
  //使用 Gps 作为位置服务提供者，获得包含位置信息的 location 对象
  location = locationmanager.getLastKnownLocation (LocationManager.GPS_PROVIDER);
  //更新 EditText 的内容
  update(location);
  //添加 LocationListener 监听器
  locationmanager.requestLocationUpdates(LocationManager.GPS_PROVIDER,
  10000, 5, new LocationListener(){
    @Override
    //当位置信息发生变化时，更新 EditText 的内容
    public void onLocationChanged(Location newLocation) {
        update(newLocation);
    }
    @Override
    //当定位设备停止服务时，更新 EditText 的内容
```

```
        public void onProviderDisabled(String provider) {
            edittext.setText("定位设备" + provider + "停止服务");;
        }
        @Override
        //当定位设备启动服务时,获取 location 对象,更新 EditText 的内容
        public void onProviderEnabled(String provider) {
            location = locationmanager.getLastKnownLocation(provider);
            update(location);
        }
        @Override
        public void onStatusChanged(String provider, int status, Bundle extras) {
        }
    });
}
```

（4）编写无返回值的方法 update()，该方法判断 Location 是否为空，如果不为空，获取 Location 对象中的经度和纬度并在 EditText 中显示，代码如下。

```
public void update(Location location){
    if (location != null){
        edittext.setText("当前位置：" + "\n");
        edittext.append("纬度：" + location.getLatitude() + "\n");
        edittext.append("经度：" + location.getLongitude());
    }
    else {
        edittext.setText("定位失败");
    }
}
```

模拟器中没有 GPS 设备,可以通过 DDMS 中的模拟器控制台向模拟器发送模拟的 GPS 数据,模拟器控制台如图 14.1 所示,启动应用程序后,由于无法得到定位信息,会出现如图 14.2 所示的状态,单击 Send 按钮,当模拟器接收到模拟数据后,文本框将显示出模拟的经度和纬度,如图 14.3 所示。

图 14.1　模拟器控制台

图 14.2　定位失败

图 14.3　定位成功

14.3　Google Map 使用

　　Google 地图是 Google 公司提供的电子地图服务，通过该服务用户可以方便查找周边商家信息、行车及公交路线，开发融合了地图服务的 Android 应用程序是程序员必须掌握的一项技术。要是用 Google 的地图服务，必须先申请经过验证 Google Map 的 API KEY。

14.3.1　申请 Map API KEY

　　申请 Map API KEY 的过程如下。
　　（1）Google 的 Map API Key 对于不同的机器有所不同，首先要取得本机的特征码，也称为指纹，获取本机指纹的可以通过 Eclipse 中的【Windows】菜单的【Preferences】选项来获得，如图 14.4 所示，这里的 SHA1 fingerprint 变量的字符串就是我们需要的指纹，在获取 API key 时需要。

图 14.4　SHA1 指纹

(2)进入 Google 的官方网站,使用注册过的账号登录,只有注册用户才能获取 API Key,登录之后在地址栏输入链接 http://code.google.com/apis/console 进入 Google 的 API 控制台,如果是第一次使用该平台,会出现新建项目的界面,如图 14.5 所示。

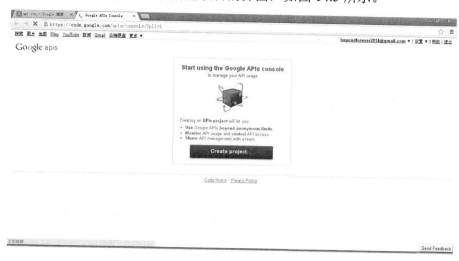

图 14.5　新建项目

(3)选择新建项目后,出现欢迎界面和用户协议选项,选择同意并单击"继续"按钮,进入 API 控制台,选择【APIs & Auth】节点下的【APIs】选项,出现 Google 提供的 API 列表,如图 14.6 所示。

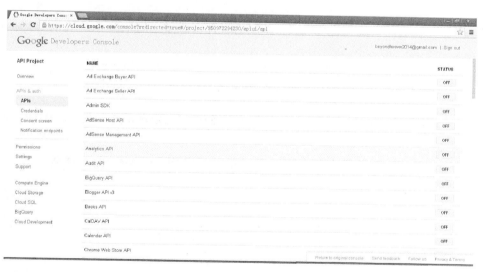

图 14.6　API 列表

(4)在列表中找到 Google Maps Android API V2,单击右侧的按钮将其状态设置为"ON",设置完毕后,该 API 会被自动置于顶部显示,如图 14.7 所示。

(5)选择【APIs & Auth】节点下的【Credential】选项,进入到创建密钥的页面,选择新建密钥的选项,出现密钥种类的选择页面,选择 Android 系统密钥,出现如图 14.8 所示的页面,该页面需要用户输入创建密钥必要的信息。

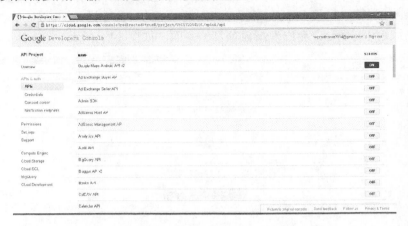

图 14.7　打开 Google Map API

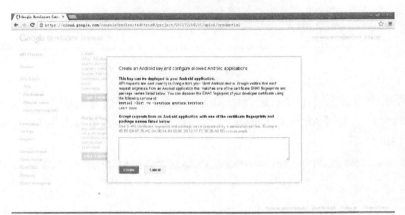

图 14.8　信息输入窗体

(6) 在文本框中,用户需要输入第一步得到 SHA1 指纹和需要使用 MAP API 的包名,两者用分号隔开,如图 14.9 所示。

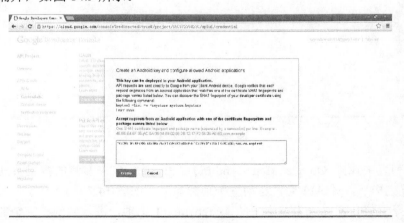

图 14.9　输入指纹和包名

（7）输入完毕后，单击创建按钮，创建 MAP API Key，如图 14.10 所示，注意，该密钥只能在指定的机器和项目中使用，如果使用其他机器，或者项目的包名发生变化，该密钥将无法使用，这种情况下，需要创建新的密钥，同一用户创建密钥的 Google 个数没有限制。

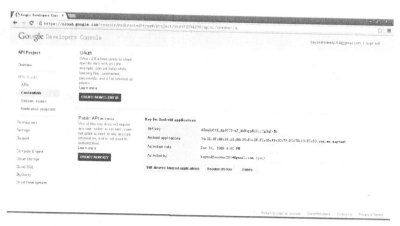

图 14.10　创建密钥

到此为止，我们已经成功地创建了 Google 的 Map API key，使用复制功能将该密钥保存好，在后面的开发中需要用到。

14.3.2　开发和测试环境搭建

取得了 API Key 后，还需要搭建能够进行开发和测试的平台。开发具有地图功能的应用程序需要 Google Play servers 的支持，打开 Android 的 SDK 管理器，查看【Extras】节点下面的 Google Play servers 是否安装，如图 14.11 所示。

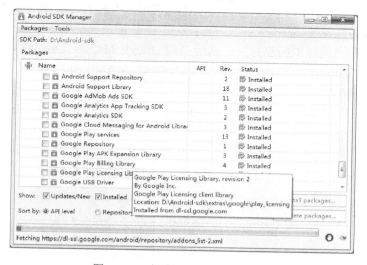

图 14.11　查看 Google Play servers

安装 Google Play servers 之后，还必须将该包导入到 Eclipse 中，选择【file】菜单的【import】选项，选择【Android】节点下的【Existing Android Code Into Workspace】选项，该包位于 Android SDK 目录下的\Extra\google\ google_play_servers\libproject 目录下，导入时要将拷贝项目到工作区的选项进行勾选，该目录下还有一些基于 Google 地图服务的示例程序，如图 14.12 所示。

导入类库之后，还需要将类库加载到对应的应用程序中，在需要添加该类库的项目上单击右键，选择【Properties】选项，在出现的窗体中选择【Android】节点，单击下方的【ADD】按钮，选中该类库，如图 14.13 所示。

图 14.12　导入 Google_Play_servers　　　　图 14.13　加载 Google_Play_servers

开发环境配置完毕之后，还需要对模拟器配置用来进行调试，模拟器的配置比较简单，从 Googel 的官方网站下载 com.google.android.vending.apk 和 com.google.android.gms.apk 两个软件并将其安装到模拟器中。

在开发环境和测试环境搭建好后，就可以将 Google Map 融合到应用程序中了。

【例 14.2】在 Eclipse 中新建一个名称为 MapTest 的 Android 的项目，该项目将一张全球地图显示在手机屏幕上。

（1）按照上一节的步骤，首先申请 API Key，加载 Google_Play_servers 类库到本项目中。

（2）修改项目的 AndroidManifest.xml 文件，在 Manifast 标签中加入下面的指令，用来设置 API Key 以及指定所使用的类库：

```
<meta-data
    android:name="com.google.android.maps.v2.API_KEY"
//此处是申请到的用于本机本项目的密钥
android:value="AIzaSyCt4rYMNzhqXNdAtI1Ygd1jpZ-eW7Q7lkw"/>
<meta-data
    //指定使用的类库
```

```xml
android:name="com.google.android.gms.version"
android:value="@integer/google_play_services_version" />
```

由于在 Google Map 中使用了 Open GL 特效，需要在 Manifast 标签中加入下面的指令。

```xml
<uses-feature android:glEsVersion="0x00020000" android:required="true"/>
```

由于地图服务需要使用网络，需要有添加网络权限等指令。

AndroidManifast.xml 的配置不当，会使程序无响应或无法实现地图服务，下面给出 AndroidManifast.xml 的全部代码，注意含有包名的代码要和申请密钥时的包名一致，代码中前面已经解释的不再注释。

```xml
<?xml version="1.0" encoding="utf-8"?>
<manifest xmlns:android="http://schemas.android.com/apk/res/android"
    package="com.ex.maptest"
    android:versionCode="1"
    android:versionName="1.0" >
    <uses-sdk
        android:minSdkVersion="9"
        android:targetSdkVersion="18" />
    <uses-permission android:name="com.ex.maptest.permission.MAPS_RECEIVE"/>
                                                        <!--允许接收地图信息-->
    <uses-permission android:name="android.permission.INTERNET" />
                                                        <!--添加网络访问权限-->
    <uses-permission android:name="android.permission.ACCESS_NETWORK_STATE" />
                                                        <!--API 自动更新权限-->
    <uses-permission android:name="android.permission.WRITE_EXTERNAL_STORAGE"/>
                                                        <!--地图缓存权限 -->
    <uses-permission android:name="com.google.android.providers.gsf.permission.
READ_GSERVICES"/>                                       <!--访问 WEB 权限-->
    <uses-permission android:name="android.permission.ACCESS_COARSE_LOCATION"/>
                                                        <!--允许粗定位-->
    <uses-permission android:name="android.permission.ACCESS_FINE_LOCATION"/>
                                                        <!--允许精确定位-->
    <uses-feature android:glEsVersion="0x00020000" android:required="true"/>
    <application
        android:allowBackup="true"
        android:icon="@drawable/ic_launcher"
        android:label="@string/app_name"
        android:theme="@style/AppTheme" >
        <activity
            android:name="com.ex.maptest.MapTestActivity"
            android:label="@string/app_name" >
            <intent-filter>
                <action android:name="android.intent.action.MAIN" />
                <category android:name="android.intent.category.LAUNCHER" />
```

```
            </intent-filter>
        </activity>
         <meta-data
           android:name="com.google.android.maps.v2.API_KEY"
           android:value="AIzaSyCt4rYMNzhqXNdAtI1Ygd1jpZ-eW7Q7lkw"/>
        <meta-data
           android:name="com.google.android.gms.version"
           android:value="@integer/google_play_services_version" />
    </application>
</manifest>
```

（3）修改项目的布局文件 activity_map_test.xml，代码如下。

```
<?xml version="1.0" encoding="utf-8"?>
<fragment xmlns:android="http://schemas.android.com/apk/res/android"
    android:id="@+id/map"
    android:layout_width="match_parent"
    android:layout_height="match_parent"
    class="com.google.android.gms.maps.SupportMapFragment"/>
```

（4）MapTestActivity 类集成 FragmentActivity 类，其他地方不需要修改，代码如下。

```
public class MapTestActivity extends FragmentActivity {
    @Override
    protected void onCreate(Bundle savedInstanceState) {
        super.onCreate(savedInstanceState);
        setContentView(R.layout.activity_map_test);
    }
}
```

运行该项目，如果开发环境、测试环境、密钥、权限、代码都正确无误的话，将会在屏幕上看到如图 14.14 所示的地图，使用鼠标左右拖曳会看到不同国家和地区的地图。

图 14.14　全球地图

14.4 地 图 定 位

上一节的实例，仅仅是一个全球地图的显示。本节利用 GPS 定位服务确定设备所在的位置，并将该位置显示在地图上，完成地图定位的功能。

【例 14.3】在 Eclipse 中新建一个名称为 LocationMap 的 Android 的项目，该项目综合使用前面介绍 GPS 服务和 Map 服务完成地图定位。

（1）以该项目的包名和 SHA1 指纹申请一个 Map Key，将 Google_Play_servers 类库到本项目中。

（2）该项目的 AndroidManifast.xml 文件和配置文件和例 14.2 一样，不再描述。

（3）在 LocationMapActivity 类中，实例化一个 LocationManager 类，使用 GPS_PROVIDER 作为位置服务提供者获取设备的经纬度信息，同时为该类绑定一个 LocationListener 监听器，当位置发生变化时，利用 GoogleMap 类的对象确定当前设备所在地图的位置，以图标的方式在地图上标明，完成地图定位，代码如下。

```java
public class LocationMapActivity extends FragmentActivity {
private LocationManager locationmanager;
private Location location;
private GoogleMap googlemap;
@Override
protected void onCreate(Bundle savedInstanceState) {
    super.onCreate(savedInstanceState);
    setContentView(R.layout.activity_location_map);
    SupportMapFragment mapfragment = (SupportMapFragment)this.getSupportFragment Manager().findFragmentById(R.id.map);
                                    //获取 SupportMapFragment 对象
    googlemap = mapfragment.getMap();    //获取 GoogleMap 对象
    googlemap.setMapType(GoogleMap.MAP_TYPE_NORMAL);    //一般地图模式
    googlemap.setTrafficEnabled(false);        //不显示导航信息
    googlemap.setMyLocationEnabled(true);        //标记当前位置
    //获取 locationmanager 对象
    locationmanager = (LocationManager)getSystemService(LOCATION_SERVICE);
    //使用 GPS 方式获取位置信息，保存在 Location 对象中
    location = locationmanager.getLastKnownLocation(LocationManager.GPS_P ROVIDER);
    //绑定监听器
    locationmanager.requestLocationUpdates(LocationManager.GPS_PROVIDER,
    10000, 5, new LocationListener(){
        @Override
        //位置变化时调用该函数，根据新坐标，刷新地图显示
        public void onLocationChanged(Location newLocation) {
            update(newLocation);
        }
```

```
            @Override
            public void onProviderDisabled(String provider) {
            }
            @Override
            //设备启动是调用此函数,在地图中标识位置
            public void onProviderEnabled(String provider) {
                location = locationmanager.getLastKnownLocation(provider);
                update(location);
            }
            @Override
            public void onStatusChanged(String provider,int status, Bundle
            extras) {
            }
        });
    }
    public void update(Location location){
        if (location != null){
          //获取 Location 对象中保存的精度和纬度
          LatLng latlng = new LatLng(location.getLatitude(),location.
          getLongitude());
          //根据当前的坐标,更新地图的显示
          googlemap.moveCamera(CameraUpdateFactory.newLatLng(latlng));
          //设置地图的缩放比例
          googlemap.animateCamera(CameraUpdateFactory.zoomTo(10));
        }
        else {
          Toast.makeText(this, "定位失败", Toast.LENGTH_SHORT).show();
        }
    }
}
```

由于模拟器不支持 GPS,使用 DMMS 的模拟控制器发送经度和纬度位置信息,如图 14.15 所示。

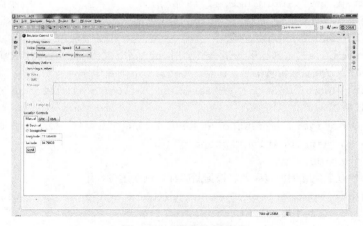

图 14.15　发送位置信息

接收到位置信息后，会触发 LocationListener 监听器的 onLocationChanged 方法，该方法定位地图的结果如图 14.16 所示。

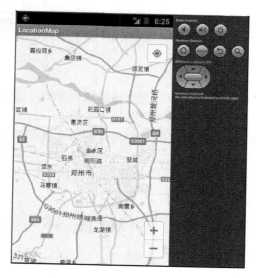

图 14.16　地图定位

14.5　本 章 小 结

本章主要介绍了 Android 平台进行定位的主要方式，以及完成定位服务常用的类和接口，还介绍了如何实现 Google 提供的地图服务，包括 Map Key 的获得、开发平台和测试平台的配置，最后结合 GPS 和 Google Map 实现了地图定位，将这两种功能融合到应用程序中，会增加 Android 系统的很多特色。

第 15 章　移 动 同 学 簿

本章介绍如何使用 Android 技术开发一个移动版同学簿。重点介绍 Android 如何获取网络数据，并进行数据的绑定，实现实时网络图片加载、Android UI 布局、UI 界面的动态更新、数据全局共享处理、界面数据交互等。通过对该实例的学习，读者可以熟悉移动网应用的开发和设计的全过程，通过本章的学习使读者能够在应用 Android 开发技术上提升一个新的台阶。

15.1　系 统 概 述

15.1.1　移动同学簿的应用背景

随着移动信息化的发展和网络信息技术的进步及移动应用产品的大面积普及，移动信息化的趋势越来越明显，移动应用开始逐渐蚕食传统互联网份额。现代网络服务提供商已经开始大规模进入移动应用战圈，网络服务早已不是守候在电脑前查找相关信息的模式。现代网络服务即时、当面、便捷。随时随地利用移动终端查询信息也成为用户使用的焦点。用户可以随时随地地浏览信息，进行网上交易，通过移动终端在线查询信息的需求也越来越迫切。手机通讯录作为手机终端最基础的功能，其质量直接影响着用户对手机的体验度。

在基于 Android 系统的众多应用中，移动同学簿是一种利用互联网或移动互联网实现通讯录信息同步更新和备份的应用和服务。通讯录是每个手机都必备的应用软件，一个人的记忆能力再好也不可能记下自己所有联系人的通讯信息，智能手机内安装一个比较好的通讯录就可以解决很多不必要的麻烦，至少不用为在关键时候自己忘了朋友的联系方式而困扰。在此背景之下，开发出一款可以切实符合用户使用习惯、安全的、同时功能齐全的同学簿可以方便用户及时地查询好友及同学的联系信息，给用户提供便捷的服务。本章通过开发移动同学簿项目为实例来说明项目分析、设计和实现的全过程。

15.1.2　移动同学簿的总体需求

移动同学簿平台主要是展示某一用户所有同学的基本信息，通过该平台可以添加人员信息、删除人员信息和搜索人员信息，并且实现传统网站信息和移动平台信息互通，从而实现数据的及时同步更新，达到数据的永久备份，使用户能够在手机更换或者丢失时，可以方便地从网络服务器获取数据信息，直接恢复原有的同学信息，使用户简单快捷地实现同学录的更新。

该系统的总体功能描述如下：系统自动检索信息，如果是新安装软件，系统会在有网络的前提下自动连接网络服务器获取数据信息，并进行列表展示。用户可以通过单击列表中某项查看同学详细信息，并可以在查看信息中拨打电话或发送短信；也可以通过"长按"在同学信息列表中删除某位同学信息；也可以在联系人详细信息页面中可以通过菜单键中的删除选项实现删除，删除联系人后主页的信息列表会自动刷新。同时该系统提供了搜索功能，可以实现根据姓名、年龄、生日、电话号码等进行查询，并且全部提供模糊查询合并产生的结果。该系统实现传统网站信息和移动平台信息互通，当数据信息通过移动终端或传统网站更改信息时，实现信息的同步更新。

15.1.3 移动同学簿的功能分析

为了方便用户的使用和操作，该系统需要提供以下核心功能。
（1）移动同学簿主要实现同学信息随时随地地进行增加、删除、修改。
（2）在使用前要对联系人进行搜索，所以该系统需要提供联系人搜索功能。搜索功能里可以根据姓名、年龄、手机号码及相应关键字查询，并给出最终合并后的查询结果。
（3）联系人信息列表需要支持无数据情况下，链接网络服务自动下载联系人信息并写入手机数据库。
（4）在删除联系人信息时，实时刷新信息列表，并清除手机数据库中相应信息，并及时同步网络服务器上的数据库信息。

根据以上功能的分析，可知该产品移动端主页面的主体功能。1）主体页面会包含其他界面启动的功能（导航条）。2）主体页面分为两个部分。首页和搜索两个展示页面。3）主页联系人信息列表显示。在无数据情况下需要链接网络服务自动下载并写入手机数据库。4）联系人查询页面。5）联系人删除。

15.1.4 移动同学簿的设计思路

该项目属于定制项目，需要配合原有网站对外提供的数据进行设计，所以真正意义上的通用是不可能的。这里所说的通用是在展示界面不变的前提下，替换数据依然可以进行正常展示，该系统首先需要现有网站提供联系人信息列表及头像地址。移动端通过网站下载该部分数据，并将数据写入到手机数据库中。除了头像半实时更新外，其余后续功能操作手机数据库中的数据。

按照这样的设计思路，可以按功能切分若干小功能模块，每个模块独立拥有一套视图界面即表示层及其相应的资源文件和一个业务处理类，通过该方法提高项目的耦合度。以下是按照功能拆分出的子功能模块：（1）联系人信息列表。（2）及时头像获取、缓存。（3）网络数据获取。（4）手机数据库写入、删除。（5）搜索功能。

15.2 系统功能模块设计

根据以上的系统功能分析和设计思路，我们可以得到该系统的模块结构图。系统框架图如图 15.1 所示。

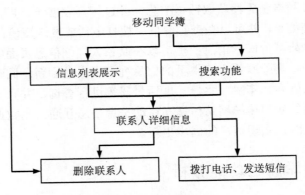

图 15.1　系统功能框架图

在设计实现时采用分层的思想，分层的好处在于代码清晰，结构分明，有利于修改、维护和复用，每个功能模块都有自己对应的页面、业务处理、数据获取、数据一系列独立的模块结构。

根据分层的思想，不同层的文件放入不同的包中，在 Layout 中存放各种布局文件，实现页面层。在 src 包中存放有根据类的不同功能把它们分到不同的包中，分别负责数据库的创建与打开，业务处理等。本次项目开发总体规划为独立模块开发，通过小模块装载完成整体功能。软件结构设计为小模块采用 页面 + 业务/流程 + 资源 + 数据 的方式。主体使用 页面 + 业务/流程（使用小模块拼装）+ 资源 + 数据的方式。

15.3　系统数据分析与设计

在 Android 应用开发中是不使用 JDBC 直接去访问网络上数据的。一般采用类似于 Ajax 的解决方案，或者采用网络编程的解决方案。

（1）Android 中的第一种数据传输方式是通过访问网络服务（网站、WebService），获取网络服务返回的 XML 文件或 Json 对象来得到数据。

（2）Android 系统本身也是基于 Linux 操作系统，所以也可以有第二种选择，使用 socket 网络编程。思路是 Android 应用程序连接服务端，发送请求，服务端进行数据库访问，并返回结果。

注意：从开发速度的角度来讲，访问网络服务的方式相对更可行，因为网络应用一般情况下都是在基于已经拥有网站或网络服务前提之下的。本次项目选择了第一种方式的 XML 获取方式。

该系统提供同学信息的快速查询，为了使数据不丢失，这里通过物理网站提供了提供数据和图像服务。由于该网站只是模拟原站的小功能，故没有进行大规模的数据库完整设计，只是提供了一张记录同学信息的表。物理网站数据库表的结构和手机客户端数据库表的结构必须保持一致，从而才能实现联系人信息列表显示时，在无数据情况下需要链接网络服务自动下载并写入手机数据库。在通过手机客户端对数据进行修改后还要返回物理网站数据库，达到数据的同步更新。所以无论是物理网站还是手机端数据库其结构相同，都仅仅包含一个同学信息表。

为了尽可能地提供联系人的详细信息，在数据库表中需要记录同学的姓名、年龄、性别、电话、地址、QQ联系方式、生日、业余爱好、梦想、紧急联系电话、所从事的工作、头像等信息。为了方便数据库的同步更新和方便存储，这里的数据类型都采用字符型。用户可以对这些信息进行查询，从而方便用户快速操作，进而获得同学的详细信息。表的结构如表15.1所示。其student表的物理设计如图15.2所示。

表 15.1 同学信息表结构

序列	名称	类型	长度	描述
1	name	varchar	255	联系人姓名
2	age	varchar	255	联系人年龄
3	sex	varchar	255	联系人性别
4	tel	varchar	255	联系人电话
5	address	varchar	255	联系人地址
6	qq	varchar	255	联系人 QQ
7	major	varchar	255	联系人专业
8	birthday	varchar	255	联系人生日
9	emergency	varchar	255	紧急联系人
10	hobby	varchar	255	联系人爱好
11	dream	varchar	255	联系人梦想
12	educational	varchar	255	联系人学历
13	wantWork	varchar	255	向往职业
14	photo	varchar	255	联系人头像

15.4 物理网站的设计与实现

该系统首先需要现有网站提供联系人信息列表及头像地址。移动端通过网站下载该部分数据，并将数据写入到手机数据库中。除了头像半实时更新外，其余后续功能操作手机数据库中的数据，所以首先创建物理网站。

该网站物理结构如图15.3所示，其中data包里的类实现数据库的连接与同学信息的查询，entity包中是同学实体类。其中各类的作用及所提供的数据如表15.2所示。

表 15.2 网站所使用的类及其作用

序列	名称	功能
1	DB.java	提供网络数据库连接对象
2	QueryStudentAll.java	获取所有联系人信息
3	StudentEntity.java	数据实体类
4	StudentMessageServlet.java	生成 XML 的 Servlet 类

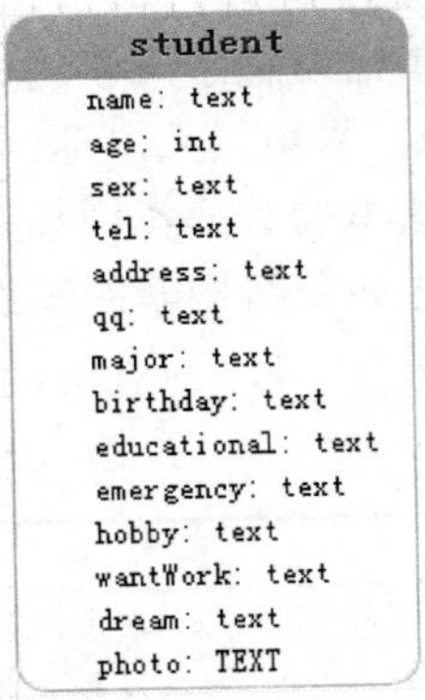

图 15.2　student 信息表 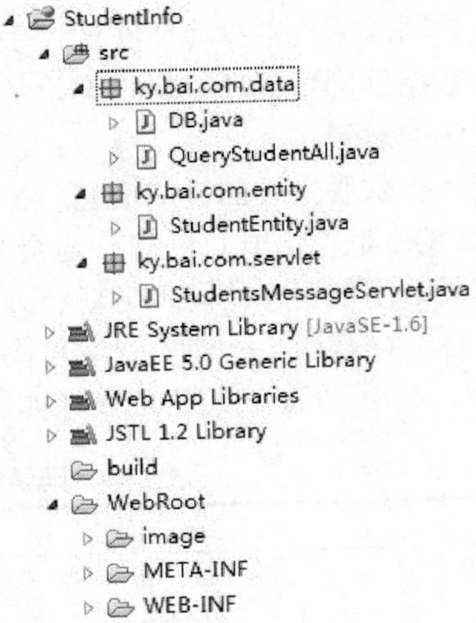　　图 15.3　网站的软件结构图

（1）数据库的连接。

由于 MySQL 数据库是一个快速、多线程、多用户的 SQL 数据库服务器，它从众多的数据库中脱颖而出，支持正规的 SQL 查询语言和采用多种数据类型，能对数据进行各种详细的查询，使用简单方便、稳定性强、占用空间少等优点。本网站选择 Mysql-5.5.24-win32 数据库。DB.java 提供网络数据库连接对象，其具体实现代码如下。

```java
package ky.bai.com.data;
import java.sql.Connection;
import java.sql.DriverManager;
import java.sql.SQLException;
//数据库连接类
public class DB {
    private static Connection conn = null;
    public static Connection getCon(){
        try {
            Class.forName("com.mysql.jdbc.Driver");
            conn = DriverManager.getConnection("jdbc:mysql://localhost:3306/cs01","root","byd");
        } catch (ClassNotFoundException e) {
            e.printStackTrace();
        } catch (SQLException e) {
            e.printStackTrace();
        }
        return conn;
    }
}
```

（2）同学信息的获取，同学簿主要实现同学信息的快速检索，当手机客户端联系人列表为空时，自动连接网络，从网站上下载联系人信息列表及头像，QueryStudentAll.java 类用来获取所有联系人信息，需要时查询全部同学信息。

```java
package ky.bai.com.data;
import java.sql.PreparedStatement;
import java.sql.ResultSet;
import java.sql.SQLException;
import java.util.ArrayList;
import java.util.List;
import ky.bai.com.entity.StudentEntity;
//获取所有学生信息类
public class QueryStudentAll extends DB {
    private PreparedStatement ps = null;
    private ResultSet rs = null;
    private static final String STUDENT_FOR_ALL = "select * from students";
    private List<StudentEntity> lt = null;
    public List<StudentEntity> getStudents() throws SQLException{
        ps = getCon().prepareStatement(STUDENT_FOR_ALL);
        rs = ps.executeQuery();
        if (!rs.wasNull()) {
            lt = new ArrayList<StudentEntity>();
            while (rs.next()) {
                StudentEntity se = new StudentEntity();
                se.setName(rs.getString(1));
                se.setAge(rs.getInt(2));
                se.setSex(rs.getString(3));
                se.setTel(rs.getString(4));
                se.setAddress(rs.getString(5));
                se.setQq(rs.getString(6));
                se.setMajor(rs.getString(7));
                se.setBirthday(rs.getString(8));
                se.setEmergency(rs.getString(9));
                se.setHobby(rs.getString(10));
                se.setDream(rs.getString(11));
                se.setEducational(rs.getString(12));
                se.setWantWork(rs.getString(13));
                se.setPhoto(rs.getString(14));
                lt.add(se);
            }
        }
        return lt;
    }
}
```

（3）同学实体类，为了持久化信息的保存，StudentEntity.java 为数据实体类，查询结果以该对象列表返回。

```java
package ky.bai.com.entity;
//学生实体类
public class StudentEntity {
    private String name = null;
    private int age = -1;
    private String sex = null;
    private String tel = null;
    private String address = null;
    private String qq = null;
    private String major = null;
    private String birthday = null;
    private String educational = null;
    private String emergency = null;
    private String wantWork = null;
    private String hobby = null;
    private String dream = null;
    private String photo = null;
    public String getPhoto() {
        return photo;
    }
    public void setPhoto(String photo) {
        this.photo = photo;
    }
    public String getName() {
        return name;
    }
    public void setName(String name) {
        this.name = name;
    }
    public int getAge() {
        return age;
    }
    public void setAge(int age) {
        this.age = age;
    }
    public String getSex() {
        return sex;
    }
    public void setSex(String sex) {
        this.sex = sex;
    }
    public String getTel() {
        return tel;
    }
    public void setTel(String tel) {
```

```java
        this.tel = tel;
    }
    public String getAddress() {
        return address;
    }
    public void setAddress(String address) {
        this.address = address;
    }
    public String getQq() {
        return qq;
    }
    public void setQq(String qq) {
        this.qq = qq;
    }
    public String getMajor() {
        return major;
    }
    public void setMajor(String major) {
        this.major = major;
    }
    public String getBirthday() {
        return birthday;
    }
    public void setBirthday(String birthday) {
        this.birthday = birthday;
    }
    public String getEducational() {
        return educational;
    }
    public void setEducational(String educational) {
        this.educational = educational;
    }
    public String getEmergency() {
        return emergency;
    }
    public void setEmergency(String emergency) {
        this.emergency = emergency;
    }
    public String getWantWork() {
        return wantWork;
    }
    public void setWantWork(String wantWork) {
        this.wantWork = wantWork;
    }
    public String getHobby() {
```

```java
        return hobby;
    }
    public void setHobby(String hobby) {
        this.hobby = hobby;
    }
    public String getDream() {
        return dream;
    }
    public void setDream(String dream) {
        this.dream = dream;
    }
}
```

（4）Servlet 类，它可以实现业务层和表现层的分离，同时又可以生成动态页面。表 15.2 中 StudentMessageServlet.java 的作用是生成查询同学信息的 XML 的 Servlet 类。

```java
package ky.bai.com.servlet;
import java.io.IOException;
import java.io.PrintWriter;
import java.sql.SQLException;
import javax.servlet.ServletException;
import javax.servlet.http.HttpServlet;
import javax.servlet.http.HttpServletRequest;
import javax.servlet.http.HttpServletResponse;
import ky.bai.com.data.QueryStudentAll;
import ky.bai.com.entity.StudentEntity;
//生成数据 XML 的 Servlet
public class StudentsMessageServlet extends HttpServlet {
    private static final long serialVersionUID = 1L;
    public StudentsMessageServlet() {
        super();
    }
    public void destroy() {
        super.destroy();
    }
    public void doGet(HttpServletRequest request, HttpServletResponse response)
            throws ServletException, IOException {
        //获取数据库数据
        QueryStudentAll all = new QueryStudentAll();
        try {
            /*
             * 设置页面编码，这里要注意与移动端的编码保持一致
             * Android 默认编码是 UTF-8
             */
            response.setCharacterEncoding("utf-8");
```

```java
            //设置文档输出格式
            response.setContentType("text/xml");
            //生成页面
            PrintWriter out = response.getWriter();
            out.println("<?xml version=\"1.0\" encoding=\"utf-8\"?>");
            out.println("<students>");
                //遍历数据生成 XML 文档内容
                for (StudentEntity se : all.getStudents()) {
                    out.println("<student>");
                    out.println("<name>" + se.getName() + "</name>");
                    out.println("<age>" + se.getAge() + "</age>");
                    out.println("<sex>" + se.getSex() + "</sex>");
                    out.println("<tel>" + se.getTel() + "</tel>");
                    out.println("<address>" + se.getAddress() + "</address>");
                    out.println("<qq>" + se.getQq() + "</qq>");
                    out.println("<major>" + se.getMajor() + "</major>");
                    out.println("<birthday>" + se.getBirthday() + "</birthday>");
                    out.println("<educational>"
                            + se.getEducational() + "</educational>");
                    out.println("<emergency>"
                            + se.getEmergency() + "</emergency>");
                    out.println("<hobby>" + se.getHobby() + "</hobby>");
                    out.println("<wantWork>" + se.getWantWork() + "</wantWork>");
                    out.println("<dream>" + se.getDream() + "</dream>");
                    out.println("<photo>" + se.getPhoto() + "</photo>");
                    out.println("</student>");
                }
            out.println("</students>");
            out.flush();
            out.close();
        } catch (SQLException e) {
            return;
        }
    }
    public void doPost(HttpServletRequest request, HttpServletResponse response)
            throws ServletException, IOException {
        doGet(request, response);
    }
    public void init() throws ServletException {}
}
```

关于网站提供的图片，这个网站直接提供了一个 image 文件夹，如图 15.4 所示。文件夹中直接存放图片，获取图片时可以直接通过 http 网址加图片文件名的形式直接读取。访问该网站的返回结果如图 15.5 所示，显示了该同学信息对应的 XML 文件。

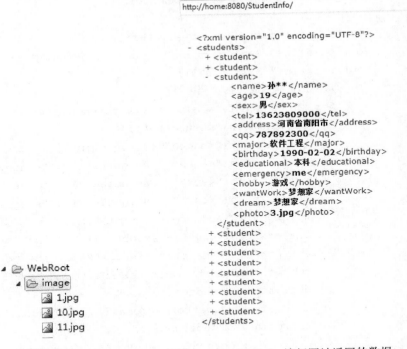

图 15.4　图片文件夹　　　　　　　　图 15.5　访问网站返回的数据

15.5　Android 移动端的设计与实现

　　整个移动应用部分基本都是以小的模块组装起来的，每个小模块都是以页面、控制、数据组成。本系统中的页面只有 5 个，其中作为结构型显示主体的页面有两个，作为主体显示页面内容填充型的页面（子页面）两个，子页面的显示结构页一个。代码部分拆分为三部分：一个主业务流程包（Activity）；一个实体包；一个工具类包。数据被拆分为两部分，第一部分动态数据获取放置在工具类包内，第二部分常量数据放置在资源目录下（values）。

15.5.1　手机端软件结构

　　代码部分拆分为三部分，分别负责传递数据、控制业务逻辑、提供组件支持。如图 15.6 所示，各类的作用如表 15.3 所示。

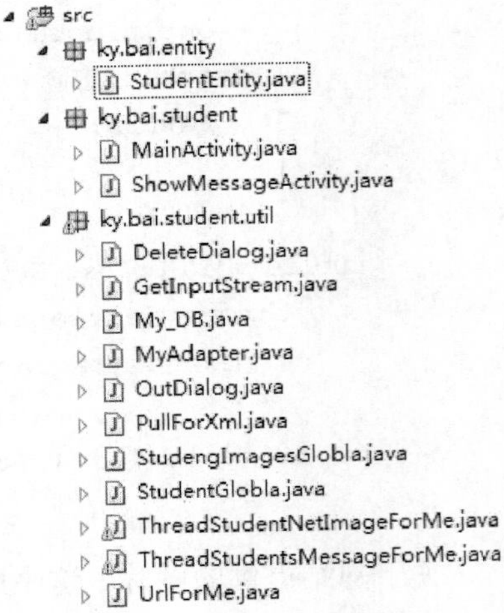

图 15.6　手机端软件结构图

表 15.3 手机端类及其功能简介

序列	名称	名称
1	StudentEntity	学生信息实体类
2	MainActivity	主页面业务、功能实现类
3	ShowMessageActivity	个人详细信息业务、功能实现类
4	DeleteDialog	删除提示框类
5	GetInputStream	提交网络访问请求并得到返回的流
6	My_DB	创建手机数据库、提供基础数据库操作功能
7	MyAdapter	自定义的列表适配器
8	OutDialog	退出提示框类
9	PullForXml	XML 解析器
10	StudengImagesGlobla	全局共享图片存储容器
11	StudentGlobla	全局共享数据列表内容容器
12	ThreadStudentNetImageForMe	联系人头像获取线程
13	ThreadStudentsMessageForMe	联系人信息获取线程
14	UrlForMe	网络地址常量类

15.5.2 移动端数据的创建与初始化

Android 手机端数据库表结构与网站所使用数据库结构相同，都仅仅包含一个同学信息表，参见 15.3 节数据库的逻辑结构设计完毕后，就可以开始在手机创建数据库和数据表了。

（1）数据库的创建与访问

本系统把数据库的创建和同学信息表的建立，还有从网站获取初始信息列表都封装到类 My_DB.java 中，该类实现数据库的创建及同学信息表的建立、数据信息的查询和将网络获取到的数据写入到数据库表中。

```
package ky.bai.student.util;

import java.util.ArrayList;
import java.util.List;
import ky.bai.entity.StudentEntity;
import android.content.ContentValues;
import android.content.Context;
import android.database.Cursor;
import android.database.sqlite.SQLiteDatabase;
import android.database.sqlite.SQLiteDatabase.CursorFactory;
import android.database.sqlite.SQLiteOpenHelper;
import android.util.Log;

public class My_DB extends SQLiteOpenHelper {
    public static final String MY_DB_NAME = "my_db";
    public static final String MY_DB_TABLE_1_NAME = "student";
```

```java
        public static final int MY_DB_VERSION = 1;
        public static final String QUERY_STUDENT_ALL
                            = "SELECT * FROM " + MY_DB_TABLE_1_NAME;
    public My_DB(Context context, String name, CursorFactory factory,
            int version) {
        super(context, name, factory, version);
    }
    @Override
    public void onCreate(SQLiteDatabase arg0) {
        arg0.execSQL("create table "
                + MY_DB_TABLE_1_NAME
                + "(name text,age int,sex text,tel text,address text,qq
                text,major  text,birthday text,educational text,emergency
                text,hobby text,wantWork text,dream text,photo text);");
    }
    @Override
    public void onUpgrade(SQLiteDatabase arg0, int arg1, int arg2) {}
    //查询student表的方法 1
    public List<StudentEntity> getStudentAll(SQLiteDatabase db){
        List<StudentEntity> lt = new ArrayList<StudentEntity>();
        StudentEntity stu = null;
        Cursor cr = db.rawQuery(QUERY_STUDENT_ALL, null);
        while (cr.moveToNext()) {
            stu = new StudentEntity();
            stu.setName(cr.getString(0));
            stu.setAge(cr.getInt(1));
            stu.setSex(cr.getString(2));
            stu.setTel(cr.getString(3));
            stu.setAddress(cr.getString(4));
            stu.setQq(cr.getString(5));
            stu.setMajor(cr.getString(6));
            stu.setBirthday(cr.getString(7));
            stu.setEmergency(cr.getString(9));
            stu.setHobby(cr.getString(10));
            stu.setDream(cr.getString(12));
            stu.setEducational(cr.getString(8));
            stu.setWantWork(cr.getString(11));
            stu.setPhoto(cr.getString(13));
            lt.add(stu);
        }
        return lt;
    }
    //将网络获取到的数据写入数据库
    public long addData(List<StudentEntity> be,SQLiteDatabase db){
        //记录操作成功次数
```

```java
long param = 0;
//开启事务
            db.beginTransaction();
            for (StudentEntity se : be) {
                ContentValues cv = new ContentValues();
                cv.put("name", se.getName());
                cv.put("age", se.getAge());
                cv.put("sex", se.getSex());
                cv.put("tel", se.getTel());
                cv.put("address", se.getAddress());
                cv.put("qq", se.getQq());
                cv.put("major", se.getMajor());
                cv.put("birthday", se.getBirthday());
                cv.put("educational", se.getEducational());
                cv.put("emergency", se.getEmergency());
                cv.put("hobby", se.getHobby());
                cv.put("wantWork",se.getWantWork());
                cv.put("dream", se.getDream());
                cv.put("photo", se.getPhoto());
                param = db.insert(MY_DB_TABLE_1_NAME, null, cv);
            }
            //确认提交,如果漏掉这一步之前所有操作自动回滚
            db.setTransactionSuccessful();
            db.endTransaction();
            return param;
    }
}
```

（2）学生实体类

由于手机端也是采用分层的思想实现各模块的,在进行数据信息的查询、信息列表的显示时最终要用到 StudentEntity 学生信息实体类。由于手机端和网站的数据表的结构完全相同,所以其对应的 JavaBean 也相同。为了方便显示同学信息的信息列表,在该类中重载了 toString()方法,实现该对象信息的字符串连接,其具体代码如下。

```java
package ky.bai.entity;
public class StudentEntity {
    private String name = null;
    private int age = -1;
    private String sex = null;
    private String tel = null;
    private String address = null;
    private String qq = null;
    private String major = null;
    private String birthday = null;
    private String educational = null;
```

```java
    private String emergency = null;
    private String wantWork = null;
    private String hobby = null;
    private String dream = null;
    private String photo = null;

    public String getName() {
        return name;
    }
    public void setName(String name) {
        this.name = name;
    }
    public int getAge() {
        return age;
    }
    public void setAge(int age) {
        this.age = age;
    }
    public String getSex() {
        return sex;
    }
    public void setSex(String sex) {
        this.sex = sex;
    }
    public String getTel() {
        return tel;
    }
    public void setTel(String tel) {
        this.tel = tel;
    }
    public String getAddress() {
        return address;
    }
    public void setAddress(String address) {
        this.address = address;
    }
    public String getQq() {
        return qq;
    }
    public void setQq(String qq) {
        this.qq = qq;
    }
    public String getMajor() {
        return major;
    }
```

```java
    public void setMajor(String major) {
        this.major = major;
    }
    public String getBirthday() {
        return birthday;
    }
    public void setBirthday(String birthday) {
        this.birthday = birthday;
    }
    public String getEducational() {
        return educational;
    }
    public void setEducational(String educational) {
        this.educational = educational;
    }
    public String getEmergency() {
        return emergency;
    }
    public void setEmergency(String emergency) {
        this.emergency = emergency;
    }
    public String getWantWork() {
        return wantWork;
    }
    public void setWantWork(String wantWork) {
        this.wantWork = wantWork;
    }
    public String getHobby() {
        return hobby;
    }
    public String getPhoto() {
        return photo;
    }
    public void setPhoto(String photo) {
        this.photo = photo;
    }
    public void setHobby(String hobby) {
        this.hobby = hobby;
    }
    public String getDream() {
        return dream;
    }
    public void setDream(String dream) {
        this.dream = dream;
    }
```

```
    @Override
    public String toString() {
        return "StudentEntity [name=" + name + ", age=" + age + ", sex=" + sex
                + ", tel=" + tel + ", address=" + address + ", qq=" + qq
                + ", major=" + major + ", birthday=" + birthday
                + ", educational=" + educational + ", emergency=" + emergency
                + ", wantWork=" + wantWork + ", hobby=" + hobby + ", dream="
                + dream + ", photo=" + photo + "]";
    }
}
```

15.5.3 首页模块的设计与实现

该系统总体分为首页、搜索页、信息列表页和删除页等几个核心模块。其中，首页模块是系统非常重要的模块，大部分操作都由此开始，它主要集成了主体信息显示、搜索两个模块。主体信息显示提供了列表显示、删除和对应跳转功能。搜索页提供了根据关键字进行搜索、对应跳转功能。详细信息页面提供了基本信息显示、更多信息显示和删除功能。

对于页面部分，使用预留位置嵌入内容页面的方式设计实现，其软件结构图如图15.7所示，每个页面都采用不同的布局格式实现的，它们被存放在 layout 包中，各文件的功能说明如表15.4所示。

图 15.7　页面结构图

表 15.4　页面文件及其功能

序列	名称	作用
1	activity_main.xml	主窗体结构页面
2	activity_sh.xml	主页子窗体2 搜索功能窗体页面
3	activity_show_message.xml	个人详细信息主窗体
4	main_page_1_list_value.xml	主页中列表的子结构页面
5	main_page_1.xml	主页子窗体1 列表显示页面

1. 首页表示层

首页的布局可以分为标题栏、主体内容栏和底部导航栏三个模块，其中，标题栏提供显示当前页面名称；主体内容部分是模块容器采用外部导入的方式进行组合，装载不同的子页面，提供不同模块的内容；底部导航负责不同模块的切换。首页的界面布局如图15.8所示。其中，linearLayout1 是头部标题容器，linearLayout2 为页面切换按钮容器，

mian_body_lin 为空白容器（子页面装载容器）。

```
▲ RelativeLayout
    ▲ linearLayout1
        ▲ LinearLayout
            Ab main_tv1 (TextView) - "首页标题"
    ▲ linearLayout2
        OK btn_show_all (Button) - "首页"
        OK btn_show_by (Button) - "搜索"
    mian_body_lin (RelativeLayout)
```

图 15.8　首页的布局结构

首页详细布局代码如下。

```xml
<!-- 页面主容器 -->
<RelativeLayout xmlns:android="http://schemas.android.com/apk/res/android"
    xmlns:tools="http://schemas.android.com/tools"
    android:layout_width="match_parent"
    android:layout_height="match_parent"
    tools:context=".MainActivity" >
<!-- 头部容器 -->
    <LinearLayout
        android:id="@+id/linearLayout1"
        android:layout_width="match_parent"
        android:layout_height="45dp"
        android:layout_alignParentLeft="true"
        android:layout_alignParentTop="true"
        android:background="@drawable/title_bg"
        android:gravity="center"
        android:baselineAligned="false">
<!-- 头部显示部分容器 -->
        <LinearLayout
            android:layout_width="wrap_content"
            android:layout_height="match_parent"
            android:layout_weight="1"
            android:gravity="center" >
<!-- 显示控件 -->
            <TextView
                android:id="@+id/main_tv1"
                android:layout_width="wrap_content"
                android:layout_height="wrap_content"
                android:text="@string/main_title" />
        </LinearLayout>
    </LinearLayout>
<!-- 底部导航容器 -->
    <LinearLayout
```

```xml
            android:id="@+id/linearLayout2"
            android:layout_width="match_parent"
            android:layout_height="45dp"
            android:layout_alignParentBottom="true"
            android:layout_alignParentLeft="true"
            android:background="@drawable/bbtm_bg" >
    <!-- 主页 -->
            <Button
                android:id="@+id/btn_show_all"
                android:layout_width="wrap_content"
                android:layout_height="match_parent"
                android:layout_weight="1"
                android:background="@drawable/btn_1_2"
                android:text="@string/user_soye"
                android:textColor="@color/black" />
    <!-- 搜索 -->
            <Button
                android:id="@+id/btn_show_by"
                android:layout_width="wrap_content"
                android:layout_height="match_parent"
                android:layout_weight="1"
                android:background="@drawable/btn_1_2"
                android:text="@string/user_sousuo"
                android:textColor="@color/black" />
        </LinearLayout>
    <!-- 中间内容部分容器 -->
        <RelativeLayout
            android:id="@+id/mian_body_lin"
            android:layout_width="wrap_content"
            android:layout_height="wrap_content"
            android:layout_above="@+id/linearLayout2"
            android:layout_alignParentLeft="true"
            android:layout_alignParentRight="true"
            android:layout_below="@+id/linearLayout1">
        </RelativeLayout>
</RelativeLayout>
```

2. 首页的业务处理

首页中的所有业务逻辑都集中在 MainActivity.java 中进行处理，其中，在 MainActivity 类进行业务处理时，会用到像删除提示框、退出提示框等自定义类。核心业务处理类如图 15.9 所示，它们放在 ky.bai.student 包中，详细代码如下。

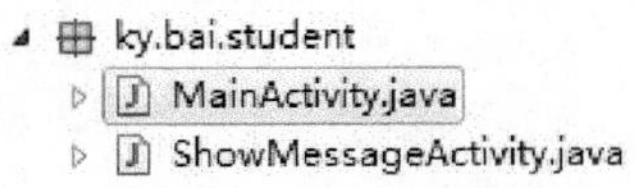

图 15.9 业务处理类

```java
package ky.bai.student;

import java.util.ArrayList;
import java.util.List;
import ky.bai.entity.StudentEntity;
import ky.bai.student.util.DeleteDialog;
import ky.bai.student.util.MyAdapter;
import ky.bai.student.util.My_DB;
import ky.bai.student.util.OutDialog;
import ky.bai.student.util.StudentGlobla;
import ky.bai.student.util.ThreadStudentsMessageForMe;
import android.os.Bundle;
import android.os.Handler;
import android.os.Message;
import android.app.Activity;
import android.app.Dialog;
import android.content.Intent;
import android.database.sqlite.SQLiteDatabase;
import android.util.Log;
import android.view.LayoutInflater;
import android.view.Menu;
import android.view.View;
import android.view.View.OnClickListener;
import android.widget.AdapterView;
import android.widget.AdapterView.OnItemClickListener;
import android.widget.AdapterView.OnItemLongClickListener;
import android.widget.Button;
import android.widget.EditText;
import android.widget.ListView;
import android.widget.RelativeLayout;
import android.widget.TextView;
import android.widget.Toast;
//主窗体类
public class MainActivity extends Activity {
    //该类ID号
    public static final int MAIN_ACTIVITY_CODE = 1000;
    private Button btn_show_all = null;
    private Button btn_show_by = null;
    private TextView main_tv1 = null;
    //获取xml布局页面的对象
    private LayoutInflater inflater = null;
    //主页面主体内容容器对象
    private RelativeLayout mian_body_lin = null;
    //主页面子页面1的列表对象
    private ListView mian_page1_list = null;
```

```java
//默认加载的第一个子页面对象(首页)
private View v = null;
//子页面一的页面数据适配器
private MyAdapter ma = null;
//专门在运行时更新界面UI的一个进程类
private Handler han = null;
//数据库创建及功能对象
My_DB db = null;
//数据库操作对象
private SQLiteDatabase job = null;
//--搜索页面内容
private Button sh_sh_btn = null;
private EditText sh_sh_et = null;
private ListView sh_sh_lv = null;
private List<StudentEntity> lt2 = null;
private MyAdapter ma2 = null;
@Override
protected void onCreate(Bundle savedInstanceState) {
    super.onCreate(savedInstanceState);
    setContentView(R.layout.activity_main);
    //声明一个界面更新进程对象
    han = new Handler() {
        @Override
        public void handleMessage(Message msg) {
            switch (msg.what) {
            case 1:
                //刷新适配器
                ma.notifyDataSetChanged();
                break;
            case 2:
                new OutDialog(MainActivity.this).getDialog("删除失败。", true)
                    .show();
                break;
            case 3:
                //将数据写入手机数据库
                long param = db.addData(StudentGlobla.lts, job);
                //写入条数与提供的数据条数相符
                if (param == StudentGlobla.lts.size()) {
                    //获取列表集合对象
                    mian_page1_list = (ListView) v.findViewById(R.id.mian_page1_list);
                    //准备适配器
                    ma = new MyAdapter(StudentGlobla.lts, MainActivity.this);
                    //适配器绑定
```

```java
            mian_page1_list.setAdapter(ma);
            //ListView 单击操作
            mian_page1_list.setOnItemClickListener(new OnItemClick
            Listener() {
                @Override
                public void onItemClick(AdapterView<?> arg0, View arg1, int arg2, long arg3) {
                    //android 中的跳转对象
                    Intent it = new Intent(MainActivity.this,ShowMessage
                    Activity.class);
                    //传值，注意类型
                    //传递实体内容
                    it.putExtra("userName", StudentGlobla.lts.get(arg2).getName());
                    it.putExtra("userAdd", StudentGlobla.lts.get(arg2).getAddress());
                    it.putExtra("userAge", StudentGlobla.lts.get(arg2).getAge());
                    it.putExtra("userBir", StudentGlobla.lts.get(arg2).getBirthday());
                    it.putExtra("userPho", StudentGlobla.lts.get(arg2).getTel());
                    it.putExtra("userSex", StudentGlobla.lts.get(arg2).getSex());
                    it.putExtra("userQQ", StudentGlobla.lts.get(arg2).getQq());
                    it.putExtra("userEmergency", StudentGlobla.lts.get(arg2)
                                                    .getEmergency());
                    it.putExtra("userMajor", StudentGlobla.lts.get(arg2).getMajor());
                    it.putExtra("userEducational", StudentGlobla.lts.get(arg2)
                                                    .getEducational());
                    it.putExtra("userHobby", StudentGlobla.lts.get(arg2).getHobby());
                    it.putExtra("userWantWork", StudentGlobla.lts.get(arg2)
                                                    .getWantWork());
                    it.putExtra("userPhoto",StudentGlobla.lts.get(arg2)
                                                    .getPhoto());
                    it.putExtra("userDre", StudentGlobla.lts.get(arg2).
                                                    getDream());
                    //传递选中项的编号
                    it.putExtra("position", arg2);
                    //启动跳转
                    startActivityForResult(it, MAIN_ACTIVITY_CODE);
                }
            });
            //ListView 长按操作
        mian_page1_list.setOnItemLongClickListener(new OnItemLongClickListener()
        {
            @Override
            public boolean onItemLongClick(AdapterView<?> arg0,
                                View arg1, final int arg2, long arg3) {
                Dialog d = new DeleteDialog(MainActivity.this, arg2,
                                han, StudentGlobla.lts, job).getDialog();
                d.show();
```

```java
                return true;
            }
        });
    }else{
        han.sendEmptyMessage(4);
    }
        break;
    case 4:
        new OutDialog(MainActivity.this).getDialog().show();
        break;
    default:
        break;
    }
  }
};
//创建数据库
db = new My_DB(MainActivity.this, My_DB.MY_DB_NAME, null,
                                    My_DB.MY_DB_VERSION);
//创建数据库操作对象
job = db.getReadableDatabase();
//----数据库查询 student 表
List<StudentEntity> lt = db.getStudentAll(job);
    if (!lt.isEmpty()) {
        StudentGlobla.lts = lt;
    } else {
    //网络数据访问线程
    ThreadStudentsMessageForMe tsmfm = new ThreadStudentsMessageForMe(han);
    new Thread(tsmfm).start();
    }
//得到获取页面用的对象
inflater = getLayoutInflater();
//获取页面中的导航按钮对象
btn_show_all = (Button) findViewById(R.id.btn_show_all);
btn_show_by = (Button) findViewById(R.id.btn_show_by);
main_tv1 = (TextView) findViewById(R.id.main_tv1);
mian_body_lin = (RelativeLayout) findViewById(R.id.mian_body_lin);
//打开时默认装载第一个页面
v = inflater.inflate(R.layout.main_page_1, null);
//获取列表集合对象
mian_page1_list = (ListView) v.findViewById(R.id.mian_page1_list);
//准备适配器
ma = new MyAdapter(StudentGlobla.lts, MainActivity.this);
//适配器绑定
mian_page1_list.setAdapter(ma);
mian_page1_list.setOnItemClickListener(new OnItemClickListener() {
```

```java
            @Override
            public void onItemClick(AdapterView<?> arg0, View arg1, int arg2, long arg3) {
                //android 中的跳转对象
                Intent it = new Intent(MainActivity.this, ShowMessageActivity.class);
                //传值，注意类型
                //传递实体内容
                it.putExtra("userName", StudentGlobla.lts.get(arg2).getName());
                it.putExtra("userAdd", StudentGlobla.lts.get(arg2).getAddress());
                it.putExtra("userAge", StudentGlobla.lts.get(arg2).getAge());
                it.putExtra("userBir", StudentGlobla.lts.get(arg2).getBirthday());
                it.putExtra("userPho", StudentGlobla.lts.get(arg2).getTel());
                it.putExtra("userSex", StudentGlobla.lts.get(arg2).getSex());
                it.putExtra("userQQ", StudentGlobla.lts.get(arg2).getQq());
                it.putExtra("userEmergency", StudentGlobla.lts.get(arg2).getEmergency());
                t.putExtra("userMajor", StudentGlobla.lts.get(arg2).getMajor());
                it.putExtra("userEducational", StudentGlobla.lts.get(arg2).getEducational());
                it.putExtra("userHobby", StudentGlobla.lts.get(arg2).getHobby());
                it.putExtra("userWantWork", StudentGlobla.lts.get(arg2).getWantWork());
                it.putExtra("userPhoto", StudentGlobla.lts.get(arg2).getPhoto());
                it.putExtra("userDre", StudentGlobla.lts.get(arg2).getDream());
                //传递选中项的编号
                it.putExtra("position", arg2);
                //启动跳转
                startActivityForResult(it, MAIN_ACTIVITY_CODE);
            }
        });
        //ListView 长按操作
        mian_page1_list.setOnItemLongClickListener(new OnItemLongClickListener() {
            @Override
            public boolean onItemLongClick(AdapterView<?> arg0,
                    View arg1, final int arg2, long arg3) {
                Dialog d = new DeleteDialog(MainActivity.this, arg2,
                                    han, StudentGlobla.lts, job).getDialog();
                d.show();
                return true;
            }
        });
        mian_body_lin.addView(v);
        //导航按钮 1 单击操作
        btn_show_all.setOnClickListener(new OnClickListener() {
            @Override
            public void onClick(View arg0) {
                main_tv1.setText(R.string.main_title);
                mian_body_lin.removeAllViews();
                mian_body_lin.addView(v);
```

```java
    }
});
//导航按钮2单击操作
    btn_show_by.setOnClickListener(new OnClickListener() {
        @Override
        public void onClick(View arg0) {
            main_tv1.setText(R.string.user_sh);
            mian_body_lin.removeAllViews();
            View v = inflater.inflate(R.layout.activity_sh, null);
            //--搜索页面内容
            sh_sh_lv = (ListView) v.findViewById(R.id.sh_sh_lv);
            sh_sh_et = (EditText) v.findViewById(R.id.sh_sh_et);
            sh_sh_btn = (Button) v.findViewById(R.id.sh_sh_btn);
            //搜索页面搜索操作
            sh_sh_btn.setOnClickListener(new OnClickListener() {
                @Override
                public void onClick(View arg0) {
                    if (StudentGlobla.lts != null) {
                        //查询结果对象集合
                        lt2 = new ArrayList<StudentEntity>();
                        //获取文本框中的关键字
                        String param = sh_sh_et.getText().toString().trim();
                        //遍历数据对象集合
                        for (StudentEntity str : StudentGlobla.lts) {
                            //如果对象名不为空且包含关键字添加到查询结果集合
                            if (str.getName() != null && str.getName().contains
                            (param)) {
                                if (!lt2.contains(str)) {
                                    lt2.add(str);
                                }
                            }
                            //如果地址名不为空且包含关键字添加到查询结果集合
                            if (str.getAddress() != null && str.getAddress()
                            .contains(param)) {
                                if (!lt2.contains(str)) {
                                    lt2.add(str);
                                }
                            }
                            //如果生日不为空且包含关键字添加到查询结果集合
                            if (str.getBirthday() != null && str.getBirthday().contains
                            (param)) {
                                if (!lt2.contains(str)) {
                                    lt2.add(str);
                                }
                            }
```

```java
        //如果年龄不为空且包含关键字添加到查询结果集合
        if ((str.getAge() + "").contains(param)) {
            if (!lt2.contains(str)) {
                lt2.add(str);
            }
        }
        //如果电话号码不为空且包含关键字添加到查询结果集合
        if (str.getTel() != null && str.getTel().contains(param)) {
            if (!lt2.contains(str)) {
                lt2.add(str);
            }
        }
    }
    //如果没有任何数据，调用弹出框提示
    if (lt2.isEmpty()) {
        new OutDialog(MainActivity.this).getDialog(true).show();
    }
    //创建搜索页面列表适配器
    ma2 = new MyAdapter(lt2, MainActivity.this);
    //列表装载适配器
    sh_sh_lv.setAdapter(ma2);
    //ListView 单击操作
    sh_sh_lv.setOnItemClickListener(new OnItemClickListener() {
        @Override
        public void onItemClick(AdapterView<?> arg0,
                        View arg1, int arg2, long arg3) {
            //android 中的跳转对象
            Intent it = new Intent(MainActivity.this,
                                ShowMessageActivity.class);
            //传值，注意类型
            //传递实体内容
            it.putExtra("userName", lt2.get(arg2).getName());
            t.putExtra("userAdd", lt2.get(arg2).getAddress());
            it.putExtra("userAge", lt2.get(arg2).getAge());
            it.putExtra("userBir", lt2.get(arg2).getBirthday());
            it.putExtra("userPho", lt2.get(arg2).getTel());
            it.putExtra("userSex", lt2.get(arg2).getSex());
            it.putExtra("userDre", lt2.get(arg2).getDream());
            it.putExtra("userEmergency", lt2.get(arg2).getEmergency());
            it.putExtra("userMajor", lt2.get(arg2).getMajor());
            it.putExtra("userEducational", lt2.get(arg2).getEducational());
            it.putExtra("userHobby", lt2.get(arg2).getHobby());
            it.putExtra("userWantWork", lt2.get(arg2).getWantWork());
            it.putExtra("userPhoto", lt2.get(arg2).getPhoto());
```

```
                    //传递选中项的编号
                    it.putExtra("position", arg2);
                    //启动跳转
                    startActivity(it);

                }
            });
        }
    }
});
//将子视图装入主页内容容器中
mian_body_lin.addView(v);
    }
  });
}
//处理带返回值的业务逻辑处理类
@Override
protected void onActivityResult(int requestCode, int resultCode, Intent data) {
    //如果是从主页跳转出去的
    if (requestCode == MAIN_ACTIVITY_CODE) {
        //如果是从搜索界面返回来的
        if (resultCode == ShowMessageActivity.SHOW_MESSAGE_CODE) {
            han.sendEmptyMessage(1);
        }
    }
  }
}
```

3．MainActivity 类中用到的自定义类

在业务类 MainActivity 中用到多个自定义的类，为了方便管理，都放入 ky.bai.student.util 包中，每个类实现不同的实用功能，从而提高代码的复用性。其中，类 My_DB 创建手机数据库、提供基础数据库操作功能和类 StudentEntity 学生信息实体类，其定义参加 15.5.2 节。其他类的定义如下。

（1）DeleteDialog 类删除提示框类用于提示用户是否确认删除。

```
package ky.bai.student.util;

import java.util.List;
import ky.bai.entity.StudentEntity;
import ky.bai.student.R;
import android.app.AlertDialog;
import android.app.Dialog;
import android.app.AlertDialog.Builder;
import android.content.Context;
```

```java
import android.content.DialogInterface;
import android.database.sqlite.SQLiteDatabase;
import android.os.Handler;
/**
 * 删除功能对应的业务类
 */
public class DeleteDialog {
    //上下文环境对象
    private Context context = null;
    //删除项编号
    private int position;
    //界面跟新进程
    private Handler han = null;
    //数据集合
    private List<StudentEntity> lts = null;
    //数据库操作对象
    private SQLiteDatabase db = null;

    public DeleteDialog(Context context, int position, Handler han,
            List<StudentEntity> lts,SQLiteDatabase db) {
        super();
        this.context = context;
        this.position = position;
        this.han = han;
        this.lts = lts;
        this.db = db;
    }

    public Dialog getDialog(){
        //创建提示框构建对象
        AlertDialog.Builder b = new Builder(context);
        //设置提示框标题
        b.setTitle("你好这是系统信息提示框!");
        //设置提示信息
        b.setMessage("您确定要删除【"
                      + lts.get(position).getName() + "】的相关数据吗? ");
        //设置提示框确认按键
        b.setPositiveButton(R.string.qr,new DialogInterface.OnClickListener() {
            @Override
            public void onClick(DialogInterface arg0, int arg1) {
                if (han != null) {
                    //设置数据库查询条件对应的参数值
                    String[] whereArgs = {lts.get(position).getName(),lts.
                    get(position).getTel()};
                    //调用数据操作实现删除功能
```

```
                    int prarm = db.delete(My_DB.MY_DB_TABLE_1_NAME
                                , "name=? and tel=?", whereArgs);
                    if (prarm > 0) {
                        //删除掉集合容器中的相应数据
                        lts.remove(position);
                        //发送界面进程更新命令
                        han.sendEmptyMessage(1);
                    }else{
                        han.sendEmptyMessage(2);
                    }
                }
            }
        });
        //设置菜单取消操作
        b.setNegativeButton(R.string.qx, new DialogInterface.OnClickListener() {
            @Override
            public void onClick(DialogInterface arg0, int arg1) {
                //--不操作
            }
        });
        return b.create();
    }
}
```

（2）MyAdapter 自定义的列表适配器，其定义如下。

```
package ky.bai.student.util;

import java.util.List;
import ky.bai.entity.StudentEntity;
import ky.bai.student.R;
import android.content.Context;
import android.view.LayoutInflater;
import android.view.View;
import android.view.ViewGroup;
import android.widget.BaseAdapter;
import android.widget.ImageView;
import android.widget.TextView;

public class MyAdapter extends BaseAdapter {
    //准备使用的数据
    private List<StudentEntity> ssa = null;
    //准备用来寻找子页面的对象
    private LayoutInflater inflater = null;
    public MyAdapter(List<StudentEntity> ssa, Context con) {
```

```java
        super();
        this.ssa = ssa;
        inflater = LayoutInflater.from(con);
    }
    //返回数据的个数：影响界面绘制的行数
    @Override
    public int getCount() {
        int param = 0;
        if (ssa != null) {
            param = ssa.size();
        }
        return param;
    }
    @Override
    public Object getItem(int arg0) {
        return null;
    }
    @Override
    public long getItemId(int arg0) {
        return 0;
    }
    //绘制界面的方法，每执行一次代表绘制了一行图像
    @Override
    public View getView(int arg0, View arg1, ViewGroup arg2) {
        //获取子页面视图
        arg1 = inflater.inflate(R.layout.main_page_1_list_value, null);
        //获取子页面视图中的控件
        TextView main_page_1_name
                = (TextView) arg1.findViewById(R.id.main_page_1_name);
        TextView main_page_1_phone
                = (TextView) arg1.findViewById(R.id.main_page_1_phone);
        ImageView main_page_1_photo
                = (ImageView) arg1.findViewById(R.id.main_page_1_photo);
        //给子页面视图中的子控件进行数据绑定
        main_page_1_name.setText(ssa.get(arg0).getName());
        main_page_1_phone.setText(ssa.get(arg0).getTel());
        //绑定照片
        if (StudengImagesGlobla.images.containsKey(
                UrlForMe.QUERY_STUDENT_IMG + ssa.get(arg0).getPhoto())) {
            main_page_1_photo.setImageBitmap(StudengImagesGlobla.images.
                    get(UrlForMe.QUERY_STUDENT_IMG + ssa.get(arg0).getPhoto()));
        }else{
            //发送网络获取图片请求，进行实时下载绑定
            ThreadStudentNetImageForMe tsnfm
                    = new ThreadStudentNetImageForMe(main_page_1_photo, ssa.get(arg0)
```

```
                .getPhoto(),StudengImagesGlobla.images);
            new Thread(tsnfm).start();
        }
        //返回绘制的当前行对象
        return arg1;
    }
}
```

（3）MyAdapter 类，定义时用到类 StudengImagesGlobla，该类用于全局共享图片存储容器的定义。

```
package ky.bai.student.util;

import java.util.HashMap;
import java.util.Map;
import android.graphics.Bitmap;
//图片集合
public class StudengImagesGlobla {
    public static Map<String,Bitmap> images = new HashMap<String,Bitmap>();
}
```

（4）ThreadStudentNetImageForMe 类，MyAdapter 类定义时用到自定义类 ThreadStudentNetImageForMe，该类主要是获取联系人头像的线程，具体定义如下。

```
package ky.bai.student.util;

import java.util.Map;

import android.graphics.Bitmap;
import android.graphics.BitmapFactory;
import android.os.Handler;
import android.os.Message;
import android.widget.ImageView;
//图片获取线程类
public class ThreadStudentNetImageForMe extends Handler implements Runnable{
    //图片控件
    private ImageView iv = null;
    //图片名称
    private String imageName = null;
    //图片对象
    private Bitmap bp = null;
    //图片缓存集合
    private Map<String,Bitmap> images = null;
    public ThreadStudentNetImageForMe(ImageView iv, String imageName
                                    ,Map<String,Bitmap> images) {
        super();
```

```java
        this.iv = iv;
        this.imageName = imageName;
        this.images = images;
    }
    @Override
    public void run() {
        //获取网络图片生成图片对象
        bp = BitmapFactory.decodeStream(GetInputStream.getInputStream(
                        UrlForMe.QUERY_STUDENT_IMG + imageName));
        //如果图片对象存在发送界面更新命令
        if (bp != null) {
            this.sendEmptyMessage(1);
        }
    }
    @Override
    public void handleMessage(Message msg) {
        super.handleMessage(msg);
        switch (msg.what) {
        case 1:
            //更新图片控件中的图片
            iv.setImageBitmap(bp);
            //将图片存入全局图片共享容器
            images.put(UrlForMe.QUERY_STUDENT_IMG + imageName, bp);
            break;
        default:
            break;
        }
    }
}
```

（5）GetInputStream 类，在 ThreadStudentNetImageForMe 类的定义中需要使用 GetInputStream.来提交网络访问请求并得到返回的流，该类的代码如下。

```java
package ky.bai.student.util;

import java.io.IOException;
import java.io.InputStream;
import java.net.HttpURLConnection;
import java.net.MalformedURLException;
import java.net.URL;

public class GetInputStream {
    public static InputStream getInputStream(String urlStr){
        //声明路径对象
        URL url = null;
        //声明HTTP访问对象
```

```java
        HttpURLConnection conn = null;
        //声明返回流接收对象
        InputStream in = null;

        try {
            //创建访问地址路径对象
            url = new URL(urlStr);
            //设置HTTP访问参数
            conn = (HttpURLConnection) url.openConnection();
            conn.setConnectTimeout(10000);
            conn.setRequestMethod("GET");
            conn.setRequestProperty("accept", "*/*");
            conn.getRequestProperty("location");
            conn.getResponseCode();
            conn.connect();
            //获得返回的流
            in = conn.getInputStream();
        } catch (MalformedURLException e) {
            e.printStackTrace();
        } catch (IOException e) {
            e.printStackTrace();
        }
        return in;
    }
}
```

(6) OutDialog 退出提示框类，也是退出功能所对应的业务类，其详细代码如下。

```java
package ky.bai.student.util;

import ky.bai.student.R;
import android.app.Activity;
import android.app.AlertDialog;
import android.app.Dialog;
import android.app.AlertDialog.Builder;
import android.content.Context;
import android.content.DialogInterface;
/**
 * 退出功能对应的业务类
 */
public class OutDialog {
    //上下文环境
    private Context context = null;
    public OutDialog(Context context) {
        super();
        this.context = context;
```

```java
    }
    //重载的退出方法1
    public Dialog getDialog(){
        AlertDialog.Builder b = new Builder(context);
        b.setTitle("你好这是系统信息提示框！");
        b.setMessage("木有数据呦，该干啥干啥去吧 ^_^");
        b.setPositiveButton(R.string.qr,new DialogInterface.OnClickListener() {
            @Override
            public void onClick(DialogInterface arg0, int arg1) {
                ((Activity)context).finish();
            }
        });
        return b.create();
    }
    //重载的退出方法2
    public Dialog getDialog(boolean flag){
        AlertDialog.Builder b = new Builder(context);
        b.setTitle("你好这是系统信息提示框！");
        b.setMessage("木有数据呦，该干啥干啥去吧 ^_^");

        if (flag) {
            b.setPositiveButton(R.string.qr,new DialogInterface.OnClickListener() {
                @Override
                public void onClick(DialogInterface arg0, int arg1) {
                    //--不操作
                }
            });
        }else{
            b.setPositiveButton(R.string.qr,new DialogInterface.OnClickListener() {
                @Override
                public void onClick(DialogInterface arg0, int arg1) {
                    ((Activity)context).finish();
                }
            });
        }
        return b.create();
    }
    //重载的退出方法3
    public Dialog getDialog(String mesage,boolean flag){
        AlertDialog.Builder b = new Builder(context);
        b.setTitle("你好这是系统信息提示框！");
        b.setMessage(mesage);
```

```java
        if (flag) {
            b.setPositiveButton(R.string.qr, new DialogInterface.OnClick
            Listener() {
                @Override
                public void onClick(DialogInterface arg0, int arg1) {
                    //--不操作
                }
            });
        }else{
            b.setPositiveButton(R.string.qr, new DialogInterface.OnClick
            Listener() {
                @Override
                public void onClick(DialogInterface arg0, int arg1) {
                    ((Activity)context).finish();
                }
            });
        }
        return b.create();
    }
}
```

（7）StudentGlobla 全局共享数据列表内容容器，定义如下。

```java
package ky.bai.student.util;

import java.util.List;
import ky.bai.entity.StudentEntity;
/**
 * 用于全局共享数据的类
 */
public class StudentGlobla {
    public static List<StudentEntity> lts = null;
}
```

（8）ThreadStudentsMessageForMe 联系人信息获取线程，用于获取联系人的信息，具体实现如下。

```java
package ky.bai.student.util;

import android.os.Handler;
//数据获取线程类
public class ThreadStudentsMessageForMe implements Runnable {
    //界面更新进程对象
    private Handler han = null;
    public ThreadStudentsMessageForMe(Handler han) {
        super();
```

```
            this.han = han;
    }
    @Override
    public void run() {
                //获取网络数据并存入全局数据共享容器
        StudentGlobla.lts=PullForXml.getStudentEntitys(GetInputStream.
                        getInputStream(UrlForMe.QUERY_STUDENT_ALL));
        if (StudentGlobla.lts != null) {
            //发送界面更新命令
            han.sendEmptyMessage(3);
        }else{
            han.sendEmptyMessage(4);
        }
    }
}
```

（9）自定义 PullForXml 类，在类 ThreadStudentsMessageForMe 定义时使用该类，该类主要实现 XML 解析，具体定义如下。

```
        package ky.bai.student.util;

import java.io.IOException;
import java.io.InputStream;
import java.util.ArrayList;
import java.util.List;
import org.xmlpull.v1.XmlPullParser;
import org.xmlpull.v1.XmlPullParserException;
import org.xmlpull.v1.XmlPullParserFactory;
import android.util.Log;
import ky.bai.entity.StudentEntity;
//XML 解析器
public class PullForXml {
    //StudentEntity 解析器
    public static List<StudentEntity> getStudentEntitys(InputStream in){
        List<StudentEntity> lt = new ArrayList<StudentEntity>();
        StudentEntity se = null;
        try {
            //创建解 Pull 解析器对象
            XmlPullParser xmlParser
                = XmlPullParserFactory.newInstance().newPullParser();
            //设置解析编码
            xmlParser.setInput(in, "utf-8");
            //获取当前事件编号
            int evtType = xmlParser.getEventType();
            //开始解析
            while (evtType != XmlPullParser.END_DOCUMENT) {
```

```java
switch (evtType) {
    //启动事件
    case XmlPullParser.START_TAG:
        //获取标签名称
        String tag = xmlParser.getName();
        //判断是否数据标签父节点
        if (tag.equalsIgnoreCase("student")) {
            //创建实体类对象
            se = new StudentEntity();
        } else if (se != null) {
            //从这里开始比对具体数据项
            if (tag.equalsIgnoreCase("name")) {
                String paramStr = xmlParser.nextText();
                se.setName(paramStr);
            } else if (tag.equalsIgnoreCase("age")) {
                String paramStr = xmlParser.nextText();
                if (paramStr != null &&
                        !"null".equalsIgnoreCase(paramStr)) {
                    se.setAge(Integer.parseInt(paramStr));
                }
            } else if (tag.equalsIgnoreCase("sex")) {
                String paramStr = xmlParser.nextText();
                se.setSex(paramStr);
            } else if (tag.equalsIgnoreCase("tel")) {
                String paramStr = xmlParser.nextText();
                se.setTel(paramStr);
            } else if (tag.equalsIgnoreCase("address")) {
                String paramStr = xmlParser.nextText();
                se.setAddress(paramStr);
            } else if (tag.equalsIgnoreCase("qq")) {
                String paramStr = xmlParser.nextText();
                se.setQq(paramStr);
            } else if (tag.equalsIgnoreCase("major")) {
                String paramStr = xmlParser.nextText();
                se.setMajor(paramStr);
            } else if (tag.equalsIgnoreCase("birthday")) {
                String paramStr = xmlParser.nextText();
                se.setBirthday(paramStr);
            } else if (tag.equalsIgnoreCase("educational")) {
                String paramStr = xmlParser.nextText();
                se.setEducational(paramStr);
            } else if (tag.equalsIgnoreCase("emergency")) {
                String paramStr = xmlParser.nextText();
                se.setEmergency(paramStr);
```

```java
                } else if (tag.equalsIgnoreCase("hobby")) {
                    String paramStr = xmlParser.nextText();
                    se.setHobby(paramStr);
                } else if (tag.equalsIgnoreCase("wantWork")) {
                    String paramStr = xmlParser.nextText();
                    se.setWantWork(paramStr);
                } else if (tag.equalsIgnoreCase("dream")) {
                    String paramStr = xmlParser.nextText();
                    se.setDream(paramStr);
                } else if (tag.equalsIgnoreCase("photo")) {
                    String paramStr = xmlParser.nextText();
                    se.setPhoto(paramStr);
                }
            }
            break;
        //标签结束事件
        case XmlPullParser.END_TAG:
            if (xmlParser.getName().equalsIgnoreCase("student")
                    && se != null && lt != null) {
                //数据对象存入数据缓存
                lt.add(se);
                se = null;
            }
            break;
        default:
            break;
        }
        //记录下一个标签编号
        evtType = xmlParser.next();
    }
} catch (XmlPullParserException e) {
    e.printStackTrace();
} catch (IOException e) {
    e.printStackTrace();
}
return lt;
}
```

4．首页运行效果图

首页的运行结果如图 15.10 所示。在手机没有用户信息列表时，在有网络的情况下，会自动连接网络搜索服务器，下载联系人信息列表到手机终端，供用户使用。主页上有页面跳转按钮，可直接进入用户信息搜索页。

15.5.4 信息列表展示页的设计与实现

（1）信息列表展示页面表示层

展示页属于首页的第一个子页面，该页面只负责展示基本信息，所以该展示页面布局很简单，只要装入列表控件即可。注意，列表中的展示结构（行）也需要一个相对独立的小布局页面，即列表本身也需要再引入一个更小单位的布局页面。其中，main_page_1.xml 是展示页面布局页面，在其中仅包含一个列表控件 listView。main_page_1_list_value.xml 为其子布局页面，包含了 TextView、ImageView 用于显示联系人的姓名、电话和照片信息。展示页详细布局代码部分如下。

图 15.10　首页运行效果

```xml
<?xml version="1.0" encoding="utf-8"?>
<LinearLayout xmlns:android="http://schemas.android.com/apk/res/android"
    android:layout_width="match_parent"
    android:layout_height="match_parent"
    android:orientation="vertical" >
<!-- 用于展示的列表 -->
    <ListView
        android:id="@+id/mian_page1_list"
        android:layout_width="match_parent"
        android:layout_height="wrap_content" >
    </ListView>
</LinearLayout>
```

展示页子页面如下。

```xml
<?xml version="1.0" encoding="utf-8"?>
<LinearLayout xmlns:android="http://schemas.android.com/apk/res/android"
    android:layout_width="match_parent"
    android:layout_height="match_parent"
    android:orientation="vertical" >
    <LinearLayout
        android:layout_width="match_parent"
        android:layout_height="wrap_content" >
<!-- 图像容器 -->
        <LinearLayout
            android:layout_width="wrap_content"
            android:layout_height="match_parent"
            android:layout_weight="0.3"
            android:gravity="center"
            android:orientation="vertical" >
            <ImageView
```

```xml
            android:id="@+id/main_page_1_photo"
            android:layout_width="75dp"
            android:layout_height="75dp"
            android:src="@drawable/ic_launcher" />
    </LinearLayout>
    <!-- 文本内容容器 -->
    <LinearLayout
        android:layout_width="wrap_content"
        android:layout_height="match_parent"
        android:layout_weight="0.7"
        android:orientation="vertical"
        android:gravity="center_vertical">

        <LinearLayout
            android:layout_width="match_parent"
            android:layout_height="wrap_content"
            android:paddingLeft="10dp" >
    <!-- 名称位置 -->
            <TextView
                android:id="@+id/main_page_1_name"
                android:layout_width="wrap_content"
                android:layout_height="wrap_content"
                android:text="Large Text"
                android:textAppearance="?android:attr/textAppearanceLarge" />

        </LinearLayout>

        <LinearLayout
            android:layout_width="match_parent"
            android:layout_height="wrap_content"
            android:gravity="left|center"
            android:paddingLeft="10dp" >

            <ImageView
                android:id="@+id/imageView2"
                android:layout_width="wrap_content"
                android:layout_height="wrap_content"
                android:src="@android:drawable/btn_star_big_on" />
    <!-- 电话号码位置 -->
            <TextView
                android:id="@+id/main_page_1_phone"
                android:layout_width="wrap_content"
                android:layout_height="wrap_content"
                android:text="Medium Text"
                android:textAppearance="?android:attr/textAppearanceMedium" />
```

```
            </LinearLayout>
        </LinearLayout>
    </LinearLayout>
</LinearLayout>
```

(2)信息列表展示页的详细业务处理

该部分所有业务代码均集成在 MainActivity.java 中,参见 15.5.3 小节首页模块业务处理类 MainActivity,其中,显示同学信息列表的部分。

(3)信息列表展示子模块的运行结果

如图 15.10 所示首页运行结果中,中间部分信息列表。

15.5.5 搜索页面的设计及实现

搜索模块是首页子页面 2,该模块主要实现由用户输入搜索关键字,根据用户的输入查询符合条件的同学信息,所以该也需要提供关键字输入和查询结果显示两部分。

(1)搜索页的表示层

该界面提供关键字输入、提供用于显示查询结果的展示列表,该页的页面布局包含两部分区域,其中,区域一 activity_sh.xml 是搜索页面布局页面,提供搜索关键字的输入。main_page_1_list_value.xml 是搜索页的子布局页面,用于显示搜索结果的显示,与 15.5.4 小节展示页的子页面布局相同。

搜索页布局,采用 LinearLayout 的布局,页面包含 EditText 接受搜索关键字的输入,Button 为搜索按钮。ListView 为查询结果的显示,其具体代码如下。

```xml
<RelativeLayout xmlns:android="http://schemas.android.com/apk/res/android"
    xmlns:tools="http://schemas.android.com/tools"
    android:layout_width="match_parent"
    android:layout_height="match_parent"
    tools:context=".ShActivity" >
    <LinearLayout
        android:id="@+id/linearLayout1"
        android:layout_width="match_parent"
        android:layout_height="60dp"
        android:gravity="center">
    <!-- 关键字输入框 -->
        <EditText
            android:id="@+id/sh_sh_et"
            android:layout_width="wrap_content"
            android:layout_height="wrap_content"
            android:layout_weight="1"
            android:ems="10" >
            <requestFocus />
        </EditText>
    <!-- 搜索按钮 -->
```

```xml
    <Button
        android:id="@+id/sh_sh_btn"
        android:layout_width="wrap_content"
        android:layout_height="wrap_content"
        android:layout_weight="1"
        android:drawableLeft="@drawable/sh"
        android:text="@string/user_sh"
        android:background="@drawable/btn_1_2"/>
</LinearLayout>
<LinearLayout
    android:layout_width="match_parent"
    android:layout_height="wrap_content"
    android:layout_alignParentLeft="true"
    android:layout_below="@+id/linearLayout1"
    android:orientation="vertical" >
<!-- 展示列表 -->
    <ListView
        android:id="@+id/sh_sh_lv"
        android:layout_width="match_parent"
        android:layout_height="wrap_content" >
    </ListView>
</LinearLayout>
</RelativeLayout>
```

搜索页子页面的布局与展示页与 15.5.4 小节展示页的子页面布局相同，请参见 15.5.4 小节 main_page_1_list_value.xml 子布局的定义部分。

（2）搜索页的业务处理

该部分所有业务代码同上节展示页面部分均集成在 MainActivity.java 中，参见 15.5.3 小节首页模块业务处理类 MainActivity 的搜索页面操作部分。

（3）搜索模块的运行结果

搜索页的运行结果如图 15.11、图 15.12 所示。图 15.11 所示为搜索关键字的输入界面，输入要搜索的内容，点击搜索按钮，进行模糊查询，查询结果如图 15.12 所示。

图 15.11　搜索页运行效果 1

图 15.12　搜索页运行效果 2

15.5.6 个人详细信息页的设计与实现

详细信息页用来展示某个联系人的详细信息，并且可以给这个人拨打电话、发送短信等。

1. 详细信息页的表示层

由于该页面提供个人详细信息的显示，可以拨打电话、发送短信等功能，该页面 activity_show_message.xml 采用 RelativeLayout 的布局，设置不同的组件，分别显示用户详细信息。个人详细信息展示页面布局如下：

```xml
<RelativeLayout xmlns:android="http://schemas.android.com/apk/res/android"
    xmlns:tools="http://schemas.android.com/tools"
    android:layout_width="match_parent"
    android:layout_height="match_parent"
    tools:context=".ShowMessageActivity" >
<!-- 垂直方向滚容器 -->
    <ScrollView
        android:id="@+id/scrollView1"
        android:layout_width="match_parent"
        android:layout_height="match_parent"
        android:layout_alignParentLeft="true"
        android:layout_alignParentTop="true" >
        <LinearLayout
            android:layout_width="match_parent"
            android:layout_height="match_parent"
            android:orientation="vertical" >
            <LinearLayout
                android:layout_width="match_parent"
                android:layout_height="wrap_content"
                android:layout_marginBottom="5dp"
                android:layout_marginTop="5dp" >
                <LinearLayout
                    android:layout_width="wrap_content"
                    android:layout_height="match_parent"
                    android:layout_weight="1"
                    android:gravity="center"
                    android:orientation="vertical" >
<!-- 头像 -->
                    <ImageView
                        android:id="@+id/user_photo"
                        android:layout_width="80dp"
                        android:layout_height="80dp"
                        android:src="@drawable/ic_launcher" />
```

```xml
            </LinearLayout>
            <LinearLayout
                android:layout_width="wrap_content"
                android:layout_height="match_parent"
                android:layout_weight="1"
                android:orientation="vertical" >
<!-- 姓名 -->
                <LinearLayout
                    android:layout_width="match_parent"
                    android:layout_height="wrap_content" >
                    <TextView
                        android:id="@+id/textView1"
                        android:layout_width="wrap_content"
                        android:layout_height="wrap_content"
                        android:text="@string/user_name" />
                    <TextView
                        android:id="@+id/user_name"
                        android:layout_width="wrap_content"
                        android:layout_height="wrap_content"
                        android:text="TextView" />
                </LinearLayout>
<!-- 年龄 -->
                <LinearLayout
                    android:layout_width="match_parent"
                    android:layout_height="wrap_content" >
                    <TextView
                        android:id="@+id/textView3"
                        android:layout_width="wrap_content"
                        android:layout_height="wrap_content"
                        android:text="@string/user_age" />
                    <TextView
                        android:id="@+id/user_age"
                        android:layout_width="wrap_content"
                        android:layout_height="wrap_content"
                        android:text="TextView" />
                </LinearLayout>
<!-- 性别 -->
                <LinearLayout
                    android:layout_width="match_parent"
                    android:layout_height="wrap_content" >
                    <TextView
                        android:id="@+id/textView5"
                        android:layout_width="wrap_content"
                        android:layout_height="wrap_content"
                        android:text="@string/user_sex" />
```

```xml
            <TextView
                android:id="@+id/user_sex"
                android:layout_width="wrap_content"
                android:layout_height="wrap_content"
                android:text="TextView" />
        </LinearLayout>
        <!-- 生日-->
        <LinearLayout
            android:layout_width="match_parent"
            android:layout_height="wrap_content" >
            <TextView
                android:id="@+id/textView7"
                android:layout_width="wrap_content"
                android:layout_height="wrap_content"
                android:text="@string/user_bir" />
            <TextView
                android:id="@+id/user_bir"
                android:layout_width="wrap_content"
                android:layout_height="wrap_content"
                android:text="TextView" />
        </LinearLayout>
        <!-- 梦想-->
        <LinearLayout
            android:layout_width="match_parent"
            android:layout_height="wrap_content" >
            <TextView
                android:id="@+id/textView14"
                android:layout_width="wrap_content"
                android:layout_height="wrap_content"
                android:text="@string/user_dre" />
            <TextView
                android:id="@+id/user_dre"
                android:layout_width="wrap_content"
                android:layout_height="wrap_content"
                android:text="TextView" />
        </LinearLayout>
    </LinearLayout>
</LinearLayout>
<!-- 显示更多-->
<LinearLayout
    android:id="@+id/user_show_message_lay"
    android:layout_width="match_parent"
    android:layout_height="wrap_content"
    android:layout_weight="1"
    android:gravity="center"
```

```xml
            android:orientation="vertical">
            <LinearLayout
                android:layout_width="match_parent"
                android:layout_height="match_parent"
                android:orientation="horizontal"
                android:gravity="right"
                android:paddingRight="15dp">
                <TextView
                    android:id="@+id/user_show_message_tv"
                    android:layout_width="wrap_content"
                    android:layout_height="wrap_content"
                    android:text="@string/user_more" />

                <ImageView
                    android:id="@+id/user_show_message_img"
                    android:layout_width="16dp"
                    android:layout_height="16dp"
                    android:layout_marginLeft="3dp"
                    android:src="@drawable/zk" />
            </LinearLayout>
<!-- 显示更多的内容容器-->
            <LinearLayout
                android:id="@+id/user_more_message_lay"
                android:layout_width="match_parent"
                android:layout_height="match_parent"
                android:orientation="vertical"
                android:visibility="gone" >
<!-- 紧急联系人-->
                <LinearLayout
                    android:layout_width="match_parent"
                    android:layout_height="wrap_content"
                    android:gravity="center" >
                    <TextView
                        android:id="@+id/textView2"
                        android:layout_width="wrap_content"
                        android:layout_height="wrap_content"
                        android:text="@string/user_emer"
                        android:textSize="18sp" />
                    <TextView
                        android:id="@+id/user_emergency"
                        android:layout_width="wrap_content"
                        android:layout_height="wrap_content"
                        android:text="TextView"
                        android:textColor="@color/red"
                        android:textSize="18sp" />
```

```xml
            </LinearLayout>
<!-- 专业-->
            <LinearLayout
                android:layout_width="match_parent"
                android:layout_height="wrap_content"
                android:orientation="vertical"
                android:gravity="center">
                <LinearLayout
                    android:layout_width="wrap_content"
                    android:layout_height="match_parent"
                    android:layout_weight="1"
                    android:gravity="center" >
                    <TextView
                        android:id="@+id/textView6"
                        android:layout_width="wrap_content"
                        android:layout_height="wrap_content"
                        android:text="@string/user_maj" />
                    <TextView
                        android:id="@+id/user_major"
                        android:layout_width="wrap_content"
                        android:layout_height="wrap_content"
                        android:text="TextView" />
                </LinearLayout>
<!-- 学历 -->
                <LinearLayout
                    android:layout_width="wrap_content"
                    android:layout_height="match_parent"
                    android:layout_weight="1"
                    android:gravity="center" >
                    <TextView
                        android:id="@+id/textView10"
                        android:layout_width="wrap_content"
                        android:layout_height="wrap_content"
                        android:text="@string/user_edu" />
                    <TextView
                        android:id="@+id/user_educational"
                        android:layout_width="wrap_content"
                        android:layout_height="wrap_content"
                        android:text="TextView" />
                </LinearLayout>
            </LinearLayout>
<!-- 爱好-->
            <LinearLayout
                android:layout_width="match_parent"
                android:layout_height="match_parent"
```

```xml
            android:gravity="center" >
            <TextView
                android:id="@+id/textView15"
                android:layout_width="wrap_content"
                android:layout_height="wrap_content"
                android:text="@string/user_hob" />
            <TextView
                android:id="@+id/user_hobby"
                android:layout_width="wrap_content"
                android:layout_height="wrap_content"
                android:text="TextView" />
        </LinearLayout>
<!-- 向往的职业-->
        <LinearLayout
            android:layout_width="match_parent"
            android:layout_height="wrap_content"
            android:gravity="center" >
            <TextView
                android:id="@+id/textView18"
                android:layout_width="wrap_content"
                android:layout_height="wrap_content"
                android:text="@string/user_wtw" />
            <TextView
                android:id="@+id/user_wantWork"
                android:layout_width="wrap_content"
                android:layout_height="wrap_content"
                android:text="TextView" />
        </LinearLayout>
        <LinearLayout
            android:layout_width="match_parent"
            android:layout_height="wrap_content" >
        </LinearLayout>
    </LinearLayout>
</LinearLayout>
<!-- 联系地址-->
    <LinearLayout
        android:layout_width="match_parent"
        android:layout_height="match_parent"
        android:layout_marginBottom="5dp"
        android:layout_marginTop="5dp"
        android:layout_weight="1"
        android:gravity="center"
        android:orientation="horizontal" >
        <TextView
```

```xml
                android:id="@+id/textView12"
                android:layout_width="wrap_content"
                android:layout_height="wrap_content"
                android:text="@string/user_add" />
            <TextView
                android:id="@+id/user_add"
                android:layout_width="wrap_content"
                android:layout_height="wrap_content"
                android:text="TextView" />
        </LinearLayout>
<!-- QQ-->
        <LinearLayout
            android:layout_width="match_parent"
            android:layout_height="wrap_content"
            android:gravity="center" >
            <TextView
                android:id="@+id/textView9"
                android:layout_width="wrap_content"
                android:layout_height="wrap_content"
                android:text="@string/user_qq" />
            <TextView
                android:id="@+id/user_qq"
                android:layout_width="wrap_content"
                android:layout_height="wrap_content"
                android:text="TextView" />
        </LinearLayout>
<!-- 联系电话-->
        <LinearLayout
            android:layout_width="match_parent"
            android:layout_height="wrap_content"
            android:layout_marginBottom="5dp"
            android:layout_marginTop="5dp"
            android:gravity="center" >
            <TextView
                android:id="@+id/textView16"
                android:layout_width="wrap_content"
                android:layout_height="wrap_content"
                android:text="@string/user_pho" />
            <TextView
                android:id="@+id/user_pho"
                android:layout_width="wrap_content"
                android:layout_height="wrap_content"
                android:text="TextView" />
        </LinearLayout>
<!-- 打电话、发短信-->
```

```xml
<LinearLayout
    android:layout_width="match_parent"
    android:layout_height="wrap_content"
    android:gravity="center" >
    <Button
        android:id="@+id/user_call"
        android:layout_width="wrap_content"
        android:layout_height="wrap_content"
        android:layout_marginRight="2dp"
        android:background="@drawable/btn_1_2"
        android:padding="8dp"
        android:text="@string/user_btn_call"
        android:textColor="@color/white"
        android:textSize="15sp" />
    <Button
        android:id="@+id/user_sms"
        android:layout_width="wrap_content"
        android:layout_height="wrap_content"
        android:layout_marginLeft="2dp"
        android:layout_marginRight="2dp"
        android:background="@drawable/btn_1_2"
        android:padding="8dp"
        android:text="@string/user_btn_sms"
        android:textColor="@color/white"
        android:textSize="15sp" />
            </LinearLayout>
        </LinearLayout>
    </ScrollView>
</RelativeLayout>
```

2. 详细信息页的业务处理

详细信息页业务、逻辑部分由 ShowMessageActivity.java 处理，该类在进行业务处理时会用到自定义的共享图片存储器 StudengImagesGlobla 类、网络地址常量类 UrlForMe 及头像获取线程类等，详细代码如下。

```java
package ky.bai.student;

import ky.bai.student.util.DeleteDialog;
import ky.bai.student.util.My_DB;
import ky.bai.student.util.StudengImagesGlobla;
import ky.bai.student.util.StudentGlobla;
import ky.bai.student.util.UrlForMe;
import android.net.Uri;
import android.os.Bundle;
```

```java
import android.os.Handler;
import android.os.Message;
import android.app.Activity;
import android.app.Dialog;
import android.content.Intent;
import android.database.sqlite.SQLiteDatabase;
import android.view.Menu;
import android.view.MenuItem;
import android.view.MenuItem.OnMenuItemClickListener;
import android.view.View;
import android.view.View.OnClickListener;
import android.widget.Button;
import android.widget.ImageView;
import android.widget.LinearLayout;
import android.widget.TextView;
//个人详细信息业务、逻辑处理类
public class ShowMessageActivity extends Activity {
    //本类ID编号
    public static final int SHOW_MESSAGE_CODE = 2000;
    //页面跳转对象
    private Intent it = null;
    //编号存储器
    private int position = -1;
    //界面UI进程对象
    private Handler han = null;
    //提示文本对象
    private TextView user_name = null;
    private TextView user_age = null;
    private TextView user_sex = null;
    private TextView user_bir = null;
    private TextView user_add = null;
    private TextView user_qq = null;
    private TextView user_pho = null;
    private TextView user_dre = null;
    private TextView user_emergency = null;
    private TextView user_major = null;
    private TextView user_educational = null;
    private TextView user_hobby = null;
    private TextView user_wantWork = null;
    private TextView user_show_message_tv = null;
    //图片对象
    private ImageView user_show_message_img = null;
    private ImageView user_photo = null;
    //详细信息开关容器对象
    private LinearLayout user_more_message_lay = null;
```

```java
        private LinearLayout user_show_message_lay = null;
        private boolean userMessageShowFlag = true;
        //拨打电话按钮
        private Button user_call = null;
        //发送短信按钮
        private Button user_sms = null;

        private String userName = null;
        private String userAdd = null;
        private String userBir = null;
        private String userPho = null;
        private String userSex = null;
        private String userDre = null;
        private String userQQ = null;
        private String userEmergency = null;
        private String userMajor = null;
        private String userEducational = null;
        private String userHobby = null;
        private String userWantWork = null;
        private String userPhoto = null;

    @Override
    protected void onCreate(Bundle savedInstanceState) {
        super.onCreate(savedInstanceState);
        setContentView(R.layout.activity_show_message);
        han = new Handler(){
        @Override
        public void handleMessage(Message msg) {
            switch (msg.what) {
                case 1:
                    Intent it = new Intent(ShowMessageActivity.this,
                    MainActivity.class);
                    setResult(SHOW_MESSAGE_CODE, it);
                    finish();
                    break;
                default:
                    break;
                }
            }
        };
        //获取界面上的控件
        user_name = (TextView) findViewById(R.id.user_name);
        user_age = (TextView) findViewById(R.id.user_age);
        user_sex = (TextView) findViewById(R.id.user_sex);
        user_bir = (TextView) findViewById(R.id.user_bir);
```

```java
user_add = (TextView) findViewById(R.id.user_add);
user_qq = (TextView) findViewById(R.id.user_qq);
user_pho = (TextView) findViewById(R.id.user_pho);
user_dre = (TextView) findViewById(R.id.user_dre);
user_emergency = (TextView) findViewById(R.id.user_emergency);
user_major = (TextView) findViewById(R.id.user_major);
user_educational = (TextView) findViewById(R.id.user_educational);
user_hobby = (TextView) findViewById(R.id.user_hobby);
user_wantWork = (TextView) findViewById(R.id.user_wantWork);
user_show_message_tv=(TextView) findViewById(R.id.user_show_message_tv);
user_more_message_lay
    = (LinearLayout) findViewById(R.id.user_more_message_lay);
user_show_message_lay
    = (LinearLayout) findViewById(R.id.user_show_message_lay);
user_show_message_img
    = (ImageView) findViewById(R.id.user_show_message_img);
user_photo = (ImageView) findViewById(R.id.user_photo);
user_call = (Button) findViewById(R.id.user_call);
user_sms = (Button) findViewById(R.id.user_sms);
//获取跳转对象
it = getIntent();
int userAge = -1;
if (it != null) {
    //接收上个页面传递过来的值
    userName = it.getStringExtra("userName");
    userAdd = it.getStringExtra("userAdd");
    userBir = it.getStringExtra("userBir");
    userPho = it.getStringExtra("userPho");
    userSex = it.getStringExtra("userSex");
    userDre = it.getStringExtra("userDre");
    userQQ = it.getStringExtra("userQQ");
    userEmergency = it.getStringExtra("userEmergency");
    userMajor = it.getStringExtra("userMajor");
    userEducational = it.getStringExtra("userEducational");
    userHobby = it.getStringExtra("userHobby");
    userWantWork = it.getStringExtra("userWantWork");
    userPhoto = it.getStringExtra("userPhoto");
    userAge = it.getIntExtra("userAge", -1);
    position = it.getIntExtra("position", -1);
    user_name.setText(userName);
    user_age.setText(userAge + "");
    user_sex.setText(userSex);
    user_bir.setText(userBir);
    user_add.setText(userAdd);
    user_qq.setText(userQQ);
```

```java
            user_pho.setText(userPho);
            user_dre.setText(userDre);
            user_emergency.setText(userEmergency);
            user_major.setText(userMajor);
            user_educational.setText(userEducational);
            user_hobby.setText(userHobby);
            user_wantWork.setText(userWantWork);
            //从全局图片共享容器中取头像
            user_photo.setImageBitmap(StudengImagesGlobla.images
                    .get(UrlForMe.QUERY_STUDENT_IMG + userPhoto));
        }
        //单击展开更多信息
        user_show_message_lay.setOnClickListener(new OnClickListener() {
            @Override
            public void onClick(View arg0) {
                if (userMessageShowFlag) {
                    user_more_message_lay.setVisibility(View.VISIBLE);
                    user_show_message_img.setImageResource(R.drawable.zk_up);
                    user_show_message_tv.setText(R.string.user_more2);
                    userMessageShowFlag = false;
                }else{
                    user_more_message_lay.setVisibility(View.GONE);
                    user_show_message_img.setImageResource(R.drawable.zk);
                    user_show_message_tv.setText(R.string.user_more);
                    userMessageShowFlag = true;
                }
            }
        });
        //调用系统拨号界面
        user_call.setOnClickListener(new OnClickListener() {
            @Override
            public void onClick(View arg0) {
                Intent it = new Intent(Intent.ACTION_CALL,Uri.parse("tel:"
                    + userPho)); startActivity(it);
            }
        });
        //调用系统短信界面
        user_sms.setOnClickListener(new OnClickListener() {
            @Override
            public void onClick(View arg0) {
                Uri smsToUri = Uri.parse("smsto:" + userPho);
                Intent intent = new Intent(Intent.ACTION_SENDTO, smsToUri);
                intent.putExtra("sms_body", "经常联系.");
                startActivity(intent);
            }
```

```java
        });
    }
    //系统菜单
    @Override
    public boolean onCreateOptionsMenu(Menu menu) {
        //加载自定义菜单页面
        getMenuInflater().inflate(R.menu.show_message, menu);
        //设置菜单单击事件
        menu.getItem(0).setOnMenuItemClickListener(new OnMenuItemClickListener() {
            @Override
            public boolean onMenuItemClick(MenuItem arg0) {
                //删除手机数据库中的数据
                SQLiteDatabase db = 
                openOrCreateDatabase(My_DB.MY_DB_NAME, MODE_PRIVATE, null);
                Dialog d = new DeleteDialog(ShowMessageActivity.this,
                        position, han, StudentGlobla.lts,db).getDialog();
                d.show();
                return true;
            }
        });
        return true;
    }
}
```

3. ShowMessageActivity 类中用到的自定义类

（1）DeleteDialog 类和 My_DB 类及 StudengGlobla、StudengImagesGlobla 等类的定义同 15.5.3 中首页业务处理类中用到的自定义类的定义相同。

（2）UrlForMe 网络地址常量类。

```java
package ky.bai.student.util;
//访问地址常量类
public class UrlForMe {
//数据访问地址
    public static final String QUERY_STUDENT_ALL
                = "http://192.168.1.104:8080/StudentInfo/";
//图片访问地址
    public static final String QUERY_STUDENT_IMG
                = "http://192.168.1.104:8080/StudentInfo/image/";
}
```

注意：本次项目中使用的网站是搭建在本机的。UrlForMe 类是移动端上的代码，所以在测试时一定要注意主机地址要保持一致，在确认网站运行正常之后，查询本机 IP 地址，将 UrlForMe 类中的地址常量 http://192.168.1.104:8080 中的 192.168.1.104 换成查询到的本机 IP，并保证手机与网站在同一网络的相同网段中才能正常运行整个案例。

4．个人详细信息页运行结果

个人详细信息页的运行结果如图 15.13、图 15.14 所示。图 15.13 所示为第一种详细信息显示，其中，显示更多详细信息收起，图 15.14 所示是详细显示所有的信息。可以单击拨打电话和发送短信和联系人进行联系。

图 15.13　联系人详细 1　　　　　　图 15.14　联系人详细 2

15.5.7　删除功能的设计与实现

删除模块实现时是通过用户查询到相应用户信息后，单击删除按钮，出现删除信息提示对话框，让用户确认是否删除的操作。

（1）删除模块的表示层

删除模块是通过对话框来提示用户是否确认删除，所以该模块没有专门的布局文件，表示层通过自定义对话框实现，其定义见 15.5.3 小节首页中 DeleteDialog 类的定义。

（2）删除模块的业务层

删除模块模块业务层的处理也集成在类 MainActivity 中完成，真正的业务处理有类 DeleteDialog 完成。

（3）删除功能的运行结果

删除用户信息运行结果如图 15.15、图 15.16 所示。图 15.15 所示单击删除按钮时，弹出的删除信息提示框，用户可以选择确认将更新首页信息列表界面，如果选择取消，将不执行任何操作。图 15.16 所示显示了在某人详细信息查询时也可以删除该联系人信息，当单击"删除该用户信息"按钮时，同样出现删除确认提示信息对话框，等待用户的输入。

图 15.15　删除功能的运行结果 1　　图 15.16　删除功能的运行结果 2

15.6　本 章 小 结

本章通过一个简单的开发实例同学簿的分析设计与实现,让读者来了解一个系统的开发流程。一个同学簿系统除了查询、信息显示、删除功能外,还应该具有添加、数据同步等更完善的功能,由于篇幅限制,文中只介绍了核心模块的设计与实现。不过,只要读者理解了这部分内容,完全有能力设计出其他功能模块和更复杂的系统的。

本章应用软件工程的设计思想,带领读者轻松地走完了一个系统的开发流程,相信读者通过本例的学习,在 Android 应用开发技术提升一个新的台阶。

将 UrlForMe 类中的地址常量 http://192.168.1.104:8080 中的 192.168.1.104 换成查询到的本机 IP，并保证手机与网站在同一网络的相同网段中才能正常运行整个案例。

4．个人详细信息页运行结果

个人详细信息页的运行结果如图 15.13、图 15.14 所示。图 15.13 所示为第一种详细信息显示，其中，显示更多详细信息收起，图 15.14 所示是详细显示所有的信息。可以单击拨打电话和发送短信和联系人进行联系。

图 15.13　联系人详细 1　　　　　图 15.14　联系人详细 2

15.5.7　删除功能的设计与实现

删除模块实现时是通过用户查询到相应用户信息后，单击删除按钮，出现删除信息提示对话框，让用户确认是否删除的操作。

（1）删除模块的表示层

删除模块是通过对话框来提示用户是否确认删除，所以该模块没有专门的布局文件，表示层通过自定义对话框实现，其定义见 15.5.3 小节首页中 DeleteDialog 类的定义。

（2）删除模块的业务层

删除模块模块业务层的处理也集成在类 MainActivity 中完成，真正的业务处理有类 DeleteDialog 完成。

（3）删除功能的运行结果

删除用户信息运行结果如图 15.15、图 15.16 所示。图 15.15 所示单击删除按钮时，弹出的删除信息提示框，用户可以选择确认将更新首页信息列表界面，如果选择取消，将不执行任何操作。图 15.16 所示显示了在某人详细信息查询时也可以删除该联系人信息，当单击"删除该用户信息"按钮时，同样出现删除确认提示信息对话框，等待用户的输入。

图 15.15　删除功能的运行结果 1　　图 15.16　删除功能的运行结果 2

15.6　本章小结

本章通过一个简单的开发实例同学簿的分析设计与实现，让读者来了解一个系统的开发流程。一个同学簿系统除了查询、信息显示、删除功能外，还应该具有添加、数据同步等更完善的功能，由于篇幅限制，文中只介绍了核心模块的设计与实现。不过，只要读者理解了这部分内容，完全有能力设计出其他功能模块和更复杂的系统的。

本章应用软件工程的设计思想，带领读者轻松地走完了一个系统的开发流程，相信读者通过本例的学习，在 Android 应用开发技术提升一个新的台阶。